Advances in Agroforestry

Volume 13

Series editor
P.K. Ramachandran Nair, Gainesville, USA

More information about this series at http://www.springer.com/series/6904

Jagdish Chander Dagar
Paramjit Singh Minhas
Editors

Agroforestry for the Management of Waterlogged Saline Soils and Poor-quality Waters

Springer

Editors
Jagdish Chander Dagar
Central Soil Salinity Research Institute
Karnal, Haryana, India

Paramjit Singh Minhas
National Institute of Abiotic Stress Management
Pune, Maharashtra, India

ISSN 1875-1199 ISSN 1875-1202 (electronic)
Advances in Agroforestry
ISBN 978-81-322-2657-4 ISBN 978-81-322-2659-8 (eBook)
DOI 10.1007/978-81-322-2659-8

Library of Congress Control Number: 2016930308

Springer New Delhi Heidelberg New York Dordrecht London

Printed on acid-free paper

Springer (India) Pvt. Ltd. is part of Springer Science+Business Media (www.springer.com)

Foreword

About two-fifths of the total food of the world is produced from 260 million hectares (Mha) of irrigated agriculture covering about 17 % of the cropped area. Though the provision and expansion of irrigation has helped in increasing food production and nutritional security, irrigation without adequate drainage is leading to waterlogging and secondary salinization especially in arid and semiarid regions. Nearly 20 % of the irrigated area is facing secondary salinization and one-third is threatened with waterlogging. The problem is severe in areas underlain with poor-quality groundwater. The intense competition from urban and industrial sectors is gradually reducing the share of freshwater for agriculture. Moreover, with low consumptive use, the nonagricultural uses are leading to the generation of huge volumes of wastewaters. Since the present sewage irrigation practices in most of the developing countries are not satisfactory, their reuse results in progressive and irreversible accumulation of salts, toxic materials, and heavy metals in soil and groundwater. Health hazards from pathogenic contaminations further multiply the complexities from their reuse.

The conventional engineering technologies of subsurface drainage for managing waterlogged saline soils and sewage treatment plants for treating wastewater are often unaffordable for resource-poor countries. Fast-growing forest tree plantations have shown promise for controlling waterlogged situations across the diverse agroecological regions. The safe and sustainable use of wastewater through high-transpiration rate plantation crops can also serve as a low-cost alternative. Thus, full and efficient utilization of this upsurge in scientific agroforestry seems both a challenge and an opportunity to the agroforestry scientific community. In order to understand the challenge and to provide sustainable solutions to these daunting problems, agroforestry scientists need access to synoptic information on multidimensional aspects of scientific knowledge of latest techniques.

The contributors to this book are experts in their relevant fields and have added their experiences to the value of this publication. I sincerely appreciate the contribution of all authors and the editors who have brought this publication to high standard. I hope the publication will be very useful for scientists working in the field of salinity, policymakers, environmentalists, educationists, and researchers for shaping this very important field of present-day science. I congratulate all editors, including series editor Dr PKR Nair, for their splendid accomplishment and express the global agroforestry community's gratitude to them for adding such valuable information to the agroforestry literature.

M.S. Swaminathan Research Foundation
Third Cross Street, Taramani Institutional Area
Chennai 600 113, India

M.S. Swaminathan

Preface

Salt-induced land degradation is a global phenomenon, afflicting millions of hectares within the sovereign borders of at least 75 countries. Besides endangering the food security, it has far-reaching and unacceptable socioeconomic consequences since a large proportion of this land is inhabited by smallholder farmers. The anthropogenic environmental changes and the climate change are further adding to the problem. Thus, dealing with the salinity in reality is becoming a highly onerous task owing to its complex nature, uncertainty, and differential temporal and spatial impacts. Nevertheless, with the need to provide more food, feed, fuel, fodder, and fiber to the expanding population and nonavailability of new productive land, there is a need for productivity enhancement of these lands. In fact, the salt-affected lands cannot be neglected since huge investments have been made in the development of irrigation and drainage infrastructure.

The social, economic, and environmental costs being high for the on- and off-farm reclamation techniques, agroforestry is now emerging as a potential tool not only for arresting salinity but for other environmental services like mitigating climate change, sequestering carbon, and restoring biodiversity. This publication attempts to address a wide range of issues related to the agroforestry principles involved in the rehabilitation of waterlogged saline soils and judicious use of marginal quality waters. Many of the site-specific case studies and those that are typical to the catchments have been described in detail. Since agroforestry is now considered as an integral part of strategies for the reversal of salinity-related land management soil problems, the concepts and practices currently available in different parts of the world for the remunerative use of these lands have been reviewed.

The authors of the various chapters come from a wide range of disciplines—agronomy, agroforestry, ecology, soil science, soil-water engineering, forestry, etc.—and are experts in their fields and have added their long-standing field experiences for the "prescriptive" aspects. We sincerely thank all the contributors and reviewers who contributed enormously on a short notice and cooperated so splendidly under strict and difficult time schedules. We also acknowledge Dr PKR Nair, Florida University, Gainesville, who is also the series editor, for his constant encouragement and help from time to time to bring out this publication in its present shape. We are also indebted to Dr MS Swaminathan who kindly wrote the wonderful Foreword for this publication. We hope that this publication will be very useful for scientists working in the fields of soil salinity, waterlogging, and poor-quality waters; policymakers; environmentalists; students; and educationists for shaping this very important field of present-day science dealing with the livelihood security of resource-poor stakeholders.

Karnal, Haryana, India Jagdish Chander Dagar
Pune, Maharashtra, India Paramjit Singh Minhas

Contents

Introduction 1
J.C. Dagar and P.S. Minhas

Global Perspectives on Agroforestry for the Management of Salt-affected Soils 5
J.C. Dagar and P.S. Minhas

Use of Tree Plantations in Water-table Drawdown and Combating Soil Salinity 33
P.S. Minhas and J.C. Dagar

Prospects for Managing Salinity in Southern Australia Using Trees on Farmland 49
Nico Marcar

Models for Estimating Evapotranspiration of Irrigated Eucalypt Plantations 73
S. Theiveyanathan, R.G. Benyon, V. Koul, R.K. Yadav, and R.I.S. Gill

Perspectives for Bio-management of Salt-affected and Waterlogged Soils in Pakistan 97
A.S. Qureshi

Combating Waterlogging in IGNP Areas in Thar Desert (India): Case Studies on Biodrainage 109
M.M. Roy, N.D. Yadava, M.L. Soni, and J.C. Tewari

Agroforestry to Rehabilitate the Indian Coastal Saline Areas 121
J.C. Dagar and P.S. Minhas

Saline Irrigation for Productive Agroforestry Systems 145
J.C. Dagar and P.S. Minhas

Potential of Wastewater Disposal Through Tree Plantations 163
R.K. Yadav, P.S. Minhas, Khajanchi-Lal, and J.C. Dagar

Tree Plantations in Saline Environments: Ecosystem Services, Carbon Sequestration and Climate Change Mitigation ... 181
S.R. Gupta, J.C. Dagar, and Mukesh Kumar

Synthesis and Way Forward: Agroforestry for Waterlogged Saline Soils and Poor-quality Waters ... 197
P.S. Minhas and J.C. Dagar

Index ... 203

Introduction

J.C. Dagar and P.S. Minhas

Abstract

Agroforestry has now emerged as an environmentally safe technology for the remediation of serious land degradation problems such as salt-affected soils, handling waterlogging particularly in canal command areas, and disposal of marginal waters. Not only the salt-affected soils in inland areas of arid and semiarid regions but also those in coastal areas can be made productive by adopting agroforestry techniques. The integrated farming systems involving multiple components such as tree plantations, fruit trees, agricultural crops, and animal and aquaculture components can be highly remunerative. Mangroves and associate vegetation play a crucial role in protecting shores and supporting wildlife and for the livelihood security of coastal people. Domestication of halophytes can be useful for food, fodder, oil, and medicinal and aromatic purposes. Salt-tolerant but high-productivity genotypes should be developed for greater viability. The latest information on these aspects has been compiled in a format that would be useful for scientists working on soil salinity, waterlogging, and poor-quality waters; policy makers; environmentalists; students; and educationists alike.

The world population is expected to be about 9.1 billion in 2050 and thereby would annually require the additional production of one billion megagrams (Mg) of cereals and 200 million Mg of livestock products. The imperative for such agricultural growth is strongest in developing countries to ensure their food and nutritional security. With little scope for the expansion of agricultural land, the increased demand will put more stress on the fragile land and water resources. The irrigated agriculture has a long

J.C. Dagar (✉)
Central Soil Salinity Research Institute,
Karnal 132 001, Haryana, India
e-mail: dagarjc@gmail.com

P.S. Minhas
National Institute of Abiotic Stress Management,
Baramati, Pune 413 115, Maharashtra, India
e-mail: minhas_54@yahoo.co.in

J.C. Dagar, P.S. Minhas (eds.), *Agroforestry for the Management of Waterlogged Saline Soils and Poor-quality Waters*, Advances in Agroforestry 13, DOI 10.1007/978-81-322-2659-8_1

and well-documented history in sustaining human populations, and even now about 40 % food is produced from 17 % of the irrigated land. Globally, the average yield of rainfed cereals in the developing world is about 1.5 Mg ha^{-1}, whereas it is 3.3 Mg ha^{-1} in irrigated areas. However, nearly one billion hectares of arid and semiarid areas of the world are salt-affected and remain barren due to salinity or water scarcity, and about 20 % of the irrigated area is facing secondary salinization and one-third is threatened with waterlogging. Additionally with the increasing demand for good-quality land and water for urbanization and industrial development projects, agriculture is bound to be pushed more and more to the marginal lands and use of poor-quality waters is inevitable.

The problems of waterlogging and salinity can be effectively tackled by conventional engineering approaches like surface and subsurface drainage (both horizontal and vertical), which have been standardized to rehabilitate the saline waterlogged lands. However, their adoption on a large scale is being constrained by high capital investment and the associated operational and maintenance problems in addition to the disposal of drainage effluents. Specifically, the problems associated with effluent disposal are widespread, and thereby, the threat of soil and groundwater salinization induced by irrigation has become a major issue for hydrologists, agronomists, soil and irrigation scientists, and policy makers. The limitations and shortcomings of the conventional techniques for salinity control call for the alternative approaches to sustain agriculture over the long term. Keeping the above in view, the agroforestry approaches for managing waterlogged salt-affected soils and low-quality waters have been included in different chapters.

Chapter "Global Perspectives on Agroforestry for the Management of Salt-affected Soils" presents information available at a global level on the agroforestry approaches in the management of waterlogged saline lands and the experiences gained on afforestation of these lands in Australia, the Indian subcontinent, and elsewhere. Since the management practices must be cost-effective, socially acceptable, and environment friendly, the available options have been discussed especially with respect to planting techniques and other management aspects for sustaining the growth of tree plantations. It emanates that the salt-affected lands can be effectively utilized for raising forest and fruit tree species, forage grasses, and nonconventional crops. Options are also available for effectively delaying the rise in water table and thus the problem of salinization through tree plantations in semiarid and arid regions. Undoubtedly, the added advantages of agroforestry systems include the production of timber, fuelwood, and other products such as oil, fruits, fiber, and pharmaceuticals. Their contribution in providing shade and shelter, function as windbreaks, soil amelioration through litter fall, enhancement of biodiversity, and other environmental services is now well established. The salt-tolerant tree species can reclaim salt-affected lands, along with increasing the size of carbon sink in the plant-soil system and improving soil microbial activity. The integration of salt-tolerant trees with grasses is a viable land use option for improving the biological productivity and fertility of highly sodic soils. Implementing agroforestry practices to build up soil carbon stocks can lead to considerable mitigation, adaptation, and development benefits in this era of climate change.

The issues related to the role of tree plantations for controlling waterlogging and salinity in different agroclimatic situations across the world have been analyzed in chapter "Use of Tree Plantations in Water-table Drawdown and Combating Soil Salinity". The potential withdrawal of soil water by trees under favorable and stress conditions and thereby inducing for water table drawdown have been compiled. The alternatives for taking up block plantation in recharge/upland discharge areas, boundary plantations, and shelterbelts and even integrating tree plantations with subsurface drainage and tree plantations have been highlighted for various saline and waterlogged situations.

Chapter "Prospects for Managing Salinity in Southern Australia Using Trees on Farmland" assesses the prospects for managing salinity in Southern Australia where trees are typically planted in salt-affected agricultural landscapes for environmental and economic benefit. Salinity is extensive in southern Australia with outbreaks typically patchy and of varying intensity at catchment, farm, and paddock scales in the east and more widespread in the west. Tree plantations targeted to appropriate landscape positions can make an important contribution to the mitigation and management of salinity, both by reducing recharge to groundwater and by lowering water tables in discharge areas. Australian work on the management of dryland salinity using tree plantations has been of great importance and followed elsewhere. Most efforts have concentrated on managing dryland (rainfed) catchments where recharge (water intake) and discharge (water outflow) locations are usually well defined. Trees are typically planted as compact plantation or shelterbelt configurations mostly in recharge areas or upslope of discharge areas and, less commonly, to stabilize discharge areas.

Eucalyptus is the most widely planted tree species in saline areas of Australia. Several case studies are available from different regions in Australia from both field experiments and process-based modelling to illustrate growth and water use responses to salinity and opportunities for maintaining water quality of rivers and streams in salinized catchments and for both the utilization and stabilization of salt-affected land in dryland and irrigated situations. Chapter "Models for Estimating Evapotranspiration of Irrigated Eucalypt Plantations" deals with models for estimating evapotranspiration of irrigated eucalypt plantations.

In continuation with country-specific problems, Pakistan has both coastal and inland salinity and suffers from waterlogging. About 4.5 million ha (Mha) irrigated area is influenced by salinity. Introduction of many species of trees and saltbushes from Australia have been tried on saline waterlogged soils. Therefore, chapter "Perspectives for Bio-management of Salt-affected and Waterlogged Soils in Pakistan" presents the results of such attempts and their role in biomanagement of waterlogged and saline soils. Bioremediation and other reclamation measures have also been compared under conventional irrigation methods.

Chapter "Combating Waterlogging in IGNP Areas in Thar Desert (India): Case Studies on Biodrainage" again includes a site-specific case study from India, i.e., *Indira Gandhi Nahar Priyojna* (IGNP) in the *Thar Desert*, which is one of the largest irrigation projects in the world. As a result of this initiative, 1.86 Mha of land has come under cultivation. Although irrigation has greatly increased the agricultural production potential, it has also rendered vast tracts of land under waterlogged category (leading to salinization as well) on account of seepage and deep percolation losses from the irrigation network and lack of a proper drainage system. The role of plantations for the removal of groundwater through evapotranspiration has been advocated as an effective way to tackle this problem. The case studies included are based on long-term observations on many tree species.

To address coastal salinity issues, chapter "Agroforestry to Rehabilitate the Indian Coastal Saline Areas" details the specific role of agroforestry in enhancing farm productivity in saline regions. The coastal and island ecosystems have a vide variability in climate, topographical, and edaphic conditions and support diverse cultivated crops as well as natural vegetation ranging from tropical rainforests to coastal mangroves. The areas are environmentally disadvantaged both in terms of anthropogenic activities and weather adversities. These factors along with the intrusion of seawater lead to salinity and waterlogging problems. The salinity problem is expected to become severe in the scenario of sea level rise with global warming. However, these ecosystems offer a vast scope of variety of agroforestry systems involving integrated farming systems, plantation crops, spices, forages, multistory plantation-based cropping systems, and aquaculture particularly in mangrove areas. The domestication of high-value halophytic crops stands a promise for enhancing farm productivity and livelihood security. All these aspects have been included in chapter "Agroforestry to Rehabilitate the Indian Coastal Saline Areas" along with strategies for the management of coastal saline areas through agroforestry systems.

The use of highly saline waters, especially those unfit for use in crop production, deserves a special attention for long-term sustainability in agroforestry. But this demands answers on how much, how long, and what methods exist for the effective utilization of high-salinity waters in agroforestry. The possibilities of diversified uses of saline waters, the suitability of planting techniques, and other management practices for the critical control of salt and water in the root zone have been examined in chapter "Saline Irrigation for Productive Agroforestry Systems". The salt-tolerant plants having potential as fuelwood, fruits, and forages with medicinal and aromatic uses are documented, and many of those can be blended with both forest and fruit trees as agroforestry systems.

Irrigation of high-transpiring forest species has been put forward for the recycling and reuse of wastewater and conservation of nutrient energy into biomass, thereby bringing multiple benefits such as fuelwood production, environmental sanitation, and ecorestoration. Nevertheless, issues such as the loading rates these plantations can carry and their dendroremediation for environmental cleanup have been contradictory so far. The chapter "Potential of Wastewater Disposal Through Tree Plantations" reviews the role agroforestry systems play in the safe disposal of wastewaters and in revenue generation and its other environmental benefits.

Biosaline agroforestry provides various ecosystem services such as improved soil fertility, carbon sequestration, improved biodiversity, and biomass production. Provisioning services relating to biomass production have been fairly well studied. These aspects have been highlighted in chapter "Tree Plantations in Saline Environments: Ecosystem Services, Carbon Sequestration, and Climate Change Mitigation", along with its role in improving soil microbial biomass and other physicochemical properties of soils. Arbuscular mycorrhizal (AM) fungi colonize the roots of grasses in silvopastoral systems on saline and sodic soils. The dominant AM fungal species are *Glomus* and *Acaulospora* which possibly play a key role in the amelioration of salt stress and carbon sequestration. This aspect has been explained clearly advocating that by integrating trees with the naturally occurring grassland systems on highly sodic soils, the soil organic carbon content is increased manyfold.

Finally, chapter "Synthesis and Way Forward: Agroforestry for Waterlogged Saline Soils and Poor-quality Waters" synthesizes the overall gist of the chapters. On the whole, different chapters are compiled in a mode to bring out the diversified role of agroforestry in tackling complex problems of salinity and waterlogging and safer management of poor-quality waters. So far, agroforestry was considered only a sustainable farming system, but its other role as a problem-solving and remunerative tool for degraded land and water resources is quite eminent now. Besides providing sustainable livelihood security to poor families, it provides for an immense scope of environmental services. The authors trust that the book will open new vistas in the versatile field of agroforestry and will be useful for different stakeholders including agricultural scientists, environmentalists, policy makers, and social scientists.

Global Perspectives on Agroforestry for the Management of Salt-affected Soils

J.C. Dagar and P.S. Minhas

Abstract

Nearly one billion hectares of arid and semiarid areas of the world are salt affected and remains barren due to salinity or water scarcity, and about 20 % of the irrigated area is facing secondary salinization and one-third threatened with waterlogging. Using appropriate planting and other management techniques, these lands can be utilized satisfactorily for agroforestry systems by integrating forest/fruit trees, forage grasses and other conventional and nonconventional crops. The process of salinization associated with rising water table can be delayed using strip plantations. Undoubtedly, the advantages of agroforestry systems include the production of timber, fuelwood and other products such as oil, honey, fruits, fibre, pharmaceuticals, etc., carbon sequestration, diminishing the wind erosion, provision of shade and shelter, function as wind breaks, soil amelioration through litter fall, enhancement of biodiversity, and improvement in general environment. The salt-tolerant tree species reclaim salt-affected lands, along with the increase in the size of carbon sink in the plant-soil system and improving soil microbial activities. The integration of salt-tolerant trees with grasses is a viable land-use option for improving the biological productivity and fertility of highly sodic soils. Implementing agroforestry practices to build up soil carbon stocks can help as an adaptation strategy for climate change and other associated developmental benefits.

Introduction

The introduction of irrigation in arid and semiarid regions without provision of adequate drainage causes rise in water table leading to waterlogging and secondary salinization. As per FAO/UNESCO Soil Map of the World (FAO/AGL 2000), the total salt-affected area in the world is 831 million

J.C. Dagar (✉)
Central Soil Salinity Research Institute,
Karnal 132 001, Haryana, India
e-mail: dagarjc@gmail.com

P.S. Minhas
National Institute of Abiotic Stress Management,
Baramati, Pune 413115, Maharashtra, India
e-mail: minhas_54@yahoo.co.in

J.C. Dagar, P.S. Minhas (eds.), *Agroforestry for the Management of Waterlogged Saline Soils and Poor-quality Waters*, Advances in Agroforestry 13, DOI 10.1007/978-81-322-2659-8_2

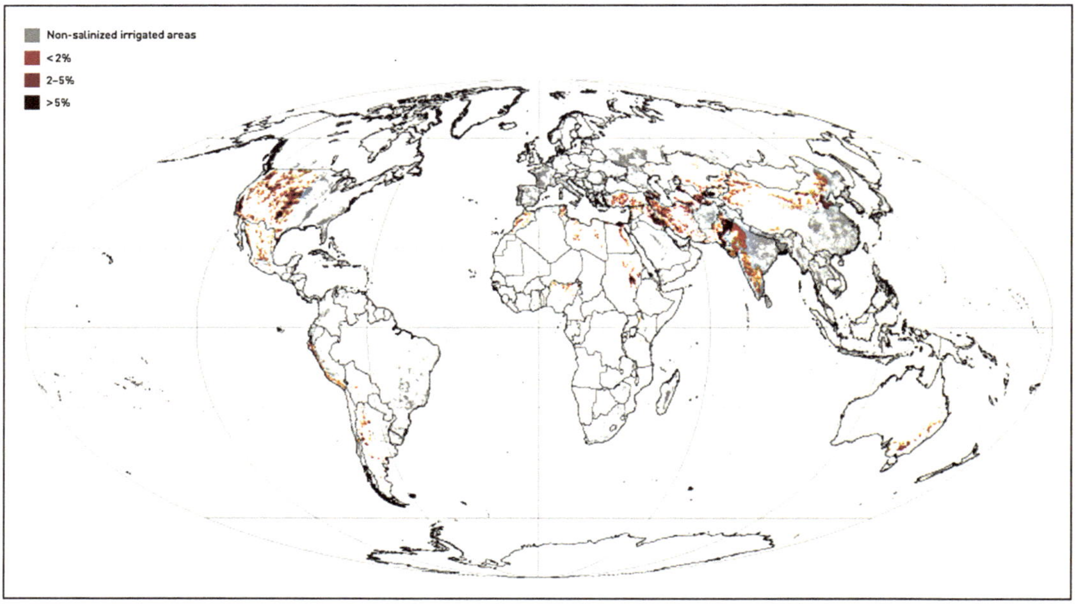

Fig. 1 Proportion of land salinized due to irrigation (Source: www.fao.org/nr/solaw)

ha (Mha), out of which 397 Mha is saline and 434 Mha sodic soils. Over one-third of the world's irrigated area faces the threat of waterlogging. On an average, it is estimated that about 100 million hectare (Mha) of land has become saline due to irrigation (Ghassemi et al. 1995; Pessarakli and Szabolcs 1999) and about 11 % of the world's irrigated areas (Fig. 1) are already affected by some degree of salinization (FAO 2012).

The problems associated with effluent disposal from sub-surface drainage are widespread. For example, discharge of drainage water from irrigated lands in the San Joaquin Valley in California into the Kesterson Reservoir resulted in problems of selenium toxicity in the biota (Cervinka et al. 1999). The Aral Sea basin faces a crisis similar to that which destroyed the Mesopotamian civilization about 4000 years ago, as the discharge of polluted and saline drainage effluent into the river systems has reached hazardous level (Heuperman et al. 2002). Similarly, the Indus basin in Pakistan, various river systems (including rivers Ganga and Yamuna and their tributaries) in India and the Murray-Darling Basin catchment in Australia are suffering the consequences of river water pollution as a result of the discharge of polluted drainage effluent from irrigation. In many countries, disposal of drainage waters into rivers is restricted as of ecological problems. As a result, the very high annual rate of installation of sub-surface drainage of the 1980s (300,000 ha per year) fell down to about 150,000 ha per year during the 1990s (Lesaffre and Zimmer 1995). In India and Pakistan, conventional engineering approaches (mainly sub-surface drainage) were tried to control waterlogging (Ritzema et al. 2008; Qureshi et al. 2008a, b) but became non-functional due to socio-economic problems and those associated with the disposal of drainage effluents.

The limitations and shortcomings of the conventional engineering-based drainage systems call for alternative approaches to keep the agriculture sustainable over the long term. The alternative approaches must be effective, affordable, socially acceptable, environment-friendly, sustainable and capable of safeguarding the precious natural resources. Agroforestry options comprising fast-growing deep-rooted vegetation with high transpiration and economic value and compatible with arable crops are now emerging as the promising alternative.

Agroforestry for Managing Waterlogging and Salinity

As an alternative, the use of woody vegetation for managing water fluxes to recharge is often referred to as bio-drainage and was earlier advocated for dryland salinity (created due to shallow water table) control (Barret-Lennard and Galloway 1996; Marcar and Khanna 1997; George et al. 1999; Stirzaker et al. 1999; Heuperman et al. 2002; Turner and Ward 2002; INCID 2003; Kapoor 2014). Efficiency of the tree plantations has also been shown for seepage control from canals and rivers (Kapoor and Denecke 2001; Manjunatha et al. 2005; Jeet-Ram et al. 2011; Sprenger et al. 2013), hence increasing agricultural production on farms. In areas where water remains stagnant creating wetlands as in Bangladesh, Ghosal et al. (2004) reported that these lands can successfully be utilized for agroforestry. Kushiev et al. (2005) reported the potential of leguminous liquorice (*Glycyrrhiza glabra*) cultivation to reclaim abandoned saline areas in the Hungry Steppes of Central Asia where elevated water tables were associated with poor irrigation management and inappropriate drainage infrastructure, while Dagar et al. (2015) have recently reported remunerative use of liquorice growing it on moderate alkali soil. The experiences gained world over for revegetation of saline lands are briefed below.

Australian Experiences

Australia has about 32 Mha of salt-affected land, the most of which is subject to primary salinity mainly in rangeland areas. Secondary salinity usually affects higher-value agricultural or irrigated land, and at least 5.7 Mha of land is at high risk of secondary salinity and 17 Mha is estimated to be at high risk by the year 2050 (Barrette-Lennard 2003). In Southern Australia, clearing of native vegetation for annual crops and pastures is recognized as a major cause of waterlogging and secondary salinity. Pastures are one of many means of obtaining productive use and rehabilitation of waterlogged salt-affected soils. These pastures include salt-tolerant fodder shrubs (species of *Atriplex, Halosarcia* and *Maireana*), perennial grasses (*Puccinellia ciliata, Thinopyrum ponticum, Distichlis spicata, Paspalum vaginatum, Sporobolus virginicus, Pennisetum clandestinum, Chloris gayana,* etc.) and some annual species. As such very limited data are available on the effects of combinations of salinity and waterlogging on plant ecological zonation particularly in pasture lands of Australia. Barrett-Lennard (2003), therefore, tried to show relative ranking of different saltland pasture species and indicator species in the salinity/waterlogging matrix (Figs. 2a and 2b). Some of the most useful species for Australian salt-affected soils have been introduced from overseas and performed well, and those include *Puccinellia ciliata* and *Thinopyrum ponticum* (Tall wheatgrass) from Turkey, *Atriplex undulata* (wavy leaf saltbush) from Argentina and *Atriplex lentiformis* (quail bush) and *Distichlis spicata* from the USA.

It is established that halophytes can accumulate salt in their tissues to high concentrations. For example, saltbush grown under rangeland conditions has leaf ash concentrations of 13–27 % (Welch 1978; Hyder 1981) and when grown in saline soils can have leaf ash concentrations up to 39 % (Malcolm et al. 1988). These attributes have led some workers to suggest that halophytes could be grown to remove salt from the soil (Chaudhri et al. 1964). Despite this optimism, the rate of 'desalting' of problem soils by halophytes would be very slow, e.g. if a saline duplex soil has chloride content of 0.17 % in surface 2 m, the total would be about 86 Mg ha^{-1} (Moore 1998). Further if halophytic crops have productivities of 2 and 10 Mg ha^{-1} per year and shoot concentrations of salt of 25 % dry weight, even after 20 years, the salt concentrations in the soil would still be at 89 and 45 % of the initial concentrations, respectively. Usually the fodder shrubs like saltbush rarely produce more than 2 Mg ha^{-1} per year under unirrigated conditions (Barrett-Lennard et al. 1990), and the salt that accumulates in leaves is returned to the soil through leaf fall. Thus, the effects of growth of halophytes on soil salinity are likely to be minimal.

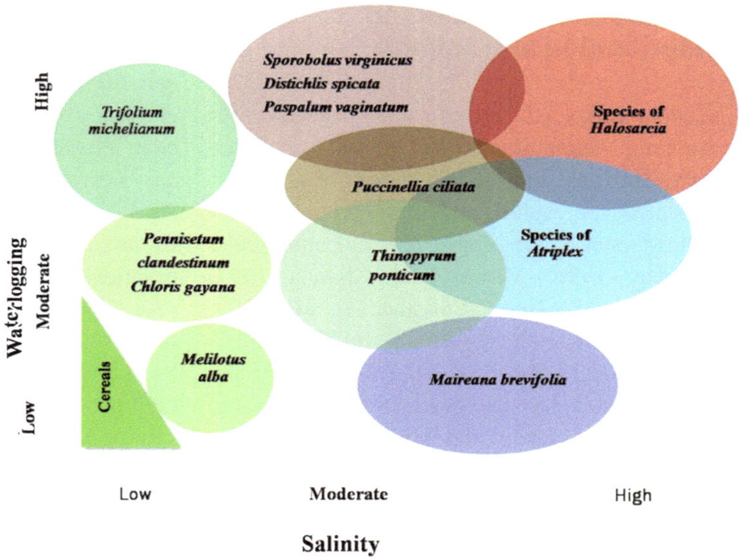

Fig. 2a Relative ranking of different saltland pasture species of Australia in salinity/waterlogging matrix (Modified from Barrett-Lennard 2003)

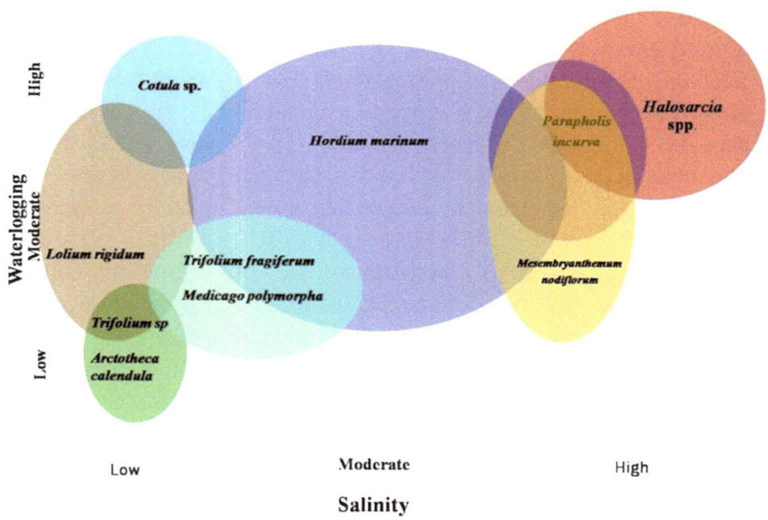

Fig. 2b Indicator species in the salinity/waterlogging matrix of saltland pastures in Australia (Modified from Barrett-Lennard 2003)

Although the long-term sustainability of tree plantations is questionable in salt-affected soils (Heuperman 1992; Thorburn 1996), it has been established that these plantations can lower water tables in soils of low to moderate salinity (Schofield et al. 1989; Schofield 1992). Various issues related to the control of water table with tree plantations have been detailed in chapter "Use of Tree Plantations in Water Table Drawdown and Combating Soil Salinity"; this section focuses primarily on some of the less accessible information obtained using stands of various species of saltbush (*Atriplex* sp.). An excellent stand of mixed saltbush species was established at three sites in North Stirlings and in the Pingaring area of Western Australia on a sandy-textured site having water table at a depth of about 1.2 m in summer. Five years later, the water table at the site had been drawn down to 2.5 m in summer, the saltbush appeared to be severely limited by moisture deficiency, and the stand had developed a self-sown understory of salt-sensitive subterranean clover (*Trifolium subterraneum*). The saltbush stand at this site had used about 25 mm of groundwater per year over the 5-year period (Barrett-Lennard and Galloway 1996). At another site with a saline shallow duplex (sand over clay) soils and a shallow (0.5–2.0 m) water table, regular measurements showed that although the water table responded to seasonal conditions, by the end of the second year, the water table at the site with the saltbush was about 0.5 m deeper than at the non-planted site (Fig. 3). Saltbush spacing trial at third site estimated the use of groundwater by plots of saltbush based on the accumulation of salt in the root zone of the plants (Malcolm et al. 1988; Barrett-Lennard and Malcolm 1999). The saline groundwater at the site was between 0.5 and 1.2 m below the surface for the duration of the experiment. Over a 2-year period, there was substantial increase in concentrations of chloride in the soil and in the salinity of the groundwater beneath the saltbush species. It was concluded that the increase in soil and groundwater salinity was primarily due to the use of groundwater by the saltbush (60–100 mm over 2 years).

The use of perennial vegetation (trees and perennial pastures) for lowering water table was reported in detail by George et al. (1999). Turner and Ward (2002) suggested that a combination of belts of trees and perennial pasture, such as saltbush (*Atriplex* spp.) and lucerne (*Medicago sativa*), can mitigate and even reverse waterlogging and secondary salinity while maintaining crop production at near-current level. Even small decrease in water table depth on salt-affected land in summer – the time of greatest evaporative demand – could decrease the accumulation of salt at the soil surface if saline water tables are less than 0.7 m deep, surface cover of the soil is generally low and accumulation of salt at the soil surface in summer is inevitable. However, if saline water table can be drawn down by 1–1.7 m using saltland pastures, the movement of salt to the soil surface in summer is decreased by at least 80 %, sufficient to change a site with a cover of patchy barley grass (*Hordeum marinum*) into one capable of growing ryegrass (*Lolium rigidum*) and balansa clover (*Trifolium michelianum*) or even a crop of barley (Barrett-Lennard 2003). Some of the experimental evidences and agroforestry approaches including silvopastoral systems have been dealt below.

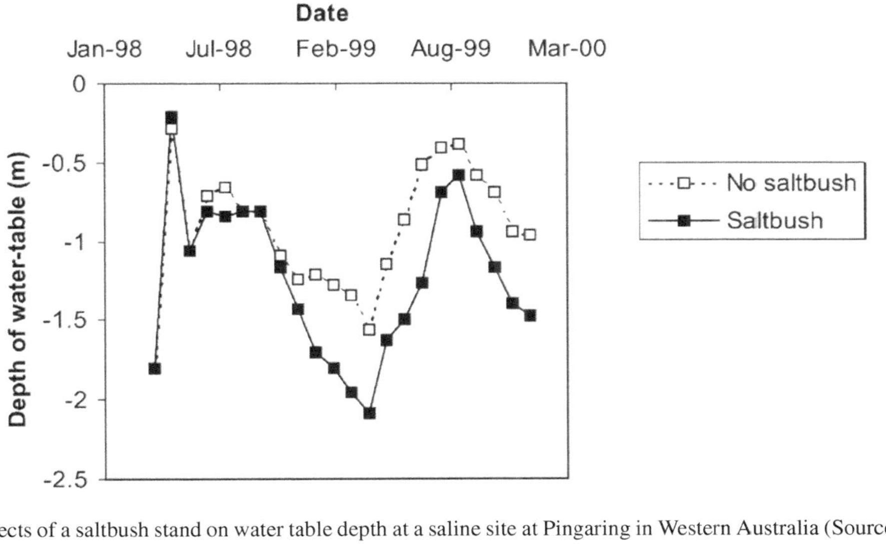

Fig. 3 Effects of a saltbush stand on water table depth at a saline site at Pingaring in Western Australia (Source: Barrett-Lennard 2002)

Suitability of Planting Techniques

In Western Australia, the process of planting the trees includes ripping, mounding, furrowing, weed control and using soil wetters (Malcolm 1989). Ripping is considered necessary on most soil types particularly clay, loam and duplex soils to allow good root penetration and their development in the subsoil. On duplex soils (having a clayey subsoil within 50 cm of the soil surface), which are common on the coastal plain, the clay subsoil should be penetrated to as great a depth as possible. Three parallel lines 0.5–1.0 m apart are ripped to at least 45 cm. Then especially on flat ground, a mound is made with a plough, road grader or three-point linkage blade on a tractor. Ripping and mounding the soil encourage leaching of salt by the rains before planting. The mound protects the seedling from flooding and helps in root growth (Negus 1988). Seedlings are planted into the niche in the ridgetop (in middle line), and the soil is pressed around the seedling to remove air pockets and fresh water is provided to the seedlings. Results are better if the site is cross ripped deeper on 1 m grid and the seedlings are planted where the lines intersected. By mounding the tree line, the trees can establish a root system above the water table allowing them to grow and develop over the first winter season. Where salinity is a serious problem, mounds should ideally be concave on the surface to collect rainfall to leach the salt to a lower level (Angell 1990). Furrow sowing followed by a thin band of surfactant sprayed in the base of the furrow will help in improving soil wetting for growth improvement of crops and pastures on water-repellent soils (Blackwell and Morrow 2006). Furrowing removes the worst of the non-wettable sand, provides weed control, makes a catchment for collecting rainfall and protects the seedlings from sand blasting. Preferably furrows should be at right angles to the prevailing summer winds. Soil wetters are used whenever furrowing is done.

The shrubs are also established on saltland by direct (niche) seeding. Niche seeders deposit saltbush fruits and a covering of vermiculite at 1–3 m intervals on a raised M-shaped mound. The shape of mound promotes leaching of salt from the soil around the fruits by rain. The vermiculite acts as mulch; it decreases evaporation, helps retain moisture around the fruits and reduces the movement of salt back into the seedbed. Elevating the seedbed above the surrounding soil reduces waterlogging. To spray the black paint or bitumen over the placement of fruits and vermiculite helps in stabilizing the placement of fruits and vermiculite (Barrrette-Lennard 2003).

Agroforestry Options

Tree-based land management strategies can contribute significantly to both the productive use of salt-affected land and minimizing the spread of salinity. Considerable scope exists in the choice of tree and shrub species for silvopastoral agroforestry schemes on salt-affected land under both rainfed and irrigated conditions. While many tree species may be unaffected by relatively low soil salt concentration (ECe up to 5 dS m^{-1}), their survival and growth will certainly be affected at higher salt concentrations. Saline drainage water, which can often have salt concentration of 10 dS m^{-1} or higher, would significantly reduce the growth and water use of pulpwood species such as *Eucalyptus grandis, E. globulus* and *E. camaldulensis*, particularly on heavy-textured soils. More salt-tolerant but less productive species such as *E. occidentalis, E. microtheca* and *Casuarina glauca* could be grown successfully. Based on observations and evaluation studies, moderately to highly salt-tolerant tree species (average root zone ECe of 15–40 dS m^{-1}) have been documented by Marcar et al. (1993), which include 38 species of *Eucalyptus*, 16 of *Melaleuca,* 13 of *Acacia* and three of *Casuarina.* Intraspecific variations between and within provenances for salt tolerance appears not to be significant in many species, but in a few species such as *E. camaldulensis, E. tereticornis* and *E. occidentalis,* the intraspecific variation is so wide that tolerance classification at specific level sometimes appears dubious. Researchers have exploited this variation and selected individual plus trees for high salt tolerance and developed clonal material for uniform growth and higher production. These include clones of *E. camaldulensis, Acacia ampliceps, A. auriculiformis, M. bracteata, M. lanceolata, M. thyoides, M. eleuterostachya, M. halmaturorum* and *C. glauca*.

For waterlogged salt-affected lands in Australia, Engel and Negus (1988) advocated to grow *Casuarina obesa* on highly saline bare areas; patchy barley grass (*Hordeum* sp.) along with trees such as *Eucalyptus occidentalis, E. sargentii* and *E. loxophleba* on moderately saline areas; thick barley grass and above species plus *E. platypus* var. *heterophylla, E. kondininensis, E. camaldulensis, E. rudis* and *E. spathulata* on slight saline areas; and sub-clover grass with species of *E. globulus* and *E. sideroxylon* on adjacent nonsaline areas. George and Frantom (1988) reported that some trees (mainly species of *Eucalyptus*) use from 10 to 100 l of water per day per tree and can reclaim or prevent sand plain seeps and can lower the water table, even where water salinity exceeds 18 dS m^{-1}. *Eucalyptus globulus, E. camaldulensis* and *E. cladocalyx* were ranked among most efficient species. It was reported by Angell (1990) that species such as *Casuarina obesa, Eucalyptus calycogona, E. grossa, E. macrandra* and *E. spathulata* are suitable for low rainfall (<500 mm); *Acacia saligna, E. astringens, E. campaspe, E. gracilis, E. griffithsii, E. occidentalis* and *E. salubris* are suitable for areas having rainfall >500 mm such as the Great Southern; and species such as *Acacia acuminata, A. saligna, E. campaspe, E. camaldulensis, E. concinna, E. gracilis, E. griffithii, E. incrassate, E. kondininensis, E. longicornis, E. loxophleba, E. macrandra* and *E. spathulata* are suitable for different wheat belt regions of Australia.

The use of trees to intercept seepage losses from irrigation supply was discussed by Webster (1984) and Sonogan and Patto (1985) with major emphasis on (i) the most suitable tree species for easy establishment, (ii) growth rates in waterlogged and salty conditions, and (iii) the impact on the water table. Among tested trees, Sargent's mallet (*Eucalyptus sargentii*) performed best from survival and growth point of view followed by *E. occidentalis* (swamp yate) and *E. kondininensis* (stocking gum). Along with these trees, perennial grass *Thinopyrum elongatum* (Tall wheatgrass) was grown in a mixed stand with *Medicago sativa* (lucerne), *Lolium perenne* (rye grass) and *Puccinellia ciliata*. These along with trees in a strip of about 50 m alongside channels in a relatively short period of 3 years could lower down water table effectively. Garrett (1993) and Bowman and Ruprecht (2000) while dealing with agroforestry for salinity control listed some useful tree species suitable for dryland salinity control. These include *Acacia implexa, A. pendula, Allocasuarina luehmannii, A. verticillata, Callistemon paludosa, Callitris columellaris, Casuarina cunninghamiana, C. glauca, Eucalyptus camaldulensis, E. microcarpa, E. occidentalis, E. polyanthemos, Melaleuca decussata, M. ericifolia, M. halmaturorum* and *M. lanceolata*. The very saline sites (with soil salinity above 18 dS m^{-1}) which do not support barley grass are best left ungrazed and be allowed to establish native salt-tolerant species. Areas having moderate salinity are suitable for raising *Atriplex* and *Casuarina glauca*. Saline areas that have a good cover of barley grass (<1 dS m^{-1}) are suitable for woodlots of *C. glauca*.

Malcolm (1993) mentioned three complementary aspects to salt-affected land management in low rainfall areas (375 mm). Those include catchment management, water control and revegetation of denuded areas. The low cost strategy involves avoiding grazing initially, introducing blue bush (*Maireana brevifolia*) and samphire (*Halosarcia* spp.) to colonize salt-affected land (spreading seeds), cropping nonsaline areas (preferably with barley), cultivating bare areas (especially if they are crusted) and allowing grazing carefully after complete establishment of saltbushes. Comprehensive revegetation involves selective sowing of seed mixtures to get a rapid, productive and well-adapted cover. It was recommended that a mixture of species (*Atriplex undulata, Maireana brevifolia, Halosarcia* spp.) be sown at a spacing of 3 m × 2 m after the sowing of *Puccinellia capillaris* grass as understory crop in the entire field. It is essential to allow planted species to become established before allowing grazing. In low recharge farming systems, tagasaste (*Chamaecytisus proliferus*) was introduced on deep sandy ridges of the farm to eliminate wind erosion and restrict groundwater recharge. Wavy leaf saltbush (*Atriplex undulata*), old man saltbush (*A. nummularia*), river saltbush (*A. amnicola*), quail brush (*A. lentiformis*) and others are suitable for low-lying salty flats and depressions, while perennial grasses such as Tall wheatgrass, Rhodes grass (*Chloris gayana*), *Phalaris aquatica* and *Puccinellia ciliata* in the highly saline areas are constituents of perennial pastures (Ryder et al. 2000). Thus, these species need to be cultivated in wider areas to establish suitable productive silvopastoral systems.

Barrett-Lennard (2003) described the establishment of pastures on waterlogged saltlands. On land of low productive potential (dominated by clay of the valley floors in lower rainfall areas and by shallow duplex soils and clays of the valley floors in higher rainfall areas), planting of pastures on these lands is unlikely to be economically feasible. These sites should be stabilized by controlling grazing and allowing natural regeneration of self-sown samphire (*Halosarcia* spp.). Land of moderate productive potential (dominated by duplex and gradational soils of lower salinity and waterlogging, salt accumulation in root zone may occur in clayey B-horizons, but not in leachable A-horizons) will be highly suited to the growth of saltland pastures. In winter-dominant rainfall regions with 300–400 mm rainfall, saltbush (*Atriplex*) species and small leaf bluebush (*Maireana brevifolia*) can be established by niche seeding or the planting of nursery-raised seedlings. These plants will have value as forages and will lower down the water table when grown with annual forage crops as understorey species. Perennial–annual partnerships are also likely to be profitable in higher rainfall (400–500 mm) areas where perennial component may be *Acacia* species and grasses such as *Puccinellia* and wheatgrass. Very high-tolerant trees may include *Casuarina* (*glauca, obesa*) and *Melaleuca* (*acuminata, eleuterostachya, glomerata, halmaturorum, lanceolata, lateriflora, thyoides*). Land of high productive potential dominated by sandplain seeps (areas where deep sands shelve out onto clays) may be planted with trees of lower tolerance to salinity and waterlogging such as species of *Eucalyptus* (*intertexta, microtheca, occidentalis, raveretiana, sargentii, socialis, striaticalyx*) as advocated by Moezel et al. (1991) and Barrett-Lennard (2003). Areas with shallow accessible groundwater of low salinity may be suited to dryland horticulture. Moore et al. (2006) advocated for perennial pastures for Western Australia because of their sustainability and profitability. The advantage of introducing lucerne in pasture (14 % of farm area) in central wheatbelt (annual rainfall 370 mm; soil deep sands, deep and shallow duplex soil with red and grey medium- to fine-textured on valley floors) was to get maximum profit, while increase in area up to 24 % could reduce groundwater recharge with farm profitability of 90 %. In the second case study, growing a mixture of perennials on the south coast (average annual rainfall 600 mm; soil-type deep sands, duplex soils and moderately deep sandy duplex soils), the pastures increased profit by manyfold.

Along with forage production, silvopastoral systems also help in the lowering down of the water table in low-lying pastures, as is evident from several examples for lowering of water table by saltland pastures. Bleby et al. (1997) reported that in Upper South East South Australia, in pasture predominated by Tall wheatgrass, water tables decreased from 0.1 to 1.4 m over 36 days in the summer when the evapotranspiration was 3.6 mm per day. In nearby areas with a cover of barley grass, water tables decreased from 0 to 1.0 m (evapotranspiration 2.8 mm per day). Barrett-Lennard and Malcolm (1999) while monitoring salt accumulation in root zone of various saltbush species (*Atriplex amnicola, A. bunburyana, A. paludosa, A. undulata* and *A. vesicaria*) over 2 years reported that evapotranspiration accounted for 60–100 mm of groundwater plus 460 mm of rainfall. The amount of groundwater used was proportional to the weight of saltbush leaf per unit soil surface area. With bladder saltbush (*A. vesicaria*), the smallest species in the experiment, evapotranspiration in summer was between 1.3 and 3.3 mm per day. With saltbush stand in North Stirlings area with initial water table in summer at 1.2 m, the drawn down was to 2.5 m in summer in 5 years (Barrett-Lennard 2003). Thus, pastures even without tall trees may also play important role in drawing down of water tables in waterlogged areas.

Valley floors within parts of the high rainfall (>500 mm) areas of the southwest of Western Australia are degraded through waterlogging and secondary salinity. Angell et al. (1994) reported some suitable salt-tolerant trees and pasture species for such areas. For highly saline region (EC 15–20 dS m^{-1}), trees such as *Casuarina obesa* and *Melaleuca cuticularis* with pasture grasses like *Puccinellia*; for moderate salinity (EC 10–15 dS m^{-1}), trees such as *Eucalyptus sargentii, E. occidentalis, E. loxphleba, E. spathulata, E. camaldulensis, E. camaldulensis, Acacia saligna* and *Casuarina cunninghamiana* associated with pasture grasses such as *Puccinellia ciliata* (Puccinellia), *Thinopyrum ponticum* (Tall wheatgrass), *Chloris gayana* (Rhodes grass) and *Phalaris* spp.; while for mildly saline (EC 5–10 dS m^{-1}) regions, pastures consisting of *Chloris gayana, Phalaris* sp., *Secale cereale, Thinopyrum ponticum, Trifolium michelianum* and *T. resupinatum* may be planted successfully with trees such as *Eucalyptus camaldulensis, E. rudis, E. robusta, E. cladocalyx, E. melliodora* and *E. microcarpa.*

Travis and Heuperman (1994) observed the effects of tree lines [*Eucalyptus globulus* (blue gum), *E. viminalis* (manna gum), *Melaleuca styphelioides* (prickly paper bark) and *M. ericifolia* (swamp paper bark)] in irrigated pastures [composed of *Lolium perenne* (rye grass), *Paspalum dilatatum* and *Trifolium repens* (white clover)] on lowering of water table. At one site, tree line section also consisted of *Acacia melanoxylon* (black wood) and small *Salix alba* (white willow). Under light soil conditions, single tree lines had no measurable impact on water table levels, while on heavier soils, there was clear localized impact. There was higher salinity (EC values) of water table under the trees on heavy soil-type site than on the lighter soil type, reflecting the higher degree of salt accumulation underneath trees growing on heavy soils. Smedema (1997), however, suggested that tree plantation could be considered for waterlogged landscape depressions and canal seepage interceptions and could be applied in parallel field drainage arrangements as an alternative to conventional field drainage systems. In Australia, it is now widely accepted that in discharge situations, enhanced evapotranspiration plantation sites will eventually succumb to salinity, unless some form of conventional drainage is installed to control salt balance to vegetation's root zone by removal of saline drainage effluent (Heuperman 2000; Heuperman et al. 2002).

Experiences from the Indian Subcontinent

Because of its fast growth rate, favourable wood properties and carbon sequestration, *Eucalyptus* is a widely planted tree species. Field trials in Indian subcontinent have shown wide variation among provenances of trees for their tolerance to frost, waterlogging and salinity; wood properties; amount of lignotubers; regeneration via coppice formation; and growth rate (Roy Chowdhury et al. 2011; Dagar 2014). Some of the experiences to rehabilitate waterlogged saline soils with tree plantations are given below.

Suitable Planting Techniques

Among the techniques, planting on high ridges was often considered beneficial for waterlogged saline soils where the most salts accumulate in surface soil and decrease with depth down the water table. Therefore, to take the advantage of low salinity and better soil moisture regimes in sub-surface layers, Tomar and Gupta (1984–1994) and Tomar (1997) tried the sub-surface planting of saplings (at a depth of 30 cm below surface) and compared it with ridge planting (40 cm high). Substantially higher salts accumulated in the ridges that resulted in poor survival and sapling growth. The greater the surface area of the ridge, the more salts accumulated in the surface 1 m root zone of ridge planted trees. Difficulty of conserving rainwater on the ridge tops and the presence of salts causing higher susceptibility to soil erosion were the other disadvantages encountered with ridge planting. In contrast, under the sub-surface planting method, roots were encountering a milder saline transmission zone and were meeting most of their water requirement from the phreatic zone. The performance of trees was better when planted with sub-surface method, but the need for spot irrigation was the main problem. This method was then improved upon by planting the saplings in the sole of furrow (60 cm wide and 20 cm deep), which was subsequently used for irrigating the tree saplings (Fig. 4). The rainwater to be accumulated in the furrows will take care of leaching down of the salts of root zone.

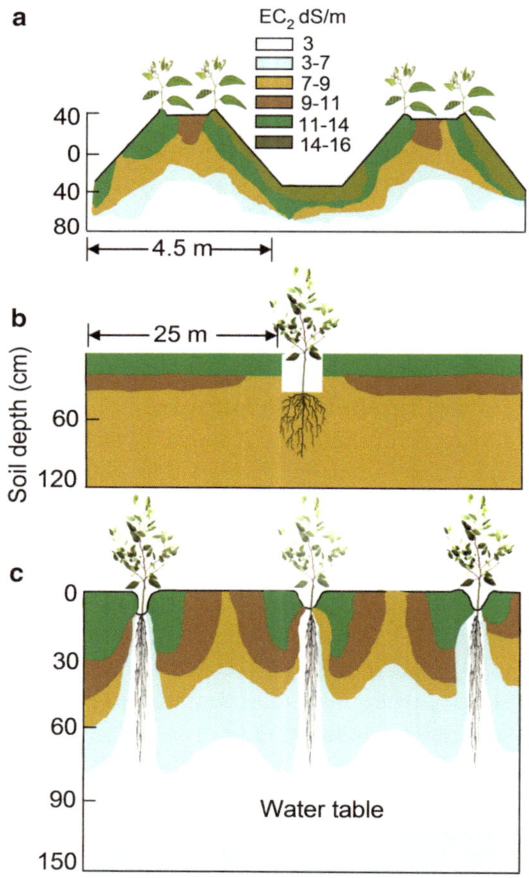

Fig. 4 Salt distribution patterns under (**a**) ridge trench, (**b**) sub-surface and (**c**) furrow planting methods in waterlogged saline soils (Source: Modified from Tomar et al. (1998) by Dagar (2014))

Salt-tolerant Plants/Halophytes

Since the pioneer attempts by Boyko (1968), Mudie (1974) and associates (Mudie et al. 1972) to cultivate halophytes with water of sea salinity, many salt-tolerant species have been evaluated for food, forage and other economic uses. It was documented by Aronson (1989) that about 1560 species in 550 genera and 117 families from saline environments have scope of economic exploitation. Dagar and Singh (2007) also reported 1140 salt-tolerant vascular species distributed in India under 541 genera and 131 families, and a number of potential woody and herbaceous halophytic genera (including coastal) of economic importance and suitable for agroforestry have been identified (Dagar 2003, 2005a, b, 2014, 2015) which include *Acacia, Arthrocnemum, Atriplex, Avicennia, Azadirachta, Borassus, Brachiaria, Bruguera, Carissa, Cassia, Casuarina, Ceriops, Chloris, Coccoloba, Cressa, Crithmum, Dichanthium, Distichlis, Eucalyptus, Juncus, Kochia, Leucaena, Leptochloa, Limonium, Lumnitzera, Maireana, Melia, Pongamia, Panicum, Paspalum, Pennisetum, Porterasia, Prosopis, Rhizophora, Salicornia, Salvadora, Seteria, Simmondsia, Sonneratia, Spergularia, Sporobolus, Suaeda, Syzygium, Tamarix, Taxodium, Terminalia, Vetiveria, Xylocarpus, Ziziphus* and *Zostera* to name a few.

About 67 salt-tolerant species (41 dicots and 26 monocots) were being grazed to various extents by sheep, goats and camels in Pakistan (Qureshi et al. 1993). Among these, 14 were legumes, 25 were grasses, and the rest belonged to other families like Amaranthaceae, Chenopodiaceae, Salvadoraceae, Tamaricaceae and Rhamnaceae. Among the salt-tolerant fodder species, *Prosopis cineraria, Acacia nilotica, Leucaena leucocephala, Sesbania* spp., *Atriplex* spp., *Leptochloa fusca, Echinochloa colonum, E. crus-galli* and *Paspalum* spp. are important, found in waterlogged salt-affected lands. Other halophytes such as *Arthrocnemum indicum, Salsola drummondii, Bienertia cycloptera, Zygophyllum simplex, Tamarix articulata, Salvadora persica* and *S. oleoides* are highly palatable to camels and are also found in India.

Based on long-term experiment on saline sodic soils in Pakistan, Qureshi et al. (1993) reported the average biomass produced after 7 years of growth varied from 100 kg per plant in *Prosopis cineraria* to 400 kg in *Eucalyptus camaldulensis,* and species like *E. camaldulensis, A. nilotica, A. lebbeck* and *L. leucocephala* were the maximum yielders of timber biomass. In another experiment on high salinity soil (ECe 38.5 dS m^{-1}), the performance of *E. camaldulensis, A. nilotica, Tamarix articulata* and *Dalbergia sissoo* was satisfactory, while *Terminalia arjuna* and *L. leucocephala* were poor performers. The survival and establishment of three species (*Azadirachta indica, Casuarina glauca* and *Grewia asiatica*) were quite good, but their growth got stunted after the cessation of irrigation. Tomar (1997) and Tomar et al. (1998) reported from experiments in semiarid regions of northwest India (rainfall 630 mm; ECe 36.4 dS m^{-1}) that biomass production after 9 years of plantation by *P. juliflora* and *Casuarina glauca* was highest among tested species (98 and 96 Mg ha^{-1}), followed by *Acacia nilotica* (67 Mg ha^{-1}), *A. tortilis* (41 Mg ha^{-1}), *Casuarina obesa* (38 Mg ha^{-1}), *Leucaena leucocephala* (30 Mg ha^{-1}) and *C. equisetifolia* and *Eucalyptus camaldulensis* (28 Mg ha^{-1} each) when planted with sub-surface or furrow technique showing their potential for waterlogged saline soils. Species like *Prosopis juliflora, Tamarix articulata, T. traupii, Acacia farnesiana, Parkinsonia aculeata, Salvadora persica* and *S. oleoides* were most tolerant to waterlogged saline soil and could be raised successfully up to salinity levels of ECe 30–40 dS m^{-1}, and species like *A. nilotica, A. tortilis, A. pennatula, Casuarina glauca, C. obesa, C. equisetifolia, Callistemon lanceolatus, Eucalyptus camaldulensis, Feronia limonia, Leucaena leucocephala* and *Ziziphus mauritiana* could be grown on sites with ECe 10–20 dS m^{-1}. Some of the important tree and shrub species suitable for saline soils have been shown in Table 1.

Table 1 Relative salt tolerance by some tree and shrub species for saline soils

ECe (dS m^{-1})	Tree species
>35 (Highly tolerant)	*Prosopis juliflora, Salvadora persica, S. oleoides, Acacia ampliceps, Tamarix ericoides, T. troupii, Arthrocnemum indicum,* species of *Atriplex, Halosarcia* and *Maireana* and all mangroves in tidal zone[a]
25–35 (Tolerant)	*Tamarix articulata, Acacia farnesiana, A. saligna, Parkinsonia aculeata, Prosopis cineraria, P. alba, Melaleuca* spp.
	([a]Associate mangrove species – *Pongamia pinnata, Terminalia catappa, Thespesia populnea, Clerodendron inerme, Manilkara littoralis, Calophyllum inophyllum, Pandanus* spp., etc.)
15–25 (Moderately tolerant)	*Casuarina (glauca, obesa, equisetifolia), Eucalyptus (occidentalis, sargentii, loxophleba), Acacia tortilis, A. nilotica, A. auriculiformis, Callistemon lanceolata, Pongamia pinnata, Eucalyptus camaldulensis, Crescentia alata, Albizia lebbeck*
10–15 (Less tolerant)	*Casuarina cunninghamiana, Eucalyptus tereticornis, E. globules, E. cladocalyx, Acacia catechu, A. eburnea, A. leucocephala, Terminalia arjuna, Samanea saman, Albizia procera, Borassus flabellifer, Prosopis cineraria, Azadirachta indica, Dendrocalamus strictus, Butea monosperma, Cassia siamea, Feronia limonia, Leucaena leucocephala, Tamarindus indica, Guazuma ulmifolia, Ailanthus excelsa, Dichrostachys cinerea, Balanites roxburghii, Maytenus emarginata, Dalbergia sissoo, Salix babylonica*

Source: Compiled from various resources, mainly Tomar et al. (1998) and Dagar (2014)

Survival will also depend on agro-ecological conditions

[a]Suitable only for coastal regions; for detailed species, see chapter "Agroforestry to Rehabilitate the Indian Coastal Saline Areas"

Agroforestry Systems on Salt-affected Lands

In Indian subcontinent, during the nineteenth century, salt-affected lands were stocked with trees in an effort to raise profitable plantations. These trees showed many beneficial side effects such as soil amelioration, environmental benefits and checking water table rise in waterlogged areas. Nevertheless, it was only during the second half of the twentieth century that tree planting became part of agricultural production and concerted reclamation strategies. Several salt-tolerant tree species have been identified in India and Pakistan (Jain et al. 1985; Sheikh 1987; Singh 1989; Aslam et al. 1993), which include highly salt tolerant species (*Prosopis juliflora, P. chilensis, P. alba, Acacia ampliceps, Salvadora persica, Tamarix articulata*) or moderately salt tolerant species (*Acacia nilotica, A. tortilis, A. cambagei, Casuarina equisetifolia, Eucalyptus camaldulensis, E. tereticornis, E. microtheca, Azadirachta indica*). Most of these results were based on short-term observations. Among shrubs, species of *Atriplex, Kochia, Suaeda, Haloxylon* and *Salsola* were reported to be highly salt tolerant, and among grasses, *Leptochloa fusca* was adjudged most tolerant to both salinity and waterlogging (Malik et al. 1986). Many species of *Paspalum, Coix, Echinochloa, Panicum, Phragmites, Brachiaria, Panicum, Atriplex, Salicornia, Juncus, Typha, Cyperus, Scirpus,* etc. have potential in waterlogged salt-affected lands. Zafar Iqbal (1990) advocated that *Eucalyptus camaldulensis* was most suitable species for waterlogged saline soils, while Altaf-Hussain and Pazir-Gul (1991), Sahibzada (1993) and Marcar and Crawford (1996) recommended local trees such as *Tamarix articulata, Acacia nilotica, A. modesta* and some exotics such as *Atriplex ampliceps, A. stenophylla* and *Casuarina obesa* for waterlogged saline conditions in Pakistan. Ahmad (1988), Ahmad and Ismail (1993) and Rashid et al. (1993) reported tolerance of different tree and forage species studied at the Nuclear Institute for Agriculture and Biology in Pakistan. Species such as *Atriplex amnicola, A. lentiformis, A. undulata, Acacia cambagei* and *Leptochloa fusca* could tolerate salinity of 20–30 dS m^{-1} (yield reduction up to 50 %), while many others such as *Sesbania aculeata, Leucaena leucocephala, Medicago sativa, Lolium multiflorum, Echinochloa colonum* and *Panicum maximum* could tolerate salinity up to ECe 10–12 dS m^{-1}.

Grassland/Pastoral Systems

In waterlogged saline areas, grasses such as *Leptochloa fusca, Aeluropus lagopoides, Eragrostis* spp., *Sporobolus helvolus, S. marginatus, S. spicatus, S. tremulus, Chloris barbata, C. gayana, C. virgata, Cynodon dactylon, Dichanthium caricosum, Echinochloa colonum, E. crus-galli, Panicum* spp., *Paspalum conjugatum, P. distichum, P. plicatulum, P. vaginatum, Brachiaria mutica, Thuarea involuta* and *Vetiveria zizanioides* can successfully be cultivated along with salt-tolerant trees and bushes such as *Atriplex, Kochia* and *Salvadora* constituting a viable and sustainable silvopastoral system to sustain live stock productivity. *Leptochloa fusca* grass is known to have special advantages in terms of both forage production and its role in soil amelioration. An associative nitrogen-fixing bacterium, *Azoarcus*, occurs as an endophyte in the roots of *L. fusca* grass – a pioneer species of alkali soils that yields 9–12 Mg ha^{-1} of dry biomass without application of any fertilizer; nearly half of the plant N of 90–120 kg ha^{-1} is derived from associative fixation (Malik and Zafar 1984; Malik et al. 1986). Among other grasses, *Aeluropus lagopoides, Sporobolus helvolus, Cynodon dactylon, Brachiaria ramosa, B. mutica, Paspalum* spp., *Echinochloa colonum, E. crus-galli, Dichanthium annulatum, Digitaria ciliaris, Vetiveria zizanioides* and *Eragrostis* spp. are important which are tolerant to both salinity and waterlogging. Species of *Atriplex, Kochia, Suaeda, Salsola, Haloxylon* and *Salvadora* are prominent forage shrubs of saline regions and relished by camel, sheep and goats.

Most of the area in the Rann of Kutch along Gujarat in India and the adjacent area in Pakistan is highly saline, and because of low rainfall and high evapotranspiration, its effects are more severe. *Prosopis juliflora, Salvadora persica, Tamarix articulata, T. troupii, Arthrocnemum indicum* and many other halophytes are the naturally growing species on these lands but with stunted growth. A silvopastoral system may be developed incorporating suitable salt-tolerant forages such as species of *Atriplex, Kochia indica, Cressa cretica, Aeluropus lagopoides, Cynodon dactylon, Dactyloctenium sindicum, Bothriochloa pertusa, Dichanthium annulatum, Leptochloa fusca* and *Sporobolus helvolus*. High value trees such as *Salvadora persica, Pongamia pinnata, Capparis decidua, Carissa carandas* and *Terminalia catappa* and fire wood trees like *P. juliflora, Acacia nilotica* and *Casuarina glauca* may be raised in furrows and above-mentioned forage species in interspaces. Oil-yielding *Salicornia bigelovii* is being grown at many places as industrial crop. Oil-yielding saltbush *Salvadora persica* performed well both in dry and waterlogged situations in saline soils. In one study, *S. persica*-based silvopastoral system was developed with *Leptochloa fusca, Eragrostis* sp. and *Dichanthium annulatum* forage grasses on clay loam saline vertisol (clay 40 %, silt 31 %, sand 29 %; pH ranging from 7.2 to 8.9; ECe from 25 to 70 dS m^{-1}) in Gujarat, India. The underground water was 0.5–2 m from the surface with EC_{iw} ranging from 55 to 60 dS m^{-1}. Based on growth pattern in terms of height, canopy spread and seed yield, a planting density of 4 m×4 m was found as optimum for *S. persica*. During the fourth year, the seed yield of *Salvadora persica* ranged from 1.84 to 2.65 Mg ha^{-1} with oil contents ranging from 576 to 868 kg ha^{-1} at different salinity levels (Rao et al. 2003). These grasses (*L. fusca, D. annulatum* and *Eragrostis* sp.) when planted on 45 cm high ridges could produce 3.17, 1.85 and 1.09 Mg ha^{-1} forage, respectively. When planted in furrows, these could yield 3.75, 1.76 and 0.54 Mg ha^{-1}, respectively, showing their potential for these highly degraded lands.

Afforestation of Coastal Saline Areas

On the coastlines wherever sulphur-containing sediments accumulate in tidal marshes or swamps or mangrove ecosystems, acid sulphate soils are formed. Rice is a major crop in the tropical and subtropical areas where acid sulphate soils are formed. However, under such conditions, it also suffers from an excess of water-soluble iron and particularly of aluminium and toxic effect of these is a consequence of the strongly acidic pH. The application of lime mitigates the adverse and harmful effects on plants, but it is a costly preposition. Therefore, alternative land-use systems need to be evolved. From the management purposes, the salinity-related problematic soils of the coastal areas may be classified (Dagar et al. 2014) as (i) land impregnated with high salinity and flooded with seawater, (ii) acid sulphate soils, (iii) land impregnated with high salinity and waterlogging but not flooded with seawater, (iv) land with low salinity and shallow water table (at 0.5–4.0 m depth) with good quality water but saline water beneath and (v) waterlogged and saline soils caused by seepage in canal command areas. Suitable agroforestry systems offer scope for increasing the income and employment generation for small farmers, meeting the local needs of fodder and fuel, conserving biodiversity, improving the coastal environment, protecting the soil from erosion and creating environment for the wildlife. Nair and Sreedharan (1986), Dagar (1991, 1994, 1995a, b, 1996, 2000, 2009, 2012), Dagar and Tomar (1998), Pandey et al. (2007), Kumar and Kunhamu (2011) and Dagar et al. (2014) have explained several agroforestry practices for coastal and island regions.

The areas lying closer to the sea are flooded regularly with seawater and, therefore, have high salinity. The tidal areas protected against high wind velocity and waves of high intensity (as in case of many creeks, lagoons and estuaries) form a suitable situation for mangroves, which are restricted in their distribution in tropics and subtropics (~between 32° N and 38° S) and encountered along the sea coasts in backwater creeks and river estuaries between approximately mid-tide to the extreme high water mark along the coasts. A wide variety of plant species are found in mangrove habitats, but of the recognized 110 species, only 54 species in 20 genera belonging to 16 families constitute the 'true mangroves' that occur exclusively in mangrove habitats and rarely elsewhere (Hogarth 1999). However, International Union for Conservation of Nature and Natural Resources (IUCN) documented 60 exclusive mangrove species (13 in the western hemisphere and 51 in the eastern hemisphere) and 23 non-exclusive mangrove species distributed over 16.9 Mha in the world (Saenger et al. 1983). Two species (*Rhizophora* x *annamalayana* and *Heritiera kanikensis*) have been documented exclusively from India (Banerjee et al. 1989; EISC 2002), making total 62 species of exclusive mangroves. The first 'World Mangrove Atlas' provided realistic assessment of existing mangrove area to be 18.23 Mha (Spalding et al. 1997). With increasing urbanization, particularly in developing countries, the mangrove habitats can judiciously be explored for meeting the demands for food, fuelwood, thatching material and increasing the productivity of coastal water for marine resources including fish and shrimp production with little disturbance to the mangrove forests (Dagar 1982, 2005a, b, 2008; Dagar et al. 1991, 1993). The coastal degraded saline areas frequently flooded with seawater may be brought under mangroves and other useful halophytes such as oil-yielding *Salvadora persica* and *Salicornia bigelovii* to augment the economy at many fronts.

Many types of fish, shrimp, crustaceans and other coastal animals inhabit mangroves and associated ecosystems, which are also utilized as food by coastal population. Besides these animals, radicles, fruits and tubers of mangroves and associated species are consumed as food items, and in many localities, people collect honey. *Nypa fruticans* is widely used for preparing alcoholic beverage. In many coastal areas where mangroves occur sporadically and there is scarcity of fodder, cattle, goats and camel are found browsing on mangroves and associate species such as *Avicennia, Bruguiera, Ceriops, Cynomitra, Ficus, Hibiscus, Pongamia, Rhizophora, Salvadora, Sonneratia, Terminalia,* etc.; the foliage of these species is also collected to feed the animals. Many halophytes and grasses are found in the Rann of Kutch area on which animals graze and these are harvested for fodder on regular bases and more palatable species can be part of well-designed silvopastoral system irrigating with saline water. Thus, mangroves are one of the most productive ecosystems and may be explored for livelihood security of coastal population and also ecological services. More details about coastal agroforestry-related systems with special reference to India are discussed in chapter "Agroforestry to Rehabilitate the Indian Coastal Saline Areas".

Experiences from Other Areas Affected with Salinity

There exist more than 250 potential staple halophyte crops in the world (Yensen et al. 1988). The question then is not if there are potential halophyte crops, but which will meet the needs of particular area and which can be grown with an economic worth. The distribution and adaptability of a species to saline and waterlogged habitats of different regions along with its economical utilization will make a species more acceptable. In many countries, halophytes have been successfully grown on saline fields to provide animal fodder (Malcolm 1982; O'Leary 1984) and have the potential for rehabilitation and even reclamation of these sites (Yensen and Yensen 1987; Watson 1987). Yuvaniyama and Arunin (1993) reported the potential of three halophytic grass species on saline soils in Thailand: *Sporobolus virginicus* have two forms suitable for high salinity and waterlogged conditions and had the highest Na/K ratio (26.5); *Spartina patens* are suitable for salt marshes and could yield 7 Mg ha^{-1} on saline soils of Egypt, with Na/K ratio of 11.1; and *Distichlis spicata* was relatively less tolerant than other two species with Na/K ratio 1.2 only. Ahmad and Ismail (1993) reported that the salt-tolerant forage and fuelwood species include *Prosopis juliflora, Cajanus cajan, Tamarix senegalensis, T. articulata, Parkinsonia aculeata, Acacia linarioides, A. nilotica, A. tortilis, Salsola passerina, Haloxylon aphyllum* and many others from woody species and various shrub and grass species.

Nearly 20 years of work with varieties of *Distichlis* grass from around the world could resulted in a number of useful cultivars, most notably a grain crop trademarked 'Wild Wheat Grain', a forage grass called 'NyPa Forage', a turf grass called 'NyPa Turf' and a reclamation grass called 'NyPa Reclamation Saltgrass' (Yensen and Bedell 1993). Malcolm (1993) and Barrett-Lennard (2003) reported that samphires (*Halosarcia pergranulata, H. lepidosperma* and *H. indica* subsp. *bideris*) and bluebush (*Maireana brevifolia*) are a group of highly salt-tolerant succulent perennial shrubs, which could be grown on waterlogged saltlands in Australia. *H. pergranulata* contains about 14 % crude protein on oven-dry basis and is better suited to sheep grazing.

Halophytes have been considered as potential crops for reuse of saline drainage water in the western portion of the San Joaquin Valley of California. Six halophytic species (*Salicornia bigelovii, Atriplex lentiformis, Distichlis spicata, Spartina gracilis, Allenrolfea occidentalis* and *Bassia hyssopifolia*) were cultivated under long-term irrigation with saline drainage water by Diaz et al. (2003). The results indicated that all species grew well under high saline-sodic soil conditions (average ECe 28.6 dS m^{-1}; SAR 39.4) with average standing biomass ranging between 3.8 and 17.4 Mg ha^{-1} dry matter depending on species. The cultivars of salt-tolerant forages (alfalfa, *Medicago sativa*; narrow leaf trefoil, *Lotus glaber*; broadleaf trefoil, *L. uliginosus*; Tall wheatgrass, *Agropyron elongatum*; alkali sacaton, *Sporobolus airoides*; kikuyu grass, *Pennisetum clandestinum*; paspalum grass, *Paspalum vaginatum*; and Bermuda grass, *Cynodon dactylon*) utilised for sequential water reuse systems were also evaluated for their biomass production and potential implications for ruminant mineral nutrition (Grattan et al. 2004a, b) and plant-ion relations (Grieve et al. 2004). The results were quite encouraging.

Zohar and Schiller (1998) tested the bio-drainage approach in Israel at five different waterlogged and salinity-affected plots representing a broad range of environmental conditions, including rainfall gradient, subsurface hydrological regime and soil salinity status. Ground water table under *E. camaldulensis* declined constantly during the 5-year monitoring period and dropped to more than 3 m below the soil surface. At the age of 3.5 years, the transpiration on summer days was between 4.5 and 5.1 mm day^{-1}. Excluding rainfall, the trees consumed 932 mm directly from underlying groundwater, thus depressing it by 1.2 m. The economic viability of growing *Eucalyptus* on abandoned lands in Israel without irrigation is considered a viable agricultural alternative that could bring higher returns than wheat, corn or cotton (Heuperman et al. 2002).

Experiences on Afforestation/Agroforestry of Alkali/Sodic Soils

The sodic soils represent about 52 % of the total salt-affected soils in the world. These soils have both high pH (>8.2) and sodicity (ESP>15) and containing soluble carbonates and bicarbonates of sodium such that $Na/Cl+SO_4>1$ (Gupta and Abrol 1990). These soils have a compact hard subsurface layer or a caliche (calcite) bed (of nodulated or amorphous $CaCO_3$) in lower depths, which imposes physical impediment to root penetration and correspondingly poor aeration when wet (due to dispersion of clay colloids by sodium); nutrition imbalances including deficiencies of zinc, calcium and magnesium (due to high pH); and toxicity of specifications (e.g. sodium and boron). Unlike soil reclamation for arable crops, where only plough layer is sought to be improved in the first instance, deep-rooted trees require reclamation of the soil to lower depths. The planting techniques should further ensure efficient utilization of rainwater and leaching of reaction products after interaction of amendments and help to root development in the soil profile and soil structural improvement for increased water retention to encourage rapid root penetration in the vertical rather than horizontal direction and minimize direct sodium toxicity hazards. Keeping this in view, several attempts were made particularly in India to develop suitable techniques for planting trees on such lands. Several forest and fruit trees, grasses, medicinal and aromatic plants and nonconventional and arable species were evaluated for their tolerance to sodicity and successful and most suitable species were identified for agroforestry.

Planting Techniques

The viable planting technique for sodic lands must ensure efficient utilization of rainwater and leaching of reaction products after interaction of amendments and help to root development in the soil profile and soil structural improvement for increased water retention to encourage rapid root penetration in the vertical rather than horizontal direction and minimize direct sodium toxicity hazards. In the past, planting methods like pits and trenches of various shapes and sizes were used for raising trees in alkali soils (Khan and Yadav 1962; Pande 1967; Yadav 1972; Yadav et al. 1975), with some intrinsic drawbacks like higher amendment needs, laborious and time-consuming operation, leaving the $CaCO_3$ layer located at about 0.5–0.7 m depth untreated and difficulties in preparing pits of larger dimension in alkali soils. The difficulties were overcome through the development of tree planting technique termed as the auger hole planting technique (Sandhu and Abrol 1981; Gill 1985; Singh and Gill 1992; Singh et al. 1993; Dagar et al. 2001b; Singh and Dagar 2005). To pierce the hard calcite layer, the auger is mounted on a tractor and used for making holes of dimensions 20–25 cm diameter and 1.2–1.8 deep. This technique recognizes that in trees, owing to their deep root systems, management of the root zone by modifying the soil environment to greater soil depths using a limited quantity of amendments (3–5 kg gypsum + 3 kg FYM + 20–30 g zinc sulphate + some insecticide per auger hole) has a vital role to play in terms of success in sapling establishment, cost of plantation and practical adaptability.

Water stagnation during rainy season is a common phenomenon. Many forests and fruit tree species can be raised on highly alkali soil, but some of the fruit trees like pomegranate (*Punica granatum*) and *bael* (*Aegle marmelos*) are unable to tolerate water stagnation. Raised and sunken bed technique of agroforestry was tested for such situations (Dagar et al. 2001a). After refilling the auger holes with soil mixture in a levelled field as mentioned in above-mentioned technique, the auger holes are marked with sticks. Parallel bunds, each of 1–2 m height and 1–2 m width, are then constructed leaving 4–5 m space between them taking soil from interspaces. The seedlings are raised on the middle of bund at marked places, and small rings are made around seedlings for initial irrigation. The interspaces are cultivated growing water-loving crops such as *kallar* grass (*Leptochloa fusca*) or rice (salt-tolerant varieties) during rainy season and Egyptian clover (*Trifolium alexandrinum*) or salt-tolerant varieties of wheat during winter. In rainfed conditions, moisture is conserved in sunken beds for raising low-water-requiring crops.

Afforestation Species

During the last three decades, numbers of studies were conducted in Indian subcontinent to establish positive benefits of the recent planting techniques and to identify most suitable species for alkali soils. Chaturvedi (1984) reported that a 30 % reduction in biomass was observed at pH 10 as compared to pH 7 in tree species like *Acacia nilotica, Terminalia arjuna* and *Pongamia pinnata,* but at pH 9.5 *Eucalyptus tereticornis, Casuarina equisetifolia* and *Acacia nilotica* could grow well in soil with ESP 30.6, whereas *Pongamia pinnata* and *Dalbergia sissoo* survived only up to pH 9.5 and ESP of 15.2 (Yadav and Singh 1986). Based on the performance of tree saplings planted in soils of different pH (7–12), relative tolerance was reported (Singh et al. 1987) in the following order: *Prosopis juliflora* > *A. nilotica* > *Haplophragma adenophyllum* > *Albizia lebbeck* > *Syzygium cuminii* (a fruit tree). Results of series of experiments conducted (Gill 1985; Gill and Abrol 1993; Singh et al. 1993; Das and Itnal 1994; Singh 1995; Dagar et al. 2001a, b; Singh and Dagar 2005; Dagar 2014) on high pH (~10) soil showed that quite good number of

Table 2 Relative tolerance of forest and fruit tree species to soil sodicity in India

pH_2 (0–1.2 m)	Fuelwood/fodder/timber species	Fruit tree species
>10	*Prosopis juliflora, Acacia nilotica, Tamarix articulata*	Not recommended
9.6–10.0	*Eucalyptus tereticornis, Capparis decidua, Pithecellobium dulce, Prosopis alba, P. cineraria, Casuarina equisetifolia*[a], *Salvadora persica, S. oleoides, Terminalia arjuna*	*Carissa carandas, Psidium guajava, Ziziphus mauritiana, Emblica officinalis*[a]
9.1–9.5	*Cordia rothii, Albizia lebbeck, Cassia siamea, Pongamia pinnata, Sesbania sesban, Parkinsonia aculeata, Dalbergia sissoo, Kigelia pinnata, Butea monosperma*	*Punica granatum*[b], *Phoenix dactylifera, Achras zapota*[a], *Tamarindus indica*[a], *Syzygium cuminii, Feronia limonia*
8.2–9.0	*Grevillea robusta, Azadirachta indica, Melia azedarach, Leucaena leucocephala, Hardwickia binata, Moringa oleifera, Populus deltoides, Tectona grandis*	*Grewia asiatica, Aegle marmelos*[b], *Prunus persica, Pyrus communis, Mangifera indica, Morus alba, Ficus* spp., *Sapindus laurifolius, Vitis vinifera*

Compiled from various sources by Dagar (2014)
[a]Frost sensitive
[b]Does not stand water stagnation, may be raised on bunds

forest and fruit trees can successfully be grown on sodic soils (Table 2) to produce suitable biomass.

Out of 30 tree species planted on highly alkali soil (pH of profile 10.1–10.6) at Saraswati in semiarid Haryana, northwest India, three species *Prosopis juliflora*, *Acacia nilotica* and *Tamarix articulata* were found economically suitable with biomass production of 51, 70 and 93 Mg ha^{-1}, respectively, in 7 years (Dagar et al. 2001b; Singh and Dagar 2005). *Tamarix articulata* ameliorated the soil by inducing the maximum reduction of exchangeable sodium percentage (ESP) and pH values followed by *P. juliflora* and *A. nilotica*. Increase in organic carbon in the surface 15 cm layer under respective species was 0.23, 0.26 and 0.10 %. At the same site, species of *Prosopis* such as *P. juliflora, P. alba, P. articulata, P. levigata, P. pallida* and *P. nigra* produced quite good biomass. All these can successfully be used as energy plantations and even in gassy fires to generate electricity in rural employment programmes. From a long-term experiment, Singh et al. (2008) reported a biomass of 56.5 Mg ha^{-1} in *Prosopis juliflora* followed by *Acacia nilotica* (50.8 Mg ha^{-1}), *Casuarina equisetifolia* (42.1 Mg ha^{-1}), *Terminalia arjuna* (41.6 Mg ha^{-1}), *Pithecellobium dulce* (32.3 Mg ha^{-1}), *Eucalyptus* (30.8 Mg ha^{-1}), *Prosopis alba* (27.8 Mg ha^{-1}), *Pongamia pinnata* (26.7 Mg ha^{-1}), *Cassia siamea* (21.7 Mg ha^{-1}) and *Azadirachta indica* (19.2 Mg ha^{-1}) when harvested after 10 years of plantation in high sodic soil of pH 10.6 in Uttar Pradesh. When harvested after 14 years of plantation, maximum biomass production per tree was achieved in *E. tereticornis, A. nilotica, P. juliflora* and *C. equisetifolia* giving 231, 217, 208 and 197 kg bole weight per plant, respectively, whereas *P. alba, P. dulce, T. arjuna, P. pinnata, A. indica* and *C. siamea* provided relatively lower bole weight of 133, 100, 97, 84, 83 and 52 kg per plant, respectively (Sharma et al. 2010). These tree species improved soil in terms of reduction of pH and exchangeable sodium percentage (ESP) and increase in organic carbon significantly. Organic carbon under *P. juliflora, P. alba, Pongamia pinnata, A. nilotica, A. indica* and *C. equisetifolia* increased to >3.5 g kg^{-1} from initial 0.8 g kg^{-1}.

Similarly, Singh and Gill (1992) and Singh (1995) from long-term studies reported that 20–36-year-old plantations ameliorated the alkali soil through litter and root decomposition in terms of lowering of soil pH and increasing organic carbon and other available nutrients (Table 3). The extent of the resulting soil enrichment differed in tree species. In the Zarifa Viran soil, *Prosopis juliflora* was most efficient in lowering the pH and enhancing organic carbon

Table 3 Effect of tree species on reclamation of highly alkali soils in India

Tree species	pH	ECe ($dS\ m^{-1}$)	Org-C ($g\ kg^{-1}$)	Avail P ($kg\ ha^{-1}$)	Avail K
Zarifa Viran (20-year-old plantation)[a]					
Acacia nilotica	8.4	0.25	8.5	59.0	499.0
Albizia lebbeck	7.9	0.32	6.2	42.6	359.0
Eucalyptus tereticornis	8.5	0.44	6.6	33.0	702.0
Prosopis juliflora	7.3	0.51	9.3	110.5	410.0
Terminalia arjuna	7.9	0.32	8.6	67.8	387.0
Original soil	10.2	0.45	1.2	20.0	278.0
Banthra (36-year-old plantation)[b]					
Albizia procera	7.7	15.0	7.2	31.0	–
Dalbergia sissoo	7.6	12.0	10.3	26.0	–
Syzygium cuminii	7.4	9.0	8.4	29.0	–
Terminalia arjuna	7.7	12.0	7.9	41.0	–
Original soil	9.8	36.0	0.7	41.0	–

Source: Data from [a]Singh and Gill (1992), [b]Singh (1996); compiled by Singh (2005)

and the available nutrients except K, which was highest under *E. tereticornis*. In Banthra soil, *Syzygium cuminii* was found to be most efficient in lowering the pH and ECe values, while organic carbon was highest under *Dalbergia sissoo*. The available P contents in this soil type had either remained unchanged or decreased after three decades. The possible reasons may include the leaching of the soluble P, its locking up as organic P in leaving or dead matter and sampling error due to the horizontal special variability so widespread in alkali soils of Indo-Gangetic alluvial plains (Singh 2005). The soils under these plantations ameliorated to the extent that arable crops could be grown successfully on these soils. These results show that sodic soils can be brought successfully under economic production at the same time to gain environmental services.

On sodic vertisols, tree species such as *Prosopis juliflora, Azadirachta indica, Salvadora persica* and *Acacia nilotica* are found to be highly successful when planted in auger holes. *Cassia siamea* and *Leucaena leucocephala* also performed satisfactorily (Sharma et al. 1992). Species such as *Acacia auriculiformis, Dalbergia sissoo, Casuarina equisetifolia, Dendrocalamus strictus* and *Hardwickia binata* did not survive for longer period. Tree species also improved clayey vertisols significantly in terms of reducing soil pH and ESP and increasing soil carbon.

Arable Crop-based Agroforestry Systems

In this land-use system, forest or fruit trees are raised in wider spaces (row to row 5–6 m, at times even more, and plant to plant 4–5 m), and the arable crops are cultivated in the interspaces on high pH soils. Singh et al. (1995) evaluated this agroforestry approach on a moderately alkali soil (pH 9.2) in irrigated condition of Haryana by planting three commercial trees, namely, poplar (*Populus deltoides*), eucalyptus (*Eucalyptus tereticornis*) and Kikar (*Acacia nilotica*) in association with rice-wheat, rice-Egyptian clover, pigeon pea/sorghum-mustard rotations and sole trees and sole crops as control. Results showed that intercrops of Egyptian clover, rice, wheat and mustard can successfully be grown for along with these trees during initial 3 years (Table 4).

Later on these, crops may be replaced with shade-loving crops such as turmeric. These intercrops help *Populus* and *Eucalyptus* grow faster but adversely affect the growth of low-water-demanding trees like *Acacia*. Soil amelioration measured in terms of decrease in pH and improvement in organic carbon and available N, P and K contents followed the following order: *Acacia-based* system > *Populus*-based system > *Eucalyptus*-based system > sole crops. The benefit/cost ratio was highest (3.30) in the case of poplar with rice-wheat followed by poplar with rice-Egyptian clover (2.95) and rice-wheat (2.79) and the lowest (1.76) in *Acacia* with rice-Egyptian clover.

Table 4 Yields (Mg ha^{-1}) of crops grown along with tree species on partially reclaimed alkali soil for 5 years in northwest India

Crops	*E. tereticornis*	*A. nilotica*	*P. deltoides*	No trees
Rice-Egyptian clover/cowpea-Egyptian clover				
Rice (G)	14.4	12.5	11.8	21.6
Egyptian clover (F)	239.7	212.7	234.6	389.1
Cowpea (F)	18.0	4.6	2.5	45.0
Rice-wheat/guinea grass-oats				
Rice (G)	13.4	11.8	10.3	21.0
Wheat (G)	9.0	8.1	8.4	16.0
Guinea grass (F)	18.1	12.3	2.5	30.0
Oats (F)	23.8	24.0	25.6	42.0
Pigeon pea-mustard/turmeric				
Pigeon pea (G)	0.7	0.8	0.2	0.7
Mustard (G)	2.3	1.8	2.0	4.0
Sorghum (F)	18.3	8.6	22.8	50.0
Turmeric (R)	22.3	5.9	8.3	22.1

Source: Singh et al. (1997)
G grain, *F* fodder, *R* rhizome

Among trees alone, poplar was most profitable followed by *Acacia* and *Eucalyptus*. Growing trees along with crops should not be viewed only as a better and economically viable in terms of food, fodder, timber and firewood production system but also as a promising option to maintain long-term sustainability and also a practical solution for sequestering C in the soil and mitigating climate change. Microbial biomass C and N were found 42 and 13 % higher in tree-based systems as compared to monocropping (Kaur et al. 2000) showing ecological sustainability of these systems for utilizing moderate alkali soils.

Singh et al. (1995), Dagar et al. (1995, 2004) and Dagar and Singh (2003, 2004) evaluated performance of arable crops in the interspaces of several forest and fruit trees such as *Tectona grandis, Ailanthus excelsa, Casuarina equisetifolia* and *Tamarindus indica* in irrigated system on reclaimed alkali soil and concluded that from irrigation and fertilizer application to crops, all the trees were benefitted showing better growth and, during initial years of establishment, normal crop yield was obtained without any yield reduction, but during later stages due to larger canopy, there was a drastic yield reduction in almost all the crops grown with all the tree species except that the remunerative yield of potato could be obtained under partial shade of *Casuarina* and there was no significant yield reduction as compared to when cultivated in open; rather there was no risk of frost under canopy.

Many of the medicinal and aromatic underexplored crops are in great demand for both internal requirements and export. But since these crops are nonconventional in nature, it is not always feasible to produce these on fertile land, which is generally used for arable crops. The marginal lands, specifically the saltlands where profitable returns are not possible from arable crops, can successfully be utilized for the cultivation of these high-value crops with marginal inputs. Results of several experiments conducted by Dagar et al. (2004, 2006) clearly indicated that aromatic grasses such as palmarosa (*Cymbopogon martini*) and lemon grass (*C. flexuosus*) could successfully be grown on moderate alkali soils up to pH 9.2, while vetiver (*Vetiveria zizanioides*) which withstands both high pH and stagnation of water could successfully be grown without significant yield reduction on highly alkali soils. Anwar et al. (1996) reported safe limit of sodicity tolerance in terms of pH and exchangeable sodium percentage (ESP) to be 9.5 and 55 for both palmarosa and vetiver, 9.0 and 50 for lemon grass and 10.0 and 55 for Jamrosa (*Cymbopogon khasans*). They also reported pH 9.5 to be the safe limit for German chamomile (*Matricaria chamomilla*) and periwinkle (*Catharanthus roseus*), while the safe limit for rye (*Secale cereale*) and for ergot (*Claviceps purpurea*) is reported to be pH 10. Medicinal psyllium (*Plantago ovata*) produced 1.47–1.58 Mg ha^{-1} unhusked grain at pH 9.2 and 1.03 to 1.12 Mg ha^{-1} at pH 9.6 showing its potential at moderate alkali soil (Dagar et al. 2006). *Matricaria chamomilla, Catharanthus roseus* and *Chrysanthemum indicum* were other interesting medicinal and flower yielding plants, which could be grown on moderate alkali soil (Dagar et al. 2009). Liquorice (*Glycyrrhiza glabra*) a leguminous crop was found one of the most suitable and highly remunerative crops for moderate alkali soils (Dagar et al. 2015), and it reclaims both alkali and waterlogged soils (Kushiev et al. 2005). All these crops can be blended suitably as intercrops in agroforestry systems on moderate alkali soils.

Silvopastoral Systems

The grazing lands of sodic soils are very poor in forage production under open grazing, but when brought under judicious management, these can be explored successfully for sustainable fodder production. Based on a series of long-term experiments (Kumar and Abrol 1983a, b, 1986; Kumar 1988a, b, 1990, 1998; Singh and Dagar 2005), it was found that kallar grass (*Leptochloa fusca*) could be rated the most tolerant grass to highly sodic soil and waterlogged conditions as compared to other grasses producing 45 Mg ha^{-1} green forage without applying any amendment. It produces more biomass in alkali soil than normal soil. It withstands prolonged stagnation of water and also ameliorates soil quickly. Another interesting grass is Rhodes grass (*Chloris gayana*), which could produce 60–65 Mg ha^{-1} green biomass and did not show any reduction in biomass up to pH 10, whereas it produced 50 Mg ha^{-1} forage at pH 10.4 (Kumar 1998). Gutton panic (*Panicum maximum*) produced 60 Mg ha^{-1} up to pH 9.6 and 45 and 35 Mg ha^{-1} at pH 10 and 10.4, respectively, and para grass (*Brachiaria mutica*) produced 90 and 70 Mg ha^{-1} at pH 9.2 and 9.6, respectively, and 50 and 40 Mg ha^{-1} at pH 10.0 and 10.4, respectively, showing their potential for sodicity tolerance. *Panicum antidotale, P. laevifolium, P. purpureum* and *Setaria anceps* were other successful grasses up to soil pH 9.6. These grasses can be grown successfully with most promising tree species such as *Prosopis juliflora, Acacia nilotica, Tamarix articulata, Casuarina equisetifolia* (susceptible for frost), *Terminalia arjuna* and *Pongamia pinnata*.

Based on 4 years' study, it was found that on an average, 15.6 Mg ha^{-1} forage of *Leptochloa fusca* could be obtained with *Prosopis juliflora;* 16.2 Mg ha^{-1} with *Acacia nilotica*, 17.0 Mg ha^{-1} with *Dalbergia sissoo* and 17.4 Mg ha^{-1} with *Casuarina equisetifolia*. However, among trees, *P. juliflora* and *A. nilotica* performed the best followed by *Dalbergia sissoo* and *Casuarina equisetifolia* (Singh and Dagar 2005). Singh et al. (1988, 1989, 1991), Dagar et al. (2001b) and Singh and Dagar (2005) evaluated several tree and grass species for their performance on highly sodic soil and mesquite (*Prosopis juliflora*), and kallar grass silvopastoral practice was adjudged the most promising for firewood and forage production and also for soil amelioration. *Leptochloa fusca* grown with *P. juliflora* produced 46.5 Mg ha^{-1} green forage in 15 cuttings over 50 months' period without application of any fertilizer or other amendments (Singh et al. 1988, 1989, 1991, 1993), and *P. juliflora* yielded 161.3 Mg ha^{-1} air-dried firewood in 6 years when planted at 2 m×2 m spacing with 55.6 Mg ha^{-1} *Leptochloa fusca* grass forage (Singh 1995). This system also ameliorated soil to a greater extent in terms of reducing soil pH, increasing organic carbon and nutrients. An associative nitrogen-fixing bacterium, *Azoarcus*, occurs as an endophyte in the roots of *kallar* grass (*L. fusca*) – a pioneer species of alkali soils that yields 9–12 Mg ha^{-1} of dry biomass without application of any nitrogen fertilizer; nearly half of the plant N of 90–120 kg ha^{-1} is derived from associative fixation (Malik and Zafar 1984; Malik et al. 1986) and helps the plants survive in adverse habitats. Growth of native non-symbiotic bacteria is improved by applying amendments (Rao 1998). Symbiotic nitrogen fixation by *Rhizobium* has been extensively investigated in salt-affected soils (Rao and Ghai 1995), and their survival is not a problem as they have considerable tolerance to high pH. This system improved the soil to such an extent that less tolerant but more palatable fodder species such as Persian clover (*Trifolium resupinatum*), Egyptian clover (*T. alexandrinum*), lucerne (*Medicago sativa*) and sweet clover (*Melilotus denticulata*) could be grown under mesquite trees after 52 months producing 23.1, 21.3, 10.3 and 8.0 Mg ha^{-1} forage, respectively (Singh et al. 1993).

The salt-affected soils of black soil zone (saline/sodic vertisols) are generally either contemporary or of secondary origin. The contemporary salty soils exist in the topographic situation having poor drainage conditions. However, the soils that might have become sodic due to unjudicious use of irrigation water can also be encountered in the irrigation command area. These lands can successfully be grown with forest and fruit trees. In one 14 years of plantation, it was found that *P. juliflora* and *Azadirachta indica* were most successful species for these soils. Among grasses, *Aeluropus lagopoides, Leptochloa fusca, Brachiaria mutica, Chloris gayana, Dichanthium annulatum, Bothriochloa pertusa* and species of *Eragrostis, Sporobolus* and *Panicum* are most successful and may form suitable silvopastoral system. In another experiment on alkaline vertisol, it was found that under 7 years of plantations of *P. juliflora* and *Azadirachta indica* forming silvopastoral system with kallar grass, the soil pH, ECe and ESP reduced from 8.8, 4 dS m^{-1} and 35 to 8.5, 1.29 dS m^{-1} and 10, respectively, under *Prosopis*-based system and 8.5, 1.3 dS m^{-1} and 14, respectively, under *Azadirachta* system. The experiments conducted in sodic vertisols with ESP 40 growing grasses like *Leptochloa fusca, Brachiaria mutica* and *Vetiveria zizanioides* showed that all these grasses performed well and the forage biomass increased during the second year. The uptake of sodium by *L. fusca* was highest followed by *B. mutica* at every stage of cutting. In 3 years, these grasses could remove 144.8, 200.0 and 63.5 kg ha^{-1} sodium from soil, respectively (AICRP 2000–2004).

A large proportion of salt-affected lands (particularly in Indian subcontinent) does not belong to individual farmers, but is either government land or in the custody of village *Panchayats*. Reclamation of such lands for crop production is not feasible because of common property rights. Raising suitable trees and grasses would appear to be a promising use of these lands. As mentioned earlier, the most promising tree species for highly alkali soils such as *Prosopis juliflora, Acacia nilotica* and *Tamarix articulata* blended with highly salt-tolerant and high biomass producing grass species like *Leptochloa fusca, Brachiaria mutica, Chloris gayana,* species of *Sporobolus* and *Panicum* form ideal silvopastoral system. The grasses can be managed through 'cut and carry' system between the interspaces of trees. Thus, we find that for highly sodic soils as well as alkaline vertisols, silvopastoral systems (especially kallar grass, *Prosopis* based) have shown promise in terms of biomass production as well as soil amelioration. Under tree cover, the bulk density of salt-affected soils decreases, and there is substantial increase in soil porosity and hence infiltration rate, water holding capacity, field capacity and permeability (Mishra et al. 2004). In India, using auger hole technique, different state forest departments have reclaimed about 60 thousand ha of highly deteriorated sodic soils through agroforestry plantations on village community lands, adjoining roads, railway lines and canals (CSSRI 2011).

Even a single intervention (e.g. protection from grazing) over a long period resulted in significant increase of soil carbon and biological amelioration of sodic soils. Integrating trees with the grasses in silvopastoral systems has been found to be effective to improve soil fertility and increase soil carbon sequestration. Organic carbon increased by 24–62 % in soils under the silvopastoral systems as compared to that in the grassland system on a sodic soil (Gupta et al. 2015). The microbial carbon, as regulated by litter and root carbon input, was found to be a good of bioamelioration of sodic soils (Table 5). Carbon sequestration also provided associated ecosystem co-benefits such as increased soil water holding capacity, better soil structure, improved soil quality and nutrient cycling. Implementing appropriate management practices to build up soil carbon stocks in grasslands could lead to considerable mitigation, adaptation and developmental benefits.

Fruit-based Agroforestry Systems

Fruit trees are comparatively less tolerant than some forest trees and, therefore, may be grown on moderate alkali soils (pH 9.0–9.6). In this land-use system, forest or fruit trees are raised in wider spaces (row to row 5–6 m, at times even more, and plant to plant 4–5 m), and the arable

Table 5 Microbial biomass carbon and soil carbon in surface 0.15 m soil under grassland systems of sodic soils in northwestern India

Grassland system	Microbial C (kg ha^{-1})	Soil carbon (kg ha^{-1})
Sporobolus marginatus[a]	85	4816
Desmostachya bipinnata[a]	112	5265
Dalbergia sissoo + *D. bipinnata*[a]	325	13,572
Acacia nilotica + *D. bipinnata*[a]	225	10,902
Prosopis juliflora + *D. bipinnata*[a]	348	14,211
Mixed grassland[c]	148	7732
D. bipinnata[c]	408	15,443
Vetiveria zizanioides[c]	475	17,088
D. bipinnata[b]	347	13,949

Source: Kaur et al. (2002); Neeraj et al. (2004); Gupta et al. (2015); Location of sites: [a]Bichhian, [b]Karnal, [c]Kurukshetra

crops are cultivated in the interspaces on high pH soils. In one trial, Singh and Singh (1990) observed that fruit trees like *Emblica officinalis, Carissa carandas, Ziziphus mauritiana, Syzygium cuminii, Grewia asiatica, Psidium guajava, Aegle marmelos* and *Vitis vinifera* when grown on different alkali soils could produce 20.5, 5.2, 15.5, 16.0, 6.0, 12.5, 6.5 and 18.3 Mg ha^{-1} fruits at different pH of 10.1, 10.0, 9.6, 9.5, 9.2, 9.0, 8.5 and 9.0, respectively. Out of ten fruit tree species tested on highly alkali soil (pH ~10) using different soil amendments (Singh et al. 1997; Dagar et al. 2001b; Singh and Dagar 2005), *Ziziphus mauritiana, Syzygium cuminii, Psidium guajava, Emblica officinalis* and *Carissa carandas* were found the most successful species showing good growth and also initiated fruit setting after 4–5 years of plantation. After 10 years, these could produce 12–25 Mg ha^{-1} fruits annually. In one trial on high pH (~10), Egyptian clover (*Trifolium alexandrinum*), wheat, rice, onion (*Allium cepa*) and garlic (*Allium sativum*) were grown successfully for 3 years in the interspaces of fruit trees *Carissa carandas, Punica granatum, Emblica officinalis, Psidium guajava, Syzygium cuminii* and *Ziziphus mauritiana*; and 10.6–16.7 Mg ha^{-1} forage from *Leptochloa fusca* grass, 1.6–3.0 Mg ha^{-1} grains from wheat, 1.8–3.4 Mg ha^{-1} onion bulb, and 2.3–4.1 Mg ha^{-1} garlic were harvested (Tomar et al. 2004) showing that during the establishment of fruit trees, suitable arable crops can successfully be harvested from interspaces of trees. As shown earlier, many forests and fruit tree species can be raised on alkali soil (pH up to 9.8), but some of these like pomegranate (*Punica granatum*) and bael (*Aegle marmelos*) are unable to tolerate water stagnation during rainy season which may be cultivated on raised bunds. Dagar et al. (2001a) planted pomegranate on bunds and water-loving kallar grass (*Leptochloa fusca*) and salt-tolerant rice (variety CSR-30) in sunken beds during rainy season. In winter season, Egyptian clover (*Trifolium alexandrinum*) and wheat (var. KRL 1–4) could be grown in sunken beds. Results showed that on average, 4.3–4.9 Mg ha^{-1} rice and 1.2–1.4 Mg ha^{-1} wheat were obtained. In second rotation, 21.3–36.8 Mg ha^{-1} fresh forage of kallar grass and 44.9–47.8 Mg ha^{-1} fresh forage of Egyptian clover were obtained. There was no yield reduction due to plantations at initial stage of growth. Another advantage was that after 2 years, soil amelioration in terms of reduction in soil pH, increase in organic matter and nitrogen contents was significant.

Fruit trees gooseberry (*Emblica officinalis*) and *ber* (*Ziziphus mauritiana*) were the most successful on alkaline vertisol (ESP 25–60) followed by sapota (*Achras zapota*). Through a series of experiments conducted on raised and sunken beds, it was concluded that both forest tree species such as *Azadirachta indica* and fruit trees like pomegranate (*Punica granatum*), Jamun (*Syzygium cuminii*) and goose berry (*Emblica officinalis*) could successfully be grown on raised bunds and rainfed rice could be grown during rainy season in sunken beds and suitable winter crops could be cultivated in residual moisture in sunken beds (CSSRI 2002–2003 to 2012–2013).

Conclusions

From adopting agroforestry especially on degraded salt-affected lads, not only the individual farmer is benefited but the entire community around the area is benefited in many ways. Soil and water conservation measures by creating the bunds across the nalas and fields and ripping along the contour result in checking the soil erosion and harvesting rainwater in situ. Creating the ponds for fish culture helps in meeting the water requirement of cattle and wildlife population in surrounding areas. The grass harvested by local people from the afforested areas helps many families to rear cattle and sustain their families. The trees influence the temperature extremes and help in percolation of rainwater into loose and porous soils (recharging of subsoil water). The live fencing helps in the protection of crop fields. Several medicinal herbs are collected by the local practitioners who sustain their families by subscribing plant-based drugs. Recently the attention is being paid towards commercial forestry, raising block plantation of commercial trees and also species yielding biodiesel such as *Jatropha curcas, Pongamia pinnata, Euphorbia antisyphilitica* and *Ricinus communis*. This approach will change the economical scenario by reducing the import of fossil fuels. As discussed above, trees play a vital role both in lowering down water table and sequestering carbon. Adopting biosaline agroforestry, the nomadic behaviour of large population will be checked in dry regions. There is a need to change policies and tilting these towards agroforestry. Agroforestry on salt-affected lands will not only bring these lands under economic utility but also will release pressure on forest resources and through environmental services should help in mitigating climate change.

References

Ahmad CN (1988) Coordinated research programme on saline agriculture. Final report, PARC, Islamabad, Pakistan

Ahmad R, Ismail S (1993) Studies on selected of salt-tolerance plants for food, fodder and fuel from world flora. In: Lieth H, Al Masoom A (eds) Towards the rational use of high salinity tolerant plants, vol 2. Kluwer Academic Publishers, Dordrecht, pp 295–304

AICRP (2000–2004) Biennial reports of the all India coordinated research project on management of salt-affected soils and use of saline water in Agriculture, CSSRI, Karnal, India, p 199

Altaf-Hussain, Pazir-Gul (1991) Selection of suitable tree species for saline and waterlogged areas. Pak J For 41(1):34–43

Angell K (1990) Tree planting on farms in high rainfall areas. Bulletin no. 4174, Western Australian Department of Agriculture, South Perth, Western Australia, p 19

Angell K, Scott P, Smith A (1994) Managing saline, high rainfall valleys and flats. Farm-note no. 25/94. Western Australian Department of Agriculture, S. Perth, Western Australia

Anwar M, Patra DD, Singh DV (1996) Influence of soil sodicity on growth, oil yield and nutrient accumulation in vetiver (*Vetiveria zizanoides*). Ann Arid Zone 35(1):49–52

Aronson JA (1989) HALOPH-A data base of salt tolerant plants of the world. Office of Arid Lands Studies, the University of Arizona, Tucson, p 75

Aslam Z, Mujtab M, Alshtar J, Waheed R, Malik KA, Naqvi M (1993) Biological methods for economically utilising salt-affected soils in Pakistan. In: Davidson N, Galloway R (eds) Productive use of saline land. ACIAR proceedings no. 12, Canberra, Australia, pp 29–31

Banerjee LK, Sastry ARK, Nayar MP (1989) Mangroves in India: identification manual. Botanical Survey of India, Calcutta, p 113

Barrett-Lennard EG (2002) Restoration of saline land through revegetation. Agric Water Manage 53(1):213–226

Barrett-Lennard EG (2003) Saltland pastures in Australia: a practical guide. State of Western Australia, Department of Agriculture, CSIRO, CRC for Plant-based management of dryland salinity, Australia, p 176

Barrett-Lennard EG, Galloway R (1996) Saltbush for water table reduction and land rehabilitation. Aust J Soil Water Conserv 9:21–24

Barrett-Lennard EG, Malcom CV (1999) Increased concentrations of chloride beneath stands of saltbushes (*Atriplex* species) suggest substantial use of groundwater. Aust J Exp Agric 39:949–955

Barrett-Lennard EG, Warren BE, Malcolm CV (1990) Agriculture on saline soils – direction for the future. In: Myers BA, West DW (eds) Revegetation of saline land. Institute for Irrigation and Salinity Research, Tatura, pp 37–45

Blackwell P, Morrow G (2006) Furrow sowing for improved crops and pastures on water repellent soils. Farmnote no. 111/96 [Reviewed on May 2006]. Department of Agriculture and Food, Western Australia, Perth

Bleby TM, Aucote M, Kennett-Smith AK, Walker GR, Schachtman DP (1997) Seasonal water use characteristics of Tall wheat grass (*Agropyron elongatum* (Host) Beauv.) in a saline environment. Plant Cell Environ 20:1361–1371

Bowman S, Ruprecht J (2000) Blackwood river catchment flood risk study. Report no. SWH 29, Surface water hydrology report series, water and rivers commission, East Perth, pp 36

Boyko H (ed) (1968) Saline irrigation for Agriculture and Forestry. W. Junk, The Hague

Cervinka V, Diener J, Erickson J, Finch C, Martin M, Menezes F, Peters D, Shelton J (1999) Integrated system for agricultural drainage management on irrigated farmland. Final report grant no. 4-FG-20-11920, US Department of the Interior Bureau of Reclamation, Westside Resource Conservation District, Fine Points, California, USA

Chaturvedi AN (1984) Firewood crops in areas of brackish water. Indian For 110(4):364–366

Chaudhri II, Shah BH, Naqvi N, Mallick IA (1964) Investigations on the role of *Suaeda fruticosa* Forsk. in the reclamation of saline and alkaline soils in West Pakistan plains. Plant Soil 21:1–7

CSSRI (2002–2003 to 2012–2013) Annual reports of Central Soil Salinity Research Institute, Karnal, India

CSSRI (2011) CSSRI vision 2030. Central Soil Salinity Research Institute, Karnal, India

Dagar JC (1982) Some ecological aspects of mangrove vegetation of Andaman and Nicobar Island. Sylvatrop Philipp For Res J 7:177–216

Dagar JC (1991) Agroforestry practices for coastal Bay Islands. J Indian Soc Coast Agric Res 9(1&2):433–439

Dagar JC (1994) Strategies for sustainable use of island resources of India. J Indian Soc Coast Agric Res 12(1&2):187–191

Dagar JC (1995a) Agroforestry systems for Andaman and Nicobar Islands in India. Intern Tree Crops J 8(2&3):107–128

Dagar JC (1995b) Ecology of halophytic vegetation in India: a review. Int J Ecol Env Sci 21:273–296

Dagar JC (1996) Rehabilitation of coastal saline lands by planting with suitable species. J Indian Soc Coast Agric Res 14:185–191

Dagar JC (2000) Agroforestry systems for coastal and island regions. Indian J Agrofor 2:59–74

Dagar JC (2003) Biodiversity of Indian saline habitats and management & utilization of high salinity tolerant plants with industrial application for rehabilitation of saline areas. In: Alsharhan AS, Wood WW, Goudie AS, Fowler A, Abdellatif EM (eds) Desertification in the third millennium. Swets & Zeitlinger Publishers, Lisse, pp 151–172

Dagar JC (2005a) Salinity resources in India: an overview. Bull Nat Inst Ecol 15:69–80

Dagar JC (2005b) Ecology, management and utilization of halophytes. Bull Nat Inst Ecol 15:81–97

Dagar JC (2008) Indian mangroves: status, management and their plausible benefits for livelihood security. J India Soc Coast Agric Res 26(2):121–128

Dagar JC (2009) Opportunities for alternate land uses in salty and water scarcity areas. Int J Ecol Env Sci 35(1):53–66

Dagar JC (2012) Utilization of degraded lands/habitats and poor quality water for livelihood security and mitigating climate change. Indian J Agrofor 14(1):1–16

Dagar JC (2014) Greening salty and waterlogged lands through agroforestry systems. In: Dagar JC, Singh AK, Arunachalam A (eds) Agroforestry systems in India: livelihood security & environmental services, vol 10, Advances in agroforestry. Springer Publishers, New Delhi, pp 333–344

Dagar JC (2015) Agroforestry for restoration, conservation and resilience of waste/degraded lands: challenges and opportunities. In: Dhyani SK, Ram-Newaj, Badri-Alam, Inder-Dev (eds) Agroforestry: present status and way forward. Biotech Books, New Delhi, pp 323–362

Dagar JC, Singh G (2003) Pasture production in degraded and problematic soil status and prospects. In: Jakhmola RC, Jain RK (eds) Sustainable animal production. Pointer Publishers, Jaipur, pp 94–119

Dagar JC, Singh G (2004) Evaluation of crops in agroforestry with *Casuarina equisetifolia* (Linn.) plantations. Indian J Agrofor 3(1):47–50

Dagar JC, Singh G (2007) Biodiversity of saline and waterlogged environments: documentation, utilization and management. National Biodiversity Authority, Chennai, p 76

Dagar JC, Tomar OS (1998) Suitable agroforestry systems for coastal saline regions: case study. J Range Manage Agrofor 19:165–172

Dagar JC, Mongia AD, Bandyopadhyay AK (1991) Mangroves of Andaman & Nicobar Islands. Oxford & IBH Publishing Co. Ltd, New Delhi, p 166

Dagar JC, Singh NT, Mongia AD (1993) Characteristics of mangrove soils and vegetation of Bay Islands in India. In: Lieth H, Al Masoom A (eds) Towards the rational use of high salinity tolerant plants, vol 1. Kluwer Academic Publishers, Dordrecht, pp 59–80

Dagar JC, Singh G, Singh NT (1995) Evaluation of crops in agroforestry with teak (*Tectona grandis* Linn-f.), maharukh (*Ailanthus excelsa* Roxb.) and tamarind (*Tamarindus indica* Linn.) on reclaimed salt-affected soil. J Trop For Sci 7(4):623–634

Dagar JC, Sharma HB, Shukla YK (2001a) Raised and sunken bed technique for agroforestry on alkali soils of northwest India. Land Degrad Dev 12:107–118

Dagar JC, Singh G, Singh NT (2001b) Evaluation of forest and fruit trees used for rehabilitation of semi-arid alkali/sodic soils in India. Arid Land Res Manage 15:115–133

Dagar JC, Tomar OS, Kumar Y, Yadav RK (2004) Growing three aromatic grasses in different alkali soils in semi-arid regions of northern India. Land Degrad Dev 15:143–151

Dagar JC, Kumar Y, Tomar OS (2006) Cultivation of medicinal Isabgol (*Plantago ovata*) in alkali soils in semiarid regions of northern India. Land Degrad Dev 17:275–283

Dagar JC, Gururaja-Rao G, Shukla YK, Sharma HB (2009) Performance of three flower yielding plants in different sodic soils. Indian J Hortic 66(3):404–409

Dagar JC, Pandey CB, Chaturvedi CS (2014) Agroforestry: a way forward for sustaining fragile coastal and island agro-ecosystems. In: Dagar JC, Singh AK, Arunachalam A (eds) Agroforestry systems in India: livelihood security & environmental services, vol 10,

Advances in agroforestry. Springer, New Delhi, pp 185–232

Dagar JC, Yadav RK, Dar SR, Ahamad-Sharif (2015) Liquorice (*Glycyrrhiza glabra*): A potential salt-tolerant highly remunerative medicinal crop which also helps in remediation of alkali soils. Curr Sci 108(9):1683–1688

Das SK, Itnal CJ (1994) Capability based land use systems: role in diversifying dryland agriculture. In: Soil management for sustainable agriculture in dryland area. Bull Indian Soc Soil Sci 16:92–100

Diaz FJ, Benes SE, Grattan SR (2003) Field performance of halophytic species under irrigation with saline drainage water in the San Joaquin Valley of California. Agric Water Manage 118:59–69

EISC (2002) Mangroves of India: state-of-art report, ENVIS publication series, 2/2002. Environmental Information System Centre (EISC), Annamalai University, Parangipettai, p 140

Engel R, Negus T (1988) Controlling saltland with trees. Farm-note no. 46/88. Western Australian Dept. of Agriculture, S. Perth, Western Australia

FAO (2012) FAO statistical year book 2012, world food and agriculture. Food and Agriculture Organization of the United Nations, Rome, p 366

FAO/AGL (2000) Extent and causes of salt-affected soils in participating countries. FAO/AGL-global network on integrated soil management for sustainable use of salt-affected lands http://www.fao.org/ag/agl/agll/spush/topic2.htm

Garrett B (1993) Agroforestry for salinity control. In: Race D (ed) Agroforestry: trees for productive farming. Daratech Pvt. Ltd, East Melbourne, pp 109–113

George R, Frantom P (1988) Reclaiming sand plain seeps with small blocks of trees. Farm-note no. 116/88. Western Australian Department of Agriculture, S. Perth, Western Australia

George RJ, Nulsen RA, Ferdowsian R, Raper GP (1999) Interaction between trees and ground waters in recharge and discharge areas- a survey of Western Australian sites. Agric Water Manage 39:91–113

Ghassemi F, Jakeman AJ, Nix HA (1995) Salinisation of land and water resources: human causes, extent, management and case studies. CABI Publishing, Wallingford, p 517

Ghosal TK, Ghosh P, Rezaul Haq AHM (2004) Cultivating wetlands in Bangladesh. LEISA Magazine 20.4, December, pp 1–6. http://www.agriculturesnetwork.org/magazines/farming-with-nature

Gill HS (1985) Studies on the evaluation of selected tree species for their tolerance to sodicity and mechanical impedance in a highly sodic soil with particular reference to root growth behaviour. Ph.D. thesis, Kurukshetra University, Kurukshetra, India

Gill HS, Abrol IP (1993) Afforestation and amelioration of salt-affected soils in India. In: Davidson N, Galloway R (eds) Productive use of saline land. ACIAR proceedings no. 42, Canberra, pp 23–27

Grattan SR, Grieve CM, Poss JA, Robinson PH, Suarez DL, Benes SE (2004a) Evaluation of salt-tolerant forages for sequential water reuse systems: I. Biomass production. Agric Water Manage 70(2):109–120

Grattan SR, Grieve CM, Poss JA, Robinson PH, Suarez DL, Benes SE (2004b) Evaluation of salt-tolerant forages for sequential water reuse systems: III. Potential implications for ruminant mineral nutrition. Agric Water Manage 70(2):137–150

Grieve CM, Poss JA, Grattan SR, Suarez DL, Benes SE, Robinson PH (2004) Evaluation of salt-tolerant forages for sequential water reuse systems: II. Plant-ion relations. Agric Water Manage 70(2):121–135

Gupta RK, Abrol IP (1990) Salt affected soils: their reclamation and management for crop production. In: Lal R, Stewart BA (eds) Advances in soil science, vol III, Soil degradation. Springer, New York, pp 223–288

Gupta SR, Jangra R, Dagar JC (2015) Carbon pools and fluxes in grassland systems on sodic soils of northern India. In: Singh AK, Dagar JC, Arunachalam A, Gopichandran R, Shelat KN (eds) Climate change modelling, planning and policy for agriculture. Springer India, New Delhi, pp 119–140

Heuperman A (1992) Trees in irrigation areas: the bio-pumping concept. Trees Nat Resour 34:20–25

Heuperman AF (2000) Biodrainage: an Australian overview and two Victorian case studies. Proceedings 8th ICID international drainage workshop, New Delhi, India

Heuperman AF, Kapoor AS, Denecke HW (2002) Biodrainage: principles, experiences and applications. International Programme for Technology & Research in Irrigation and Drainage. IPTRID Secretariat, FAO, Rome, p 78

Hogarth PJ (1999) The biology of mangroves. Oxford University Press, Oxford

Hyder SZ (1981) Preliminary observations on the performance of some exotic species of Atriplex in Saudi Arabia. J Range Manage 34:208–210

INCID (2003) Biodrainage status in India and other countries. Indian National Committee on Irrigation and Drainage (INCID), New Delhi, p 40

Jain BL, Muthana KD, Goyal RS (1985) Performance of tree species in salt-affected soils in arid regions. J Indian Soc Soil Sci 33:221–224

Jeet-Ram DJC, Khajanchi L, Singh G, Toky OP, Tanwar VS, Dar SR, Chauhan MK (2011) Bio-drainage to combat waterlogging, increase farm productivity and sequester carbon in canal command areas of northwest India. Curr Sci 100(11):1673–1680

Kapoor AS (2014) Managing groundwater for irrigated agriculture – the relevance of biodrainage model. Rawat Publications, Jaipur, p 341

Kapoor AS, Denecke HW (2001) Biodraiange and biodisposal: the Rajasthan experience. In: GRID, IPTRID's network magazine no. 17, IPTRID, FAO, Rome

Kaur B, Gupta SR, Singh G (2000) Soil carbon, microbial activity and nitrogen availability in agroforestry systems on moderately alkaline soils in northern India. Appl Soil Ecol 15(3):283–294

Kaur B, Gupta SR, Singh G (2002) Bioamelioration of a sodic soil by silvopastoral system in northwestern India. Agrofor Syst 54:13–20

Khan MAW, Yadav JSP (1962) Characteristics and afforestation problems of saline alkali soils. Indian For 88(4):259–271

Kumar A (1988a) Long term forage yields of five tropical grasses on an extremely sodic soil and resultant soil amelioration. Expl Agric 24:89–96

Kumar A (1988b) Performance of forage grasses in saline soils. Indian J Agron 33:26–30

Kumar A (1990) Effect of gypsum compared with that of grasses on the yield of forage crops on a highly sodic soil. Exp Agric 24:89–96

Kumar A (1998) Forage in slat-affected soils. In: Tyagi NK, Minhas PS (eds) Agricultural salinity management in India. CSSRI, Karnal, pp 489–498

Kumar A, Abrol IP (1983a) Effect of periods of submergence on the performance of selected grasses. Indian J Agric Sci 53:694–698

Kumar A, Abrol IP (1983b) Effect of gypsum on tropical grasses grown in normal and extremely sodic soil. Exp Agric 19:167–177

Kumar A, Abrol IP (1986) Grasses on alkali soils. Bulletin no. 11, CSSRI, Karnal, India

Kumar BM, Kunhamu TK (eds) (2011) Quarter century of agroforestry research in Kerala: a compendium of research publications. Kerala Agriculture University, Kerala, p 404

Kushiev H, Noble AD, Abdullaev I, Toshbekov U (2005) Remediation of abandoned saline soils using *Glycyrrhiza glabra*: a study from the Hungry steppes of central Asia. Int J Agric Sustain 3(2):102–112

Lesaffre B, Zimmer D (1995) Review of Western European experience in sub-surface drainage. Keynote address in national seminar on sub-surface drainage, Jaipur, Rajasthan, India, p 52

Malcolm CV (1982) Wheatbelt salinity. review of the salt land problems in South Western Australia. Western Australian technical bulletin no. 52, S. Perth, Western Australia

Malcolm C (1989) Grazing and management of saltland shrubs. Farm-note no. 87/89. Western Australian Department of Agriculture, S. Perth, Western Australia

Malcolm CV (1993) Saltland management-revegetation. Farm-note no. 44/86, Western Australian Department of Agriculture, S. Perth, Western Australia

Malcolm CV, Clarke AJ, D'Antuono MF, Swaan TC (1988) Effects of plant spacing and soil conditions on the growth of five *Atriplex species*. Agric Ecosyst Environ 21:265–279

Malik KA, Zafar Y (1984) Use of ^{15}N isotope dilution technique for quantification of associative BNF in kallar grass. In: Veegar C, Newton WE (eds) Advances in nitrogen fixation research. Nijhoff Junk, The Hague, p 58

Malik KA, Aslam Z, Naqvi M (1986) Kallar grass-a plant for saline lands. Nuclear Institute for Agriculture and Biology, Faisalabad, p 93

Manjunatha MV, Hebbara M, Patil SG, Kuligod VB, Minhas PS (2005) Effect of trees alone or with grasses on halting canal seepage and shallow water table control in saline vertisols. J Indian Soc Soil Sci 53(2):254–257

Marcar NE, Crawford D (1996) Tree-growing strategies for productive use of saline land. Aust J Soil Water Conserv 9:34–40

Marcar NE, Khanna PK (1997) Reforestation of salt-affected and acid soils. In: Nambier EKS, Brown AG (eds) Management of soil, nutrients and water in tropical plantation forests, ACIAR monograph 43. ACIAR, Canberra, pp 481–524

Marcar NE, Crawford DF, Leppert PM (1993) The potential of trees for utilization and management of salt-affected land. In: Davidson N and Galloway R (eds) Productive use of saline land. ACIAR proceedings no. 42, Canberra, pp 17–22

Mishra A, Sharma SD, Pandey R, Mishra L (2004) Amelioration of highly alkaline soil by trees in northern India. Soil Use Manage 20:325–332

Moore G (1998) Soilguide: a handbook for understanding and managing agricultural soils, Bulletin 4343. Agriculture Western Australia, South Perth, p 328

Moore G, Sanford P, Wiley T (2006) Perennial pastures for Western Australia. Bulletin 4690. Department of Agriculture and Food, Western Australia, Perth

Mudie PJ (1974) The potential economic uses of halophytes. In: Reimold RJ, Queen WH (eds) Ecology of halophytes. Academic, New York, pp 565–597

Mudie PJ, Shmitt WR, Luard EJ, Rutherford JW, Wilfson FH (1972) Preliminary studies on seawater irrigation. Halophyte research project, Scripps Institute of Oceanography, Publication no. 1

Nair MA, Sreedharan C (1986) Agroforestry farming systems in the homesteads of Kerala, southern India. Agrofor Syst 4:339–363

Neeraj, Gupta SR, Malik V, Kaur B, Neelam (2004) Plant diversity, carbon dynamics and soil biological activity in tropical successional grassland systems at Kurukshetra Int J Ecol Environ Sci 30:285–298

Negus T (1988) Tree for saltland. Farm-note no. 110/88. Western Australian Department of Agriculture, S. Perth, Western Australia

O'Leary JW (1984) The role of halophytes in irrigated agriculture. In: Salinity tolerance in plants: strategies for crop improvement. Wiley, New York, pp 285–300

Pande GC (1967) Afforestation of usar lands. In: Proceedings of 11th all India silviculture conference, FRI, Dehra Dun

Pandey CB, Rai BR, Singh L, Singh A (2007) Homegardens of Andaman and Nicobar, India. Agric Syst 92:1–22

Pessarakli M, Szabolcs I (1999) Soil salinity and sodicity as particular plant/crop stress factors. In: Pessarakli M (ed) Handbook of plant and crop stress, 2nd edn. Marcel Dekker, New York, pp 1–16

Qureshi RH, Nawaz S, Mahmood T (1993) Performance of selected tree species under saline-sodic field conditions in Pakistan. In: Lieth H, Al Masoom A (eds) Towards the rational use of high salinity tolerant plants, vol 2. Kluwer Academic Publishers, Dordrecht, pp 259–269

Qureshi AS, McCornick PG, Qadir M, Aslam Z (2008a) Managing salinity and waterlogging in the Indus Basin of Pakistan. Agric Water Manage 95(1):1–10

Qureshi AS, Gill MA, Sarwar A (2008b) Sustainable groundwater management in Pakistan: challenges and opportunities. Irrig Drain 57:1–10

Rao DLN (1998) Biological amelioration of salt-affected soils. In: Subba Rao NS, Donimergues YR (eds) Microbial interactions in agriculture and forestry, vol I. Oxford & IBH Publication Co, New Delhi, pp 219–238

Rao DLN, Ghai SK (1995) Predicting nitrogen fixation and N accumulation in field grown annual *Sesbania* spp. Proc Indian Nat Sci Acad(B) 61:57–62

Rao GG, Nayak AK, Chinchmalatpure AR (2003) *Salvadora persica*: a life support species for salt affected black soils. Technical bulletin 1/2003. CSSRI, Karnal, India, p 44

Rashid A, Khattak JK, Khan MZ, Iqbal MJ, Akbar F, Khan P (1993) Selection of halophytic forage shrubs for the Peshawar valley, Pakistan. In: Davidson N, Galloway R (eds) Productive use of saline land. ACIAR proceedings no. 42, Canberra, pp 56–61

Ritzema HP, Satyanaryana T, Raman S, Boonstra J (2008) Subsurface drainage to combat waterlogging and salinity in irrigated lands in India. Lesson learnt in farmers' fields. Agril Water Manage 95:179–189

Roy Chowdhury S, Kumar A, Brahmanand PS, Ghosh S, Mohanty RK, Jena SK, Sahoo N, Panda GC (2011) Application of Bio-drainage for reclamation of waterlogged situations in deltaic Orissa, Research bulletin 53. Directorate of Water Management (ICAR), Bhubaneswar, p 32

Ryder A, Bowyer J, Skinner G, McCarron C (2000) Low recharging farming systems: case studies on the South Coast. Agriculture Western Australia, Miscellaneous publication 22, Perth

Saenger P, Hegerl EJ, Davie JDS (1983) Global status of mangrove ecosystems. Working groups on mangrove ecosystems, IUCN commission on ecology, UNEP/WWF, Environmentalist 3, Supplement no. 3, p 88

Sahibzada MH (1993) Identification of fast growing salt tolerant tree species. Pak J For 43(4):216–220

Sandhu SS, Abrol IP (1981) Growth responses of *Eucalyptus tereticornis* and *Acacia nilotica* to selected cultural treatments in a highly sodic soil. Indian J Agric Sci 51:437–443

Schofield NJ (1992) Tree planting for dryland salinity control in Australia. Agrofor Syst 20:1–23

Schofield NJ, Loh IC, Scott PR, Bartle JR, Ritson P, Bell RW, Borg H, Anson B, Moore R (1989) Vegetation strategies to reduce salinities of water resource catchments in south west Western Australia. Report no. WS 33. Water Authority of Western Australia, Leederville p 81

Sharma OP, Verma SK, Kool YM, Desnonki SC (1992) Progress report of thr project evaluation of tree plantations for the control of salinity and water table. Jawaharlal Nehru Krishi Vishava Vidyalaya, Indore, p 26

Sharma SK, Dagar JC, Singh Gurbachan (2010) BIOSAFOR: biosaline (Agro) forestry: remediation of saline wastelands through production of renewable energy, biomaterial and fodder. Technical bulletin no. 4/2010. Central Soil Salinity Research Institute, Karnal, p 26

Sheikh MI (1987) Energy plantations for marginal and problematic lands, Pakistan. GCP/RAS/III/NET Fd. Document 5, Bangkok, FAO, p 81

Singh NT (1989) Salt affected soils in India. In: Khushoo TN, Deekshtalu BI (eds) Land and soils. Har Anand Publications, New Delhi, pp 65–102

Singh G (1995) An agroforestry practice for the development of salt lands using *Prosopis juliflora* and *Leptochloa fusca*. Agrofor Syst 29:61–75

Singh B (1996) Influence of forest litter on reclamation of semi-arid sodic soils. Arid Soil Res Rehab 10:201–211

Singh NT (2005) Irrigation and soil salinity in the Indian subcontinent – past & present. Lehigh University Press, Bethlehem, p 404

Singh G, Dagar JC (2005) Greening sodic soils: Bichhian model. Technical bulletin no. 2/2005. Central Soil Salinity Research Institute, Karnal, p 51

Singh G, Gill HS (1992) Amelicrative effect of tree species on characteristic of sodic soils at Karnal. Indian J Agri Sci 62:142–146

Singh IS, Singh RK (1990) Fruits as a component of agroforestry. In: Agroforestry: present status and scope for future development in farming systems. Narendra Dev University of Agriculture & Technology, Faizabad

Singh UN, Bhatt DN, Yadav JSP (1987) Growth and biomass production of certain forest species as influenced by varying pH levels. In: Proceedings of international symposium on afforestation of salt affected soils, vol 2. CSSRI, Karnal, pp 51–62

Singh G, Abrol IP, Cheema SS (1988) Agroforestry on alkali soil-effect of planting methods and amendments on initial growth, biomass accumulation and chemical composition of mesquite (*Prosopis juliflora* (SW) DC) with inter-space planted with and without Karnal grass (*Diplachne fusca* (Linn) P. Beauv.). Agrofor Syst 7:135–160

Singh G, Abrol IP, Cheema SS (1989) Effects of gypsum application on mesquite (*Prosopis juliflora*) and soil properties is an abandoned sodic soil. For Ecol Manage 29:1–14

Singh G, Gill HS, Abrol IP, Cheema SS (1991) Forage yield, mineral composition, nutrient cycling and ameliorating effects of Karnal grass (*Leptochloa fusca*) grown with *Prosopis juliflora* in a highly alkali soil. Field Crops Res 26:45–55

Singh G, Singh NT, Tomar OS (1993) Agroforestry in salt-affected soils. Technical bulletin no. 17, CSSRI, Karnal, India, 65 p

Singh G, Singh NT, Dagar JC, Singh H, Sharma VP (1995) An evaluation of agriculture, forestry and agroforestry practices in moderately alkali soil in northwestern India. Agrofor Syst 37:279–295

Singh G, Dagar JC, Singh NT (1997) Growing fruit trees in a highly alkali soils – a case study. Land Degrad Dev 8:257–268

Singh YP, Sharma DK, Singh G, Nayak AK, Mishra VK, Singh R (2008) Alternate land use management for sodic soils. CSSRI technical bull no. 2/2008, CSSRI, Karnal, p 16

Smedema LK (1997) Biological drainage: myth or opportunity? GRID Magazine of the IPTRID network, issue 11, p 3

Sonogan RMC, Patto PM (1985) The use of trees and other vegetation in the management of seepage from earthen channels in the Wimmera-Mallee region. Internal report, Department of Conservation, Forests and Land and Rural water Commission of Victoria, Australia

Spalding M, Blasco F, Field C (1997) World mangrove atlas. International Society for Mangrove Ecosystems, Okinawa, p 180

Sprenger M, Oelmann Y, Weihermuller L, Wolf S, Wilcke W, Potvin C (2013) Tree species and diversity effects on soil water seepage in a tropical plantation. For Ecol Manage 309:76–86

Stirzaker RJ, Cook FJ, Knight JH (1999) Where to plant trees on cropping land for control of dryland salinity: some approximate solutions. J Agric Water Manage 39:115–133

Thorburn P (1996) Can shallow water tables be controlled by the revegetation of saline lands? Aust J Soil Water Conserv 9(3):45–50

Tomar OS (1997) Technologies of afforestation of salt-affected soils. Intnl Tree Crops J 9(2):131–158

Tomar OS, Gupta RK (1984–1994) Performance of forest tree species in saline soils. Annual reports. CSSRI, Karnal

Tomar OS, Gupta RK, Dagar JC (1998) Afforestation techniques and evaluation of different tree species for waterlogged saline soils in semiarid tropics. Arid Soil Res Rehab 12(4):301–316

Tomar OS, Dagar JC, Singh YP (2004) Forest and fruit trees for alkali soils. Indian Farming 53(11):44–47

Travis KA, Heuperman AF (1994) Agroforestry-checkbank plantings; the interaction of checkbank plantings and irrigated pastures under shallow water table conditions. Final report to the Murray-Darling Basin Commission. Victoria Department of Agriculture technical report series no. 215

Turner NC, Ward PR (2002) The role of agroforestry and perennial pasture in mitigating waterlogging and secondary salinity, summary. Agric Water Manage 53:271–275

van der Moezel PG, Pearce-Pinto GVN, Bell DT (1991) Screening for salt and waterlogging tolerance in *Eucalyptus* and *Melaleuca* species. For Ecol Manage 40:27–37

Watson C (1987) Preliminary assessment of halophytes for salt-affected lands of Thailand. STDP, Bangkok, p 199

Webster A (1984) Intercepting channel seepage in the Wimmera-Mallee; the use of trees and tall wheat grass. In: Water talk, vol 52. Rural Water Commission of Victoria, Australia

Welch BL (1978) Relationships of soil salinity, ash, and crude protein in *Atriplex canescens*. J Range Manage 31:132–133

Yadav JSP (1972) Improvement of saline and alkali soils through biological methods. Indian For 101:385–395

Yadav JSP, Singh K (1986) Response of *Casuarina equisetifolia* to soil salinity and sodicity. J Indian Soc of Coast Agric Res 4:1–8

Yadav JSP, Bhumbla DR, Sharma OP (1975) Performance of certain forest species on a saline sodic soil. In: Proceedings of international symposium on new development in the field of salt affected soils of sub-commission of salt affected soils. International Society of Soil Science at Cairon, pp 683–690

Yensen NP, Bedell JL (1993) Consideration for the selection, adoption and application of halophyte crops to highly saline desert environments as exemplified by the long-term development of cereal and forage cultivars of *Distichlis* spp. (Poaceae). In: Lieth H, Al Masoom A (eds) Towards the rational use of high salinity tolerant plants, vol 2. Kluwer Academic Publishers, Dordrecht, pp 305–313

Yensen NP, de Yensen SB (1987) Development of halophyte grain: prospects for reclamation of salt ruined land. J Wash Acad Sci 77(4):209–214

Yensen NP, Bedell JL, de Yensen SB (1988) The history of the development of *Distichlis* cultivars for grain, forage and sad. In: Proceedings of the Mexican national conference on halophytes. Ciudad Obregon, Sonora, Mexico

Yuvaniyama A, Arunin S (1993) Growth of three halophytic grasses on salt-affected soil in Northeast Thailand. In: Davidson N, Galloway R (eds) Productive use of saline land. ACIAR proceedings no. 42, Canberra, Australia, pp 32–35

Zafar I (1990) Effect of soil salinity on the early development of three important species and two provenances of Acacia nilotica. In: Proceedings of the workshop on tree production from saline waterlogged soils, held in Faisalabad, Pakistan, pp 6–23

Zohar Y, Schiller G (1998) Growth and water use by selected seed sources of eucalyptus under high water table and saline conditions. Agric Ecosys Environ 69:265–277

Use of Tree Plantations in Water-table Drawdown and Combating Soil Salinity

P.S. Minhas and J.C. Dagar

Abstract

The prime requirement for rehabilitating saline waterlogged soils is reverting the flux of water for leaching salts beyond active root zone. Though engineering approaches like surface and sub-surface drainage have been standardised for controlling salt and water balance, their adoption is constrained by high capital investment and associated operational and maintenance problems in addition to drainage water disposal. As an alternative, use of tree plantations (often referred as 'biodrainage') has been advocated without long-term verification. The main force behind this notion is water profligate nature of some tree species and their deep root systems. However, it is now clear that the water use by trees varies with specific site conditions defining soil type, evaporative demands and even the depth to groundwater and its salinity. Under favourable conditions (sandy and deep soils, shallow water table of good quality, cooler climate), trees may draw soil water at about 0.8 × Epan, but these may reduce to about 0.2 × Epan under less optimal conditions (clayey and shallow soils, saline/deeper water table, hot and dry summer, etc.). Nevertheless, the major advantage of tree plantations in waterlogging control can be viewed in terms of year-round water with drawls and that too from rain-recharged soil profiles/the shallow water tables. The tree plantations, especially of *Eucalyptus* species, have been shown to drawdown water table, of course their spatial extent being governed by tree transpiration rates and hydraulic characters of soils. But for trees to be effective, land requirements have been estimated to be very high (10–50 % of the total land), the other issue being the

P.S. Minhas (✉)
National Institute of Abiotic Stress Management,
Baramati, Pune 413115, Maharashtra, India
e-mail: minhas_54@yahoo.co.in

J.C. Dagar
Central Soil Salinity Research Institute,
Karnal 132 001, Haryana, India
e-mail: dagarjc@gmail.com

J.C. Dagar, P.S. Minhas (eds.), *Agroforestry for the Management of Waterlogged Saline Soils and Poor-quality Waters*, Advances in Agroforestry 13, DOI 10.1007/978-81-322-2659-8_3

salt accumulation, once the deeper-rooted trees skim out the water. All these factors indicate towards the myths being created for biodrainage without sufficient data and experimentation to support the notion. The alternatives being proposed are shifting shelterbelts instead of block plantation in recharge/upland discharge area, boundary plantations in flat land and even integrating sub-surface drainage and tree plantations. These issues call for GIS/remote sensing-based prognosis of hot spot areas to be covered under tree plantations. Further, the modelling efforts to predict salinity with proposed plantations should help in afforestation designs and highlight other management options so as to promote the bio-management of waterlogged and saline soils.

Introduction

One of the major processes degrading the soil productivity in irrigated commands of arid and semiarid regions of the world, especially where irrigation is practiced without natural or artificial drainage, is the rising trend in water table and resultant accumulation of salts in excess for practical and normal production of crops. Additional cause of rising water tables is large-scale clearing of native vegetation from landscapes to allow for increased amounts of rainfall to permeate through soils (recharge) and resulting in dryland salinity. It is estimated that about 20 % of the global irrigated area across 75 countries is salt affected with far-reaching and unacceptable socio-economic and environmental consequences (Ghassem et al. 1995). The well-known salinity-afflicted regions include Murray-Darling basin in Australia, the Aral Sea (Amu and Syr Darya) basin in Central Asia, the Yellow River basin in China, the Indo-Gangetic basin in India, the Indus basin in Pakistan, the Euphrates basin in Syria and Iraq and the San Joaquin valley in USA. The anthropogenic activities and global warming are further likely to increase salt-induced soil degradation.

An estimated land area of 3.45 Mha is afflicted with secondary salinisation, while 4.52 Mha is prone to waterlogging problems in India (Singh 1992). These areas mostly exist adjacent to canals especially where drainage facilities are poor and canal levels are higher than ground level and where groundwater is of poor quality and is not pumped at rates sufficient enough to arrest rise in water table due to seepage. Even in unirrigated areas, low-lying parts, which act as discharge sites, are prone to waterlogging and saline seep problems. The prime requirement of rehabilitating the afflicted lands is that of reverting flux of water for salt leaching. To achieve this end and to control water table deep enough to provide for adequate root development, aeration and trafficability, provision for natural or artificial drainage seems essential. Though the engineering approaches like surface and sub-surface drainage have been standardised to rehabilitate the saline waterlogged lands, their adoption on large scale is being hindered by very high capital investment and associated operational and maintenance problems in addition to suitable alternatives for disposing drainage waters. As an alternative, the use of vegetation for managing water fluxes to recharge, often referred to as biodrainage, has been advocated for dryland salinity control (Heuperman et al. 2002; Turner and Ward 2002; Morris and Benyon 2005; Marcar 2009). Efficiency of the tree plantations has also been shown for seepage control from canals (Kapoor and Denecke 2001; Manjunatha et al. 2005). In fact, a traditional practice in north-western India is to have tree plantations in areas adjoining the canal embankments, but there has been a lack of information on the role of these trees and also if large-scale plantations are taken in the adjoining areas, in controlling waterlogging and salinity in flat-irrigated areas. Such information is required as some are even sceptical that tree plantations like *Eucalyptus* (Vandana-Shiva 1991), that are being recommended for biodrainage, have been one of the major culprits for excessive depletion and thus fast lowering of groundwater levels in many irrigated areas.

Moreover, farm plantations have been shown as an important and economical option for salt-degraded soils while improving physico-chemical properties through organic matter additions and also the ecological benefits (Tomar and Minhas 1998; Tomar et al. 2003), but their use for salinity control or biodrainage has been debated a lot (Minhas 2006). This is because issues related to accurate estimates of tree water use and salt accumulation, i.e. once the deep-rooted trees distil out the salts and these may affect the long-term sustainability of tree plants growing under saline conditions, remain unanswered (Morris and Collopy 1999; Silberstein et al. 1999; Mahmood et al. 2001; INCID 2003). In fact, the overall salt dynamics vis-à-vis performance of the agroforestrysystems would be governed by a combination of site-specific factors like soil type, rainfall amount and distribution, groundwater level and quality, uptake patterns and salinity tolerance of tree plantations and also the irrigation management practices in adjoining recharge areas. Keeping above in view, some of the hydrological and other information on water use by trees and also the options for better utilisation of tree-based systems for reducing the degradation of lands during rising water tables are discussed below.

Plantation Water Use

In natural environments, vegetation plays an important role in evapotranspiration and soil water storage component of the hydrological balance of an area. Tree plantations are often expected to use water at higher rates than the shorter vegetation. This is because of higher aerodynamic roughness of tree plantations, clothesline effect in tree rows and deeper rooting system, leading to access to water down to several metres. Therefore, it has been generalised that water use by trees is more than the agricultural crops. *Eucalyptus* has been the principle species, because of its importance for waterlogging control and reforestation, for which comprehensive and detailed studies on its water use have been conducted in Australia. However, in general, the transpiration demands for the most of vegetation types are determined principally by the following:

1. Climatic demand as determined by factors like humidity, solar radiation, air temperature, wind, etc.
2. Availability of soil water to roots as influenced by water supplies with rainfall/ irrigation to build up soil water storage and its movement as affected by soil texture and structure for controls over soil hydraulic properties and water-holding capacity
3. Tree factors which control physiological response mechanisms like canopy structure (leaf area index, sapwood area) and capacity of hydraulic lift to transport water, rooting depth and the response of stomata and xylem conductivity to moisture stress.

All of these are further influenced by nature (species/genotype) and age of plantation as well as agroclimatic conditions.

Though the actual water use by trees is difficult to measure, several reports have appeared employing the methods like ventilated chambers, micrometeorological methods using Bowen ratio and Eddy correlation, neutron probes, isotopes (deuterium water) and of late the sap flow/heat flux sensors for monitoring the water use by tree plantations. These are summarised in Table 1. Earlier measurements using ventilated chambers on pastures and more promising species of 8-year-old *Eucalyptus* (Greenwood et al. 1985) showed that crops and pastures discharged about 400 mm year^{-1} of rainfall, whereas the best *Eucalyptus* evaporated about 2500 mm year^{-1}. This sixfold difference was ascribed to the deep phreatophytic root system of *Eucalyptus* and to greater leaf area duration (evergreen) and advective energy. The advective energy expressed itself in measurements of rainfall interception,

Table 1 Water use by tree plantations from shallow water table sites

Authors	Method	Species	Age (years)	Rainfall (mm)	OPE (mm)	Water use (mm year^{-1})	
						Total	T/Epan
Salama et al. (1994)	Sap flow	*Eucalyptus camaldulensis*	10		2025	416	0.21
	ECw 9.0 dS m^{-1}						
Khanzada et al. (1998)	-do- ECw 20	*Acacia nilotica*	3–5	200	2650	1248	0.47
	1.5 dS m^{-1}					2225	0.84
Cramer et al. (1999)	Sap flow/isotopes	*E. camaldulensis*	6–8	738	1623	1–3[a]	~0.35
		Casuarina glauca				1.5–3[a]	~0.37
Morris and Collopy (1999)	Sap flow	*E. camaldulensis*	5–7	480	1350	339	0.25
	ECe ~ 10 dS m^{-1}	*C. cunninghamia*				359	0.27
Heuperman et al. (2002)	Salt build-up	*E. grandis*	6–16	480	1403	730	0.52
Mahmood et al. (2001)	Sap flow	*E. camaldulensis*	4–5	500	1618	1393	0.86
		E. microtheca				1048	0.65
		E. camaldulensis	6–7	250	1618	1179	0.73
		Acacia ampliceps				624	0.38
		Prosopis juliflora				235	0.15
Kapoor and Denecke (2001)	Water balance	*E. camaldulensis*	4–10	335	2971	3446	1.2
		A. nilotica					
Jeet-Ram et al. (2007)	Cl modelling	*E. tereticornis*	18	490	1715	1302	0.76
Forrester et al. (2010)	Sap flow	*E. globulus*	14	569	882	358	0.40
Macfarlane et al. (2010)	Sap flow	*Eucalyptus* spp.	3–16	320–2620			0.21–0.73
Hubbard et al. (2010)	Sap flow	*E. grandis*	5	1180		1102	0.80
Jeet-Ram et al. (2011)	Sap flow	*E. tereticornis*	5–6	212	~1700	268	~0.16
Minhas et al. (2015)	Sap flow Irrigated	*E. tereticornis*	3–10	700	1544	1087	0.70
		"(4×density)"				1371	0.88

[a]mm d^{-1}

which ranged between 19 and 34 % with an average of about 25 % of rainfall on *Eucalyptus* trees. This is determined by interception capacity of leaves and has been reported to be still higher (27–34 %) in the case of *Pinus*, Shola forest and *Acacia* (Cladder 1986). Greenwood et al. (1981) also measured total evaporation from an agroforestrysystem with high-pruned *Pinus radiata* (16 years old and 16 m high) and *Trifolium* subterranean pasture grazed by sheep to be 900 mm on a site where no evidence of perched water table was found.

Sharma (1984) utilised the water balance approach and monitored the evapotranspiration patterns of a natural Jarrah-marri catchment (*E. marginata*, *E. calophylla*, crown area 28 m^3 ha^{-1}, 30 m high) of high rainfall (>1000 mm). Water balance components for winter (April–September) and summer (October–March) averaged over 5 years (1974–1978) indicated that ratio of evapotranspiration to pan evaporation averaged 0.35 and 1.15 for respective periods. Stream flow (~112 mm) mainly occurred during winter. Soil water storage was usually the major component of water balance amounting to 200 mm of 1000 mm rainfall. Evaporation during summer was mainly derived from soil water where the monthly depletion up to 90 mm was monitored. Similarly, Tomar et al. (2003) monitored that at the time of ceasing supplemental irrigation after about 3 years of planting of 31 tree species prevalent in arid and semiarid part of north-western India, the carry-over water from rainfall (532 mm) and irrigation (324 mm) ranged between 17 and 277 mm only, and the water use between 2.5 and 6 years of planting averaged 1.8–3.6 mm d^{-1} with the better extracting species being *Acacia nilotica,Azadirachta indica* and *Eucalyptus tereticornis.*

Periodic groundwater table depths within and adjacent to plantation area have been utilised to compute water uptake by trees utilising the hydraulic water table drawdowns and even chloride modelling. Based on estimates of seepage from IGNP canal, the recharge from rainfall and irrigation and water table drawdown, Kapoor and Denecke (2001) calculated that the annual rate of evaporation by a mixed plantation of *Eucalyptus* and *Acacia* (4–10 years) in an arid climate was 3446 mm which was 1.2 times the open pan evaporation (2971 mm). Similarly based upon chloride modelling, Jeet-Ram et al. (2007) have computed that the average annual recharge during 18 years of *Eucalyptus* plantation was 1547 cm, and considering the effective rainfall, the annual evapotranspiration (ET) averaged 1302 mm. Further, computations using Dupuit-Forchheimer formula showed that the trees were still actively transpiring water and ET between April 2004 and March 2005 (19th year) equalled about 1052 mm.

Though each of the above methods has its own advantages and limitations, the sap flow sensors which monitor transpiration on the basis of thermoelectric heat pulse are considered to be a remarkable improvement with automatic and extraordinary accurate measurements of transpiration. Hunt and Beadle (1998) in a study on *Eucalyptus nitens* mixed with *Acacia dealbata* in Tasmania recorded mean daily sap flux ranging from 1.4 to 103.6 l d^{-1}. Plantation transpiration varied from 1.4 to 2.8 mm d^{-1} in mixed 8-year-old plots and was 0.85 mm d^{-1} in a mixed 4-year-old plot. However, from regression models, it was predicted that, in the absence of *Acacia* competition, plantation water use for the 8-year-old stand would approach 5–6 mm day^{-1} during the growing season. Morris and Collopy (1999) reported average single tree water use to vary from less than 10 l d^{-1} in winter and over 30 l d^{-1} in summer. Stand water use averaged 0.9 ± 1.0 mm d^{-1} over 2 years and was evidently limited by soil water availability. The leaf area index (LAI) of the *E. camaldulensis* stand was estimated as 2.07. Morris et al. (2004) monitored tree growth, water use, climate and soil water conditions over 12 months in two 3–4-year-old *Eucalyptus urophylla* plantations on the Leizhou Peninsula of southern China. One of the plantations (Hetou) was established on a sandy soil of sedimentary origin with low water storage capacity, and the other (Jijia) plantation was established on a clay soil formed on basalt. Sapwood area was ~50 % higher at latter than at former because of differences in plant spacing (1994 versus 1356 stems ha^{-1}). Annual water use was 542 mm at former and 559 mm at latter, with mean sap flux densities of 2772 and 1839 l m^{-2} d^{-1}, respectively. Limitations to water use, imposed by climatic and soil factors, were quantified by analysis of daily canopy conductance in relation to daytime vapour pressure deficit (VPD) and soil water content. Similar annual water use at the two sites was a result of higher VPD and soil water availability at Hetou compensating for the greater sapwood area at Jijia. Potential annual water use in the absence of soil water limitation was estimated at 916 mm at Jijia and 815 mm at Hetou. Higher water availability during the dry season and early wet season at Hetou than at Jijia was the result of deep root systems.

In Australia, mean annual ET is usually less than 650 mm in grasslands but can exceed 1300 mm in forests (Vertessy 1999; Cornish and Vertessy 2001). Relationship between mean annual rainfall in Victorian catchment with different proportions of grassland and forest cover were developed for 19 large catchments (Fig. 1). ET tended to increase with rainfall due to its higher interception as well as higher productivity,

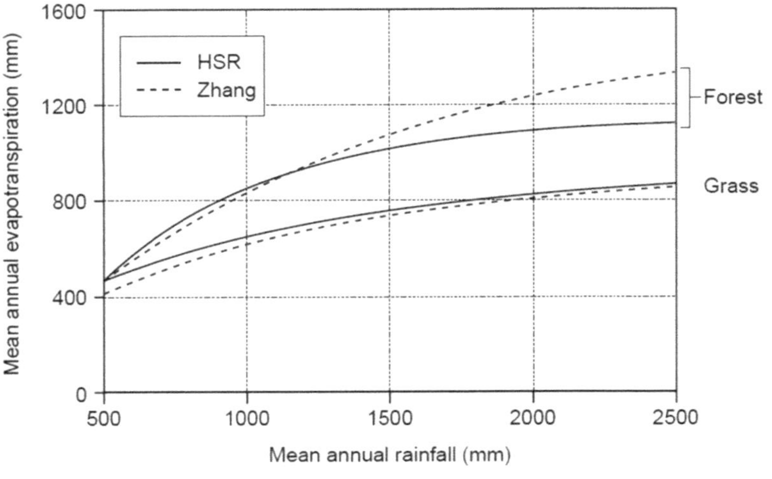

Fig. 1 Relationship between land cover, mean annual rainfall and mean annual evapotranspiration as predicted by the Holmes and Sinclair (1986) relationship and the Zhang et al. (2001) model (Source: Vertessy 2001)

while Zhang et al. (2001) predicted that the mean annual ET should rise to about 1400 mm in wet sites. However, the transpiration of trees monitored in Australia rarely exceeded 0.5 × Epan (Cramer et al. 1999; Morris and Collopy 1999), while much higher values have been reported from Pakistan even on some saline sites (Khanzada et al. 1998; Mahmood et al. 2001). It thus emerged from the latter studies that though *Eucalyptus* may have high water use, it appears unlikely to be greater than of alternative species with similar growth rate as *Acacia nilotica*. Later, Morris and Benyon (2005) observed the transpiration (T; mm $year^{-1}$) in North Victoria to depend upon the total water supply (TWS; mm $year^{-1}$), i.e. rain plus irrigation and groundwater, and the defining function ($r^2 = 0.94$) was T = 0.87 TWS – 153.96. Thus, results demonstrated that once the canopy has closed (tree crowns shading the most of ground beneath), the stand water use is largely a function of water availability up to a maximum imposed by the prevailing climate.

Dye (1996) estimated annual transpiration in *Eucalyptus grandis* × *camaldulensis* clones at two sites in South Africa (MAP ~1100 mm and 1390 mm) and reported rates ranging from 900 to 1400 mm per year. Almeida et al. (2007) estimated lower annual transpiration at the Aracruz site in Brazil in a catchment dominated by *E. grandis* hybrid plantations using water balance approach. They estimated that annual transpiration ranged from 635 mm to 1092 mm over a 6-year period (annual precipitation averaged 1147 mm at the same period). The percentage of the precipitation removed by transpiration was 77 %. In Brazil, Hubbard et al. (2010) also compared the effects of irrigation on water use and water use efficiency in clonal *Eucalyptus* (*E. grandis* x *urophylla*) plantations. Annual water use at Aracruz was 1394 mm in rainfed treatments versus 1779 mm in irrigated treatments and accounted for approximately 67 % and 58 % of annual precipitation and irrigation inputs, respectively; and the increased water use in the irrigated stands was associated with higher sapwood area, leaf area index and transpiration per unit leaf area. The water use efficiency, however, was not influenced by irrigation. Stand water use at the Veracel site totalled 975 mm and 1102 mm in rainfed and irrigated treatments, respectively, and the trends were similar to other sites. Under north-Indian conditions, the annual sap flow values for a 10-year rotation of irrigated *Eucalyptus* plantations increased from 53 to 106 cm (increment of about 14 cm $year^{-1}$) between 2 and 6 years and stabilised (138–141 cm) thereafter (Minhas et al. 2015; Fig. 2). This indicated the advantages of *Eucalyptus* plantation over annual crops (rice/maize/ cotton-wheat) in terms of evapotranspiration losses which would occur

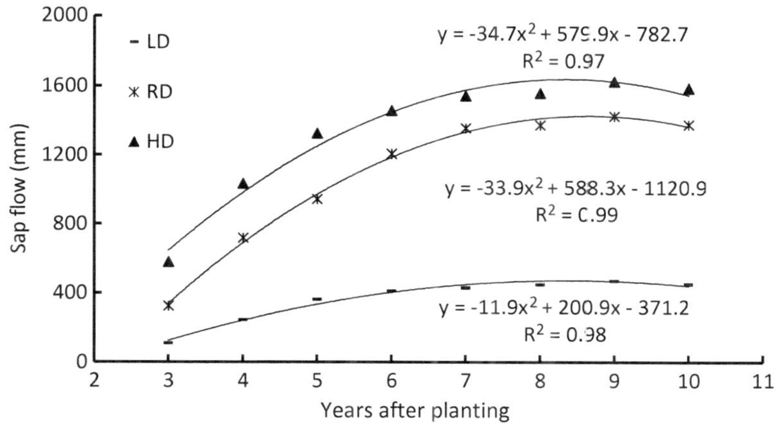

Fig. 2 Annual sap flow as a function of *Eucalyptus* plantation age (years) under varying stocking densities (LD (163), RD (553) and HD (1993 trees ha^{-1}) (Source: Minhas et al. 2015)

only after about 6 years of their planting. There were slight improvements (~11 %) with high-density (four times) plantation (~160 cm $year^{-1}$).

Tree water use is however also sensitive to local conditions including soil water and groundwater salinity, e.g. in a study with the use of saline drainage water (EC 10 dS m^{-1}), the crop coefficient of *Eucalyptus* trees was 0.83 rather than the anticipated value of 1.1 or 1.2 (Tanji and Karajeh 1993). However, for border planted (300 ha^{-1}) cloned *Eucalyptus tereticornis* of about 5 years of age, Jeet-Ram et al. (2011) reported the lowest annual sap flow values of 268 mm only. Thus, the overall water use by trees seems to vary a lot with the method used, and also the specific site conditions defining soil type, evaporative demands and even the depth to groundwater and its salinity determine the actual water use by the species under consideration. Nevertheless, the most of above studies show that under favourable conditions (sandy and deep soils, shallow water table of good quality, cooler climate), trees may draw soil water at about 0.8×Epan, but the drawls may reduce to about 0.2×Epan under less optimal conditions (clayey and shallow soils, saline/deeper water table, hot and dry summer, etc.). Nevertheless, their major advantage in waterlogging control can be viewed in terms of year-round water with drawls unless being deciduous and that too from rain-recharged soil profiles and even the shallow water tables since plantations being mostly the rainfed/nonirrigated trees are mostly deep rooted.

Groundwater Use and Drawdown with Trees

Strong impacts of agricultural development in Australia were monitored on water tables vis-à-vis stream salinity since the magnitude of increase in stream salinity was related to proportion and location of agricultural clearing in addition to annual rainfall at the sites (Schofiled and Bari 1991). Therefore, a number of experimental reforestation sites were established to unravel the contributions of recharge reduction and groundwater use in lowering water tables. Since the water table drawdowns with tree plantations would occur only when these trees use groundwater over and above the recharge through lateral flows from surrounding areas, a number of reforestation strategies were tested. These included (a) lower slope and discharge zone planting, (b) wide-spaced planting covering most of the cleared areas, (c) strips or small blocks strategically placed but covering a small proportion of the cleared area and (d) dense plantations covering a high proportion (>50 %) of the cleared area. Field measurements in Western Australia have shown that reforestation does help lower the water tables, but the response of groundwater levels beneath reforestations was found to depend primarily on area planted and crown cover. Relative to groundwater levels beneath the pastures, groundwater levels beneath the reforestation decreased proportionately with area planted (for similar crown covers) and with total tree cover (as represented by-product of area planted and crown cover). Generally groundwater levels began to decline beneath reforestation relative to pasture 3–4 years after planting. The strategy with low-percentage planting (c) had little effect on groundwater levels and was not recommended in the rainfall zone of 600–900 mm. The strategy with high-percentage planting in dense stands (d) achieved groundwater reductions of the order of 5 m in 9 years. George et al. (1999) reviewed the effects of tree planting on groundwater at 47 discharge sites in Western Australia, and conclusions were as follows:

1. Changes in water table levels ranged from increases of 1 m to decreases of 2.5 m, with the majority of sites showing a decrease.
2. Tree planting is more effective in lowering groundwater if salinity of water is low and where local flow systems, including perched water tables, predominate. This is because trees are able to take up less water with salinity above EC 5–10 dS m^{-1} and tree growth is impaired at such salinity levels.
3. Trees had little or no impact on the water tables more than 10–30 m away from the area planted.

In India, the use of plantations has mostly been viewed for halting seepage control from canals. In a study to evaluate the effects of different tree species in a saline vertisol in Tung Bhadra Project (TBP) area, *Acacia nilotica* intercepted 90 % of seepage from a canal (1991–993), followed by *Dalbergia sissoo* (84 %), *Sesbania grandiflora* and *Casuarina equisetifolia* (72 %), while *Hardwickia binata* and *Azadirachta indica* were less than 50 % (Manjunatha et al. 2005). Water tables monitored between the years 1995 and 1999 ranged between 1.3–1.8 and 1.7–2.1 m during canal running and closure periods under plantation area, while it showed a decreasing trend (1.0–0.7 m) outside the plantation. Similarly, Jeet-Ram et al. (2007) monitored groundwater table fluctuations and salinity build-up underneath an 18-year-old plantation of *Eucalyptus tereticornis* (320 m × 40 m, density 118 plants ha^{-1}, height 24.1 m, DBH 0.32 m) on a sandy-loam soil and the external irrigated fields in the vicinity of 0.55–0.73 km from this plantation. The transacts of water table depth across the plantation showed consistent depression of about 0.8–1.08 m in water table beneath the plantation compared with external irrigated fields extending up to 0.73 km, indicating groundwater uptake and flow towards plantation (Fig. 3). Seasonal variations could also be visualised with minimum drawdown occurring during monsoon season (September) when plants are able to take rain-recharged water from soil, while these were maximum during summer (April) with high climatic demand. Similar patterns have later been reported by Jeet-Ram et al. (2011) from ridge-planted *Eucalyptus* on the boundary of rice-wheat fields. The effects are expected to be still higher in arid climates and still lighter soils. For example, Kapoor and Denecke (2001) reported the effects of a 25 ha plantation (1524 m × 261 m strip) along the IGNP Phase II, mainly consisting of *Eucalyptus* and *A. nilotica* (crown density > 40 %) established during 1997 and 1994 that was initially inundated with water. The groundwater under tree plantation fell by 15.7 m in about 6 years. At 100 and 500 m from the edge of the plantation, the level of groundwater was about 9 and 7.6 m higher than at the edge, a drawdown of about 6.7 and 8.1 m, respectively, while along the cross section of canal, the drawdown was 3.6 and 7.8 m at 100 and 300 m distance. Thus, the overall groundwater dynamics underneath the plantations do reveal that water table be significantly lowered relative to adjacent irrigated areas. Therefore, it may be possible to utilise tree plantations to lower water table where these can use water faster from that of recharge from rainfall and seepage from adjoining areas.

The Spatial Extents of Water Table Drawdown

The drawdown zone beneath a plantation is a local phenomenon resulting from tree transpiration rate and hydraulic characteristics of the soil, while the water table level in the surrounding catchments is a result of the balance between

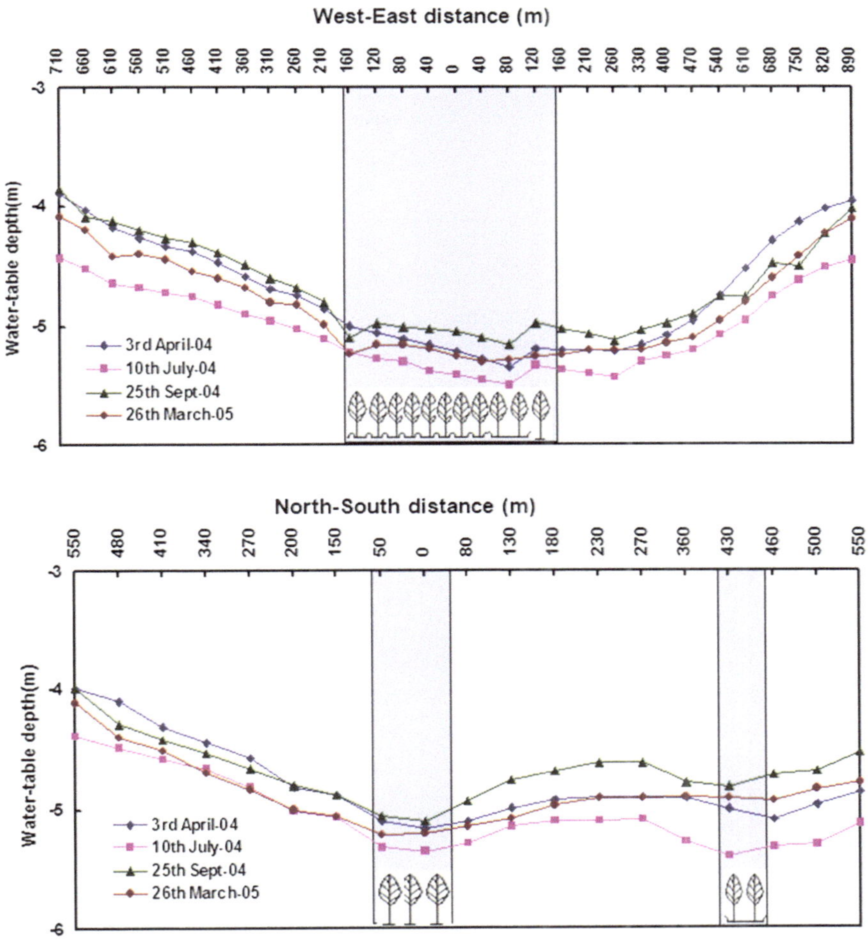

Fig. 3 Trend of groundwater table levels in the east–west transacts under 18-year-old *Eucalyptus tereticornis* plantation at Dabwali, Rohtak, in north-western India. (Source: Jeet-Ram et al. 2007)

recharge and discharge processes in the whole catchment. As such the "regional" water table is affected by groundwater removal by trees, and this effect would become evident when plantations are established on enough large scale within a single locality. The key factors influencing groundwater use by trees are depth to groundwater and its salinity, root growth characteristics, salt tolerance of tree species and soil properties that influence water availability. Evidence for use of saline groundwater by trees has often been obtained or inferred from reductions in groundwater levels, higher evapotranspiration than rainfall by plantations and more direct measurements such as stable isotope discrimination (Thorburn 1996) and chloride mass balance. Groundwater uptake by trees tends to decrease as the water table falls, as a result of lower gravitational potential of the water, greater resistance through longer root xylem transport and limits on the capacity of the tree to supply respiratory substrate to a deep root system. Some tree species which grow in seasonally or perennially wet soil adapt to hypoxic (low oxygen) conditions (e.g. by developing aerenchymatous roots as in *E. camaldulensis*); this is essential for survival in waterlogged soils, but is not required for the uptake of groundwater from the capillary zone.

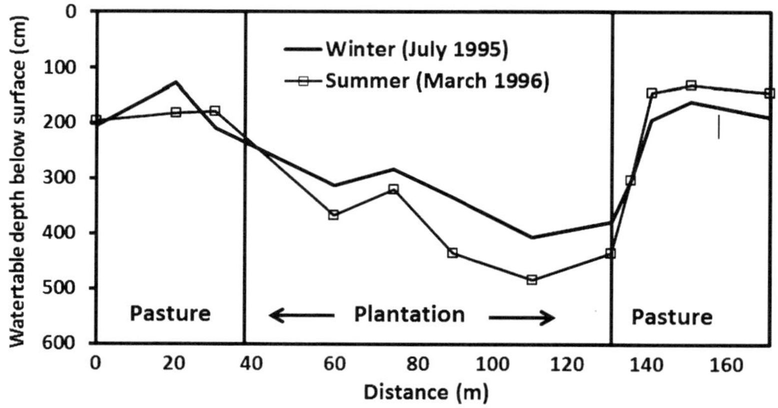

Fig. 4 Transacts of water table depth across 20-year-old rainfed *Eucalyptus* plantation at Kyabram, Australia (Source: Heuperman 2000)

Soil texture has a significant impact on groundwater use by trees, although the confounding effects of site and stand management make interpretations from studies difficult (Barrett-Lennard et al. 1999). The rate of water uptake per unit of root length will be lower from clay than from loamy and sandy soils, mainly because of reduced rate of delivery of water to tree roots. If clay subsoils with low hydraulic conductivity are present, the water table will fall as the water used by trees is replaced more slowly by restricted lateral and upward groundwater flow. Further, extension of the root system is then required to follow the source of water to greater depth (Vertessy et al. 2000).

Heuperman (2000) in a long-term experiment on a medium-textured red-brown duplex soil observed a distinct local drawdown effect on water table underneath the *Eucalyptus* plantations (2.4 ha, 16 years old). Water table levels in the adjoining paddocks were lowered, but the impact only extended to a distance of 50 m into irrigated pasture. However, when strip plantations were attempted (Travis and Heuperman 1994), the drawdown effect of tree lines on light, permeable soils subjected to flood irrigation was too small due to replenishment from adjoining pastures, whereas on heavier, less permeable soils, distinct drawdown occurred and it was confined to 5–10 m on both sides of the tree lines (Fig. 4). However, very drastic drawdown of water tables, down to 8 m to a distance of 500 m, was monitored in a desert soil during 6 years by Kapoor and Denecke (2001), while Jeet-Ram et al. (2007) observed that due to water uptake by tree roots, the influxes of water occurring from the adjoining area to a distance beyond 0.7 km were unable to compensate even during the monsoon rains. Chaudhary et al. (2000) also monitored that on an average water table (1.99 m) under 6-year-old *Eucalyptus* plantation in a 4 ha area was lower by 0.35 and 0.46 cm than in the adjoining area at about 66 and 130 m away from the edge of plantation. Thus, the spatial extent of drawdown of water seems to be strongly affected by soil texture in addition to tree characters.

Salt Accumulations Vis-à-Vis Plantation Sustainability

The long-term sustainability of tree growing in shallow water table will depend upon salt balance mechanisms underneath trees. If trees are irrigated with saline water or take up groundwater from saline water table, a typical process of "skimming/distilling" off of water from the capillary zone and its concentration through uptake by plantation roots is expected to occur. Evidence for this is presented in Fig. 5 for a groundwater-dependent *E. grandis* plantation near Kyabram (Vertessy et al. 2000). Fine roots were most abundant in usually dry soil near the surface, but high root density was also observed lower in the profile above the water table, associated with a zone of high salt concentration (measured as Cl) at

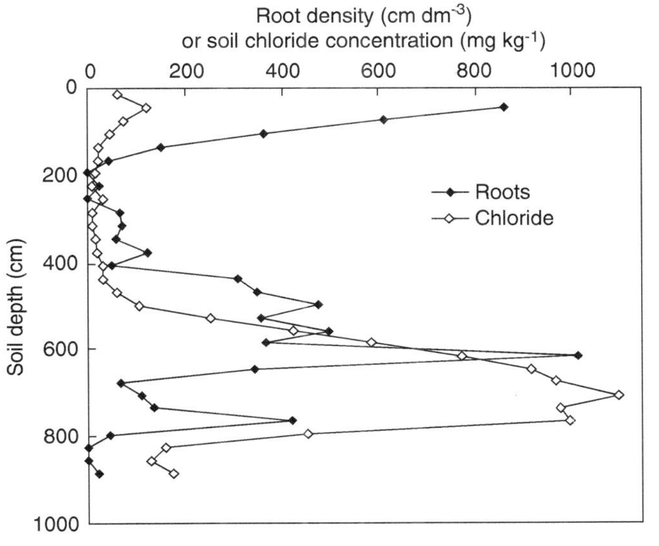

Fig. 5 Depth distribution of root density and soil chloride concentration beneath an 18-year-old *E. grandis* plantation on heavy clay soil underlain with shallow saline groundwater (Source: Marcar and Morris 2005)

600–800 cm soil depth. However, large seasonal changes in root zone salinity indicated that salt accumulation in the soil as a result of groundwater uptake is being balanced to some extent by salt removal.

Thus, there is potential for salt accumulations in soil beneath plantations, and this must raise doubts about long-term effectiveness of plantations for salinity control. The movement of salts in soil is a complex phenomenon that occurs in response to leaching flows, root water uptake and water table movements. Since the lateral movement of water in the unsaturated zone above the water table, from where the most of water uptake vis-à-vis salt accumulations occur, may be negligible, this aspect has been studied at large. The results of the several studies (Tomar et al. 1994; Heuperman et al. 2002; Morris and Collopy 1999; Silberstein et al. 1999; Jeet-Ram et al. 2007) corroborate that trees growing in area with shallow water table result in accumulation of salts in the active root zone. But the values of salinity observed underneath the *Eucalyptus* plantations were often not excessively high and could be tolerated by the most of tree plantations advocated for the seepage control (Jeet-Ram et al. 2007; Fig. 6). Thus, it seems that one commercial tree cycle should be possible for the conditions similar to the site under study, but the tree growth may be eventually limited by salt build-up where the groundwater table water is simultaneously saline. In a simulation study, Silberstein et al. (1999) predicted that plantations grown over with lower salinities may certainly continue beyond 30 years, but sensitivity of plantations to salt concentrations would have the significant influence on growth if the groundwater has salinity above 3 dS m^{-1}. Soil texture and episodic events of rainfall would also have impacts on salt build-up, e.g. under the monsoon climate, the concentration of rainfall and its episodic events during July–September usually result in leaching and deep displacement of accumulated salts especially in lighter textural soils (Minhas and Gupta 1992; Minhas 2012).

The salts accumulated in root zone are more likely to be leached if rainfall is high. The long-term growth of groundwater-dependent plantations may, therefore, be at risk when rainfall is less than 500 mm per year. However, both the rate of accumulation and the maximum salt concentration that can develop are restrained by a

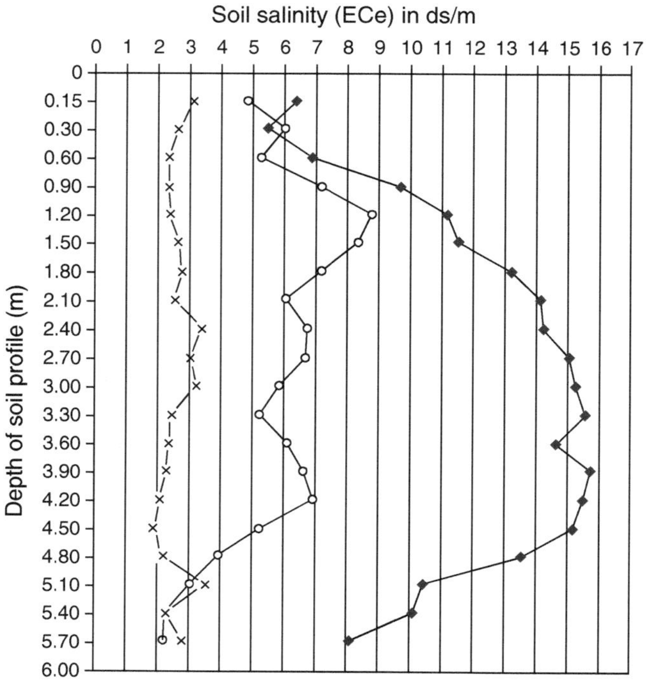

Fig. 6 Distribution of salts in the soil profiles at Dabwali, Rohtak, in north-western India. (-x-) Soil salinity in adjacent fields located in eastern side of plantation, (-o-) soil salinity inside the *Eucalyptus tereticornis* plantation and (-◆-) soil salinity in adjacent fields located in western side of plantation (Source: Jeet-Ram et al. 2007)

reduction in groundwater use that tends to occur as water tables move deeper and become more saline (Morris and Collopy 1999). Even the salt tolerance and the pattern of root growth of trees would also influence the salt dynamics and redistribution of salts. However, plantations growing over shallow, saline water tables may continue to grow provided that equilibrium salt concentrations do not exceed the maximum salinity tolerated by tree species. The typical examples of plantations growing over shallow water tables as cited by Marcar and Morris (2005) are:

- *E. grandis* (slightly salt tolerant) plantations have continued growth over 20 years over a shallow water table site near Kyabram at about 20 m^3 ha^{-1} $year^{-1}$ (Vertessy et al. 2000).
- *E. occidentalis* (highly salt tolerant) shows continued healthy growth at age 14 on a moderately to highly saline site near Stanhope (Bandara et al. 2002).
- *Melaleuca halmaturorum* (extremely salt tolerant) in natural stands draws on saline groundwater in extremely saline swamps of the lower southeast of South Australia (Mensforth and Walker 1996).

Considerations for Taking-up Tree Plantations

For taking-up plantations to control salinity, the challenge is to maintain high level of water use by trees while minimising the salt build-up in capillarity zone from which the most of the roots draw water, either through salt export or by maintaining the salts at deeper depth. Thus, for getting the best solutions under different situations, the factors like soil depth and its hydraulic properties, geology and topography, salt content, nature and distribution of remnant vegetation, annual rainfall and agricultural practices including quantities and qualities of irrigation water are important. The more often criteria used for selecting the criteria tree species include adoption to site conditions, water use and multiple uses of the selected species. When objectives are to halt seepage from canals and create hydrological balance in the adjoining landscapes, woodlots/ blocks of *Eucalyptus* (typically 100–1200 tree ha^{-1} in high-rainfall less-saline zones and 500–600 tree ha^{-1} in low-rainfall high-salinity zone) seem to be the best species especially when with micropropagation techniques, and its faster-growing vis-à-vis high-transpiring clones are now available. Measurements on water use by heat pulse method (Khanzada et al. 1998; Morris and Collopy 1999; Mahmood et al. 2001) further corroborate that the values do not vary greatly among the trees of widely differing species at the same location with good access to soil water or groundwater, even in the presence of substantial salinity variations. Other studies (Heuperman et al. 2002) also indicate that trees are not constrained for water use under low salinity conditions (ECe<4 dS m^{-1}). Otherwise, to overcome such situations where the accumulated salts may inhibit the growth of trees, a better option is to go in for salt-tolerant trees (e.g. *E. camaldulensis,E. loxophleba,E. occidentalis,E. sargentii* and *E. spathulata*), those which have had succeeded in lowering highly saline (EC up to 30 dS m^{-1}) water table (Lefroy and Scott 1994). The other higher salt-tolerant trees include species of *Tamarix, Prosopis, Acacia*, etc., but unfortunately. the most of tolerant trees with halophytic character have very low leaf areas and more often slow growth rates, which raise doubts about their suitability for waterlogging control.

To overcome such situations where the accumulated salts may inhibit the tree growth, a concept of "walking" plantations, where trees, after harvesting, are replaced with shallow-rooted irrigated crops and water table control is provided through new plantations on the adjacent sites, has also been advocated (Heuperman 2000). The other option may be to go in for row/strip plantations on the boundaries of fields to create hydraulic gradients towards these rows and also the conducive zones of salt accumulation (Jeet-Ram et al. 2011), but the optimal area of influence of these plantations and salt dynamics under such agroforestrysystems would have to be worked out. Also for the long-term sustainability of trees under such a system, some form of engineering drainage input for the removal of salts from their root zone can be integrated so that a favourable balance between leaching and salt accumulation processes could be established. Thus, various options should be tested to find out workable solutions to halt water table rise and secondary salinisation through the use of tree plantations, as these will also have regional-scale ecological and economic impacts by providing much needed fuel wood, timber, shade and shelter, function as wind breaks and other rural livelihood advantages.

Although numerous examples of groundwater-level lowering by trees have been reported, there has been little consensus on how much area is required to be put under afforestation. The actual area to be planted would be governed by the objective of such plantation, i.e. whether for halting seepage from canals or other recharge sites in undulating topographies or for water table control in irrigated areas. Thus, answers have to be site-specific and cannot be extrapolated to areas with different conditions. The most extensive data sets have been generated from Western Australia in 700 mm annual rainfall regions, where very close relationship existed between groundwater levels beneath tree plantings and amount of tree cover (Schofield 1992). However, for plantings of less than 15 % of the cleared area, no such groundwater level reduction occurred. For the normal rainfall, about 52 % of the cleared area was predicted to be reforested so as to annually lower the water table at 200 mm relative to the ground surface. Kapoor and Denecke (2001) predicted that 10 % of the area should be planted under trees to transpire the recharge volumes from IGNP. If the required halt of annual rise in water table is 1.0 m, the ratio of area to be covered under plantation should be about 25 % of the land area (Jeet-Ram et al. 2007). Even if the calculations are based on irrigated (equivalent to Epan) *Eucalyptus* plantations with annual transpiration averaging about 902 mm for a 10-year rotation (Minhas et al. 2015), the ratio of area to be covered would be around 15 %. In fact high land requirements for effective waterlogging control through tree plantations may become a constraint and, thus, raise doubts about their suitability in the most canal commands where present area under forestry is too low and land is in short supply and so expansive that it may be very difficult to spare the additional area for forestry.

Conclusions

With increasing environmental concerns, it is axiomatic that tree plantations should form an integral part of landscapes to revive their hydrological balances. Thus, the general challenges in devising plantation programmes are to specify the new vegetation regimes; those should be viable, productive and sustainable. Extensive experiences have been gained the world over on the role that the deep-rooted tree plantations can play for managing waterlogging and salinity. The results have been mixed, and their effectiveness is usually defined by factors like hydrogeology and soil type, water table depth and its salinity, climate (rainfall and evaporation), location of planting and the proportion of land revegetated, the species, their density and the configuration. Trees have generally shown potential in lowering water table under the following situations:

- The selected trees are tolerant to both salinity and waterlogging and can extend roots into phreatic zones. Choice of tree species suited to soils of different levels of salinity and waterlogging is listed in chapter "Global Perspectives on Agroforestry for the Management of Salt-affected Soils".
- The groundwaters are relatively fresh, e.g. intercepting of seepage from canal or other laterally flowing groundwater, about 2–6 m deep, and have low salinity preferably <8 dS m^{-1} especially under dryland salinity conditions.
- Soils are deep and have high transmission properties, e.g. loam sand to sandy loam. Effectiveness is less in heavy-textured soils with low water transmission.
- Reintroduction of tree plantations should be strategically located, e.g. in recharge area, and have dense configuration, e.g. interception belts over perched water tables, boundary plantations, etc.
- Effectiveness of tree plantations to lower water table is more in 400–600 mm rainfall zones due to higher transpiration to rainfall input ratio and decreases with rainfall above this range.
- When planned over block/*taluka*/regional scale, extensive areas, to the extent of 30–50 %, will have to be covered under strategically located tree plantations; otherwise, these must be combined with other agronomic/engineering measures.

In view of the above, the experiences gained can be gainfully employed to mitigate and even reverse waterlogging and secondary salinisation problems, but for these to be realistic under the present farm settings, some of the specific requirements include:

1. Utilising the GIS and remote-sensing techniques for prognosis of hot spot areas to be put under plantations. This should help in prioritisations of action plans for developing integrated command framework to control waterlogging and salinity.
2. Generating temporal information on growth satge dependent transpiration capacity and hydrological effectiveness with high-density plantations of selected salt-tolerant species and even on the role of fast-growing clonal trees, e.g. *Eucalyptus*.
3. Process-based models (like 3-G) to predict salinity within the basin under the present and afforestated conditions. Those should help in afforestation design and highlight management options and priorities.

Thus, the overall knowledge of the equilibrium root zone salinity that is likely to develop and the plant types that can grow at the acceptable rates under the prevailing conditions should help better match between specificity of site, tree species and other silvicultural and agroforestry-practices for better control over salinity.

References

Almeida AC, Soares JV, Landberg JJ, Rezende GD (2007) Growth and water balance of *Eucalyptus grandis* hybrid plantation in Brazil during rotation for pulp production. Forest Ecol Manage 251:10–12

Bandara G, Morris J, Stackpole D, Collopy J (2002) Soil profile monitoring at irrigated plantation sites in northern Victoria: progress report 2001. CFTT Report No. 2002/025, Forest Science Centre, Melbourne

Barrett-Lennard EG, Speijers J, Morris J, Marcar N (1999) Transpiration by trees on land with shallow water-tables: a survey of the literature suggests that transpiration is affected by soil texture. In: Redesign of Australian plant production systems- deep drainage and nitrate losses in the Mediterranean climate region. Land and Water Resources Research and Development Corporation, Canberra, Australia

Chaudhary MR, Chaudhary MA, Subhani KM (2000) Biological control of waterlogging and impact on soil and environment. In: Proceedings of the 8th ICID international drainage workshop, vol II. New Delhi, pp 209–222

Cladder IR (1986) Water use of eucalyptus – a review with special reference to South India. Agric Water Manage 11:333–342

Cornish PM, Vertessy RA (2001) Forest age-induced changes in evapotranspiration and water yield in a eucalypt forest. J Hydrol 242:43–63

Cramer VA, Thorburn PJ, Fraser GW (1999) Transpiration and groundwater uptake from farm forest plots of *Casuarina glauca* and *Eucalyptus camaldulensis* in saline areas of southeast Queensland, Australia. Agric Water Manage 39:187–204

Dye PJ (1996) Response of *Eucalyptus grandis* trees to soil water deficits. Tree Physiol 16:233–238

Forrester DI, Collopy JJ, Moris J (2010) Transpiration along an age series of *Eucalyptus globulus* plantations in southeastern Australia. Forest Ecol Manage 259:1754–1760

George RJ, Nulsen RA, Ferdowsian R, Raper GP (1999) Interactions between trees and groundwaters in recharge and discharge areas — a survey of western Australian sites. Agric Water Manage 39:91–113

Ghassem F, Jakeman AJ, Nix HA (1995) Salinisation of land and water resources: human causes, extent, management and case studies. CABI Publishing, Wallingford

Greenwood EAN, Beresford JD, Bartle JR (1981) Evaporation from vegetation in landscapes developing secondary salinity using the ventilated-chamber technique: III. Evaporation from a *Pinus radiata* tree and the surrounding pasture in an agroforestry plantation. J Hydrol 50:155–166

Greenwood EAN, Klien G, Beresford JD, Watson GD (1985) Differences in annual evaporation between grazed pasture and Eucalyptus species plantation on a saline farm catchment. J Hydrol 78:261–278

Heuperman AF (2000) Biodrainage: an Australian overview and two Victorian case studies. In: Proceedings of the 8th ICID international drainage workshop, New Delhi, India

Heuperman AF, Kapoor AS, Denecke HW (2002) Biodrainage: principles, experiences and applications. International Programme for Technology & Research in Irrigation and Drainage, IPTRID Secretariat, FAO, Rome, p 78

Holmes JW, Sinclair JA (1986) Water yield for some afforested catchments in Victoria. In: Hydrology and water resources symposium. Institute of Engineers, Brisbane/Barton, pp 214–218

Hubbard RM, Stape JL, Ryan MG, Almieida AC, Rojas J (2010) Effects of irrigation on water use and water use efficiency in two fast growing *Eucalyptus* plantations. Forest Ecol Manage 259:1714–1721

Hunt MA, Beadle CL (1998) Whole-tree transpiration and water-use partitioning between *Eucalyptus nitens* and *Acacia dealbata* weeds in a short-rotation plantation in northeastern Tasmania. Tree Physiol 18:557–563

INCID (2003) Biodrainage status in India and other countries. Indian National Committee on Irrigation and Drainage (INCID), New Delhi, p 40

Jeet-Ram, Garg VK, Toky OP, Minhas PS, Tomar OS, Dagar JC, Kamra SK (2007) Biodrainage potential of *Eucalyptus tereticornis* for reclamation of shallow water table areas in north-west India. Agrofor Syst 69:147–165

Jeet-Ram, Dagar JC, Khajanchi L, Singh G, Toky OP, Tanwar VS, Dar SR, Chauhan MK (2011) Bio-drainage to combat waterlogging, increase farm productivity and sequester carbon in canal command areas of north-west India. Curr Sci 100(11):1673–1680

Kapoor AS, Denecke HW (2001) Biodrainage and biodisposal: the Rajasthan Experience. In: GRID, IPTRID's Network Magazine No. 17

Khanzada AN, Morris JD, Ansari R, Salvich PG, Collopy JJ (1998) Groundwater uptake and sustainability of *Acacia* and *Prosopis* plantations in southern Pakistan. Agric Water Manage 36:121–139

Lefroy EC, Scott P (1994) Alley farming: new vision for Western Australian farmland. J Agric 35(4):119–127

Macfarlane C, Bonda C, White DA, Girg AH, Ogden GH, Silberstein RS (2010) Transpiration and hydraulic traits of old and regrowth Eucalypt forest in south-west Australia. Forest Ecol Manage 260:96–105

Mahmood K, Morris J, Collopy J, Slavich P (2001) Groundwater uptake and sustainability of farm plantations on saline sites in Punjab province, Pakistan. Agric Water Manage 48:1–20

Manjunatha MV, Hebbara M, Patil SG, Kuligod VB, Minhas PS (2005) Effect of trees alone or with grasses on halting canal seepage and shallow water table control in saline vertisols. J Indian Soc Soil Sci 53(2):254–257

Marcar NE (2009) Productive use and rehabilitation of saline land using trees. In: Nuberg I, George B, Ried R (eds) Agroforestry for natural resource management. CSIRO, Collingwood, pp 251–266

Marcar NE, Morris J (2005) Plantation productivity in saline landscapes. In: Nambiar EKS, Ferguson I (eds) New forests: wood production and environmental services. CSIRO, Collingwood, pp 51–74

Mensforth LJ, Walker GR (1996) Root dynamics of *Melaleuca halmaturorum* in response to fluctuating saline groundwater. Plant Soil 184:75–84

Minhas PS (2006) Use of tree plantations for waterlogging and salinity control: an overview and research needs. J Water Manage 13:169–179

Minhas PS (2012) Sustainable management of brackish-water agriculture. In: Lal R, Stewart BA (eds) Soil water and agronomic productivity, vol 19, Adv Soil Sci., pp 289–324

Minhas PS, Gupta RK (1992) Quality of irrigation water – assessment and management. Information and Publication Section, Indian Council of Agricultural Research, New Delhi, 123 p

Minhas PS, Yadav RK, Lal K, Chaturvedi RK (2015) Effect of long-term irrigation with wastewater on growth, biomass production and water use by *Eucalyptus tereticornis* planted at variable stocking density. Agric Water Manage 152(4):151–160

Morris JD, Benyon R (2005) Plantation water use. In: Nambiar S, Ferguson I (eds) New forests: wood production and environmental services. CSIRO Publishing, Collingwood, pp 75–112

Morris J, Collopy J (1999) Water use and salt accumulation by *Eucalyptus camaldulensis* and *Casuarina cunninghamiana* on a site with shallow saline groundwater. Agric Water Manage 39:205–227

Morris J, Zhang NN, Yang ZJ, Collopy J, Xu DP (2004) Water use by fast-growing *Eucalyptus urophylla* plantations in southern China. Tree Physiol 24:1035–1044

Salama RB, Bartle GA, Farrington P (1994) Water use of plantation *Eucalyptus camaldulensis* estimated by groundwater hydrograph separation techniques and heat pulse method. J Hydrol 22:307–323

Schofield NJ (1992) Tree planting for dryland salinity control in Australia. Agrofor Syst 20:1–23

Schofiled NJ, Bari MA (1991) Valley reforestation to lower saline ground water tables: results of Stene's farm, Western Australia. Aust J Soil Res 29:635–650

Sharma ML (1984) Evapotranspiration from *Eucalyptus* community. Agric Water Manage 8:41–56

Silberstein RP, Vertessy RA, Morris JD, Feikema PM (1999) Modelling the effects of soil moisture and solute conditions on long-term tree growth and water use: a case study from the Shepparton irrigation area, Australia. Agric Water Manage 39:283–315

Singh NT (1992) Salt affected soils in India. In: Khoshoo TN, Deekshatulu BL (eds) Land and soils. Har-Anand Publications, New Delhi, pp 65–102

Tanji KK, Karajeh FF (1993) Saline drain water re-use in agroforestry system. J Irrig Drain Eng 119(1):170–180

Thorburn PJ (1996) Can shallow water tables be controlled by the revegetation of saline lands? Aust J Soil Water Cons 9:45–49

Tomar OS, Minhas PS (1998) Afforestation of salt-affected soils. In: Tyagi NK, Minhas PS (eds) Agricultural salinity management in India. Central Soil Salinity Research Institute, Karnal, pp 453–472

Tomar OS, Minhas PS, Gupta RK (1994) Potentialities of afforestation of water-logged soils. In: Singh P, Pathak PS, Roy MM (eds) Agroforestry systems for degraded lands, vol 1. Oxford/IBH Publishing Co. Pvt. Ltd., New Delhi, pp 111–120

Tomar OS, Minhas PS, Sharma VK, Singh YP, Gupta Raj K (2003) Performance of 31 tree species and soil conditions in a plantation established with saline irrigation. Forest Ecol Manage 177(1–3):333–346

Travis KA, Heuperman AF (1994) Agroforestry – checkbank plantings; the interaction of checkbank plantings and irrigated pastures under shallow watertable conditions. Final Report to the Murray-Darling Basin Commission, Victoria Department of Agriculture Technical Report Series. No. 215

Turner NC, Ward PR (2002) The role of agroforestry and perennial pasture in mitigating waterlogging and secondary salinization, summary. Agric Water Manage 53:271–275

Vandana-Shiva (1991) Afforestation programmes and land use conflicts. Conflicts over natural resources in India. Sage, New Delhi, p 365

Vertessy RA (1999) The impacts of forestry on stream flows: a review. In: Croke J, Lane P (eds) Forest management for the protection of water quality and quantity. Proceedings of the 2nd erosion in forests meeting, Warburton, 4–6 May 1999. Cooperative Research Centre for Catchment Hydrology, Report 99/6, pp 93–109

Vertessy RA (2001) Impacts of plantation forestry on catchment runoff, pp 9–19. In: Plantation, farm forestry and water. Workshop Proceedings Publication 1/20, Water and Salinity Issues in Agroforestry No. 7, Rural Industries Research and Development Corporation, Canberra, Australia

Vertessy RA, Connell L, Morris J, Silberstein R, Heuperman A, Feikema P, Mann L, Komarzynski M, Collopy J, Stackpole D (2000) Sustainable hardwood production in shallow water table areas. RIRDC Publication No. 07/163. Rural Industries Research and Development Corporation, Canberra, Australia

Zhang L, Dawes WR, Walker GR (2001) Response of mean annual evapotranspiration to vegetation changes at catchment scale. Water Resour Res 37:701–708

Prospects for Managing Salinity in Southern Australia Using Trees on Farmland

Nico Marcar

Abstract

Trees are typically planted in salt-affected agricultural landscapes of southern Australia for environmental and economic benefit. Environmental benefits include land reclamation and containing groundwater rise, reducing salt loads in streams, carbon sequestration, and enhanced biodiversity. Most efforts have concentrated on managing dryland (rainfed) catchments where recharge (water intake) and discharge (water outflow) locations are usually well defined. Trees are typically planted in agroforestry, compact plantation, or shelterbelt configurations mostly in recharge areas or upslope of discharge areas and, less commonly, to stabilize discharge areas. In saline discharge areas, the amount and concentration of salt to which tree roots may be exposed varies with landscape position, salt load in the soil regolith, management practices, the extent of lateral subsurface flow of water, and the degree to which saline soil water and groundwater are used by trees. Tree survival and growth are progressively decreased at higher soil and groundwater salinities, with the extent of this reduction depending on tree species' (genotype) tolerance to salinity and associated stresses, such as waterlogging and sodicity. Several Australian native tree species, including *Eucalyptus camaldulensis, E. occidentalis,* and *E. spathulata* perform consistently well on saline soils. The use of selected eucalypt hybrid clones may enhance prospects for more economically productive farm forests on saline land. Saline water tables are more likely to be lowered by planting trees in saline areas if trees are planted at an appropriate scale and can reduce recharge and/or use groundwater directly. Opportunities to lower the water table beneath a plantation are greater if lateral flows from surrounding areas are relatively small.

N. Marcar (✉)
Formerly of Commonwealth Scientific and Industrial Research Organisation (CSIRO), Australia
e-mail: nico.marcar@gmail.com

J.C. Dagar, P.S. Minhas (eds.), *Agroforestry for the Management of Waterlogged Saline Soils and Poor-quality Waters*, Advances in Agroforestry 13, DOI 10.1007/978-81-322-2659-8_4

However, if the aim is to maximize the use of groundwater by trees for improved tree growth and environmental outcomes, then lateral flow from the surroundings could be an advantage. Case studies are provided from several regions in southern Australia from field experiments and process-based modeling to illustrate growth and water use responses to salinity, opportunities for maintaining water quality of rivers and streams in salinized catchments, and for both utilising and stabilizing salt-affected land in dryland and irrigated situations.

Causes, Extent, and Nature of Salinity in Southern Australia

Causes

The expression of dryland salinity in southern Australia may have several causes, and these appear to vary among and within regions. There is ongoing debate about the relevance of different processes operating in specific regions and the consequence for assessing the extent of salinity and its management. In order to prescribe management options for addressing dryland salinity, a thorough understanding is required of groundwater (either the local, intermediate, or regional) and surface water flow systems at farm to regional scales (Coram et al. 2000).

The main cause of dryland salinity in southern Australia has been extensive land clearing for agriculture and grazing, especially since the mid-nineteenth century, and the ensuing reduction in annual evapotranspiration due to a change from deep-rooted woody vegetation to shallow-rooted annual crops and pastures. This has permitted larger volumes of rainfall than previously to drain through the soil profile and recharge groundwater. As a consequence, water tables become shallower in discharge areas increasing the opportunity for dissolved salts to be carried closer to the soil surface, resulting in dryland salinity outbreaks (Marcar and Crawford 2004). This groundwater-driven model applies particularly to many parts of southwest Western Australia (George et al. 1999), the southeast of South Australia (Jolly et al. 2000), and more western parts of the Murray–Darling Basin (MDB) (Summerell et al. 2005).

However, not all dryland salinity results from groundwater-driven processes. Local lateral flows of surface and near-surface water containing salts from upslope to downslope positions can predominate especially in upland areas of the MDB and in southwest Victoria (Fawcett 2004; Bann and Field 2006, 2010). Soil structural decline, loss of organic matter, increased sodicity, and presence of impermeable subsoils can result in restricted water movement and increasing salinity which may be transient (Rengasamy 2002). Proponents of the importance of surface and near-surface hydrology as causes of salinization argue for the importance of land management (e.g., reduced grazing), improving soil structure (e.g., through improved grass establishment), and localized tree planting to foster leaching of surface salts rather than a focus on broad-scale planting of trees and perennial shrubs (Jones 2000; Bann and Field 2010). In some regions, such as parts of southwestern Victoria, salinity is proposed to have always been a part of the natural environment and not brought about by rising groundwater in response to land clearing (Dahlhaus et al. 2008).

Extent

Soil salinity, shallow water tables, and stream salinization are significant problems predominantly in low to medium (c. 350–700 mm) rainfall zones of the MDB, southwest Western Australia, the southeast of South Australia, and the southwest of Victoria as well as other regions in Australia (National Land and Water

Resources Audit 2001; Fig. 1). Salinity degrades both land and water resources, causing loss of agricultural productivity and biodiversity. Salinization of streams and rivers affects the supply and quality of water for drinking, irrigation, and industrial use, with serious economic, social, and environmental consequences for rural and urban communities (Oliver et al. 1996). Salinity has serious adverse impacts on the ecology of wetlands, rivers, and their associated habitats. Soils in discharge zones are likely to experience a range of salinity and water regimes (dry to seasonally waterlogged) that vary in extent and degree, both temporally and spatially.

Murray-Darling Basin (MDB)

The MDB covers about 1,060,000 km^2, about 14 % of Australia's total area. Salt, principally derived from rainfall, has accumulated over geological time and is a natural feature. Its flat terrain (especially in the western parts), relatively low

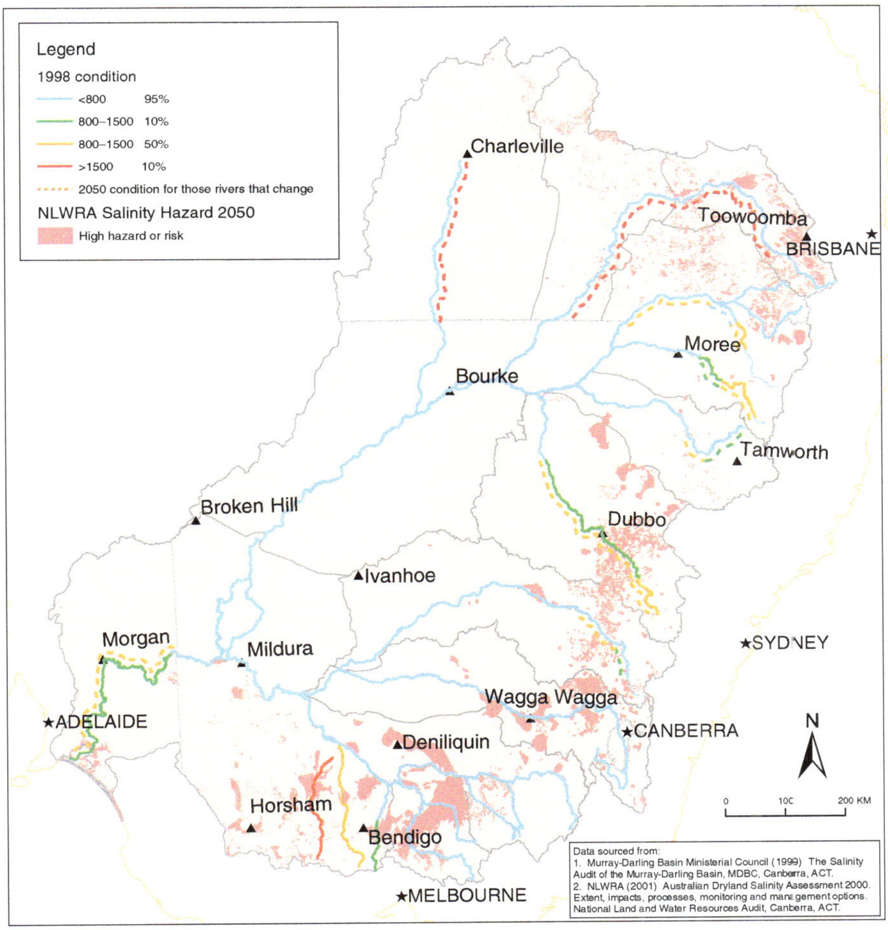

Fig. 1 Land and rivers affected by and at future risk from salinity in the the Murray-Darling Basin (Murray Darling Basin Commission 2004)

annual rainfall, and high potential evaporation have combined to concentrate salt in the soil profile and groundwater. It was estimated that in 1996 there were about 300,000 ha of salt-affected land in the MDB, although much larger areas were underlain by shallow saline groundwater, threatening further land and water degradation (Blackmore et al. 1999; Murray–Darling Basin Commission 1999). This is a consequence of increasing storage of salt in the soil regolith that is not exported to rivers and increasing stream salinity. Outbreaks of dryland salinity are typically patchy with small proportions (usually less than 5 %) of farmland affected, and salinity tends to concentrate around drainage lines and lower slopes (Australian Bureau of Statistics 2002), with mainly low to intermediate groundwater flow systems.

Irrigation salinity occurs mainly in southern parts of the MDB with more intermediate to regional groundwater flow systems and is caused by inappropriate irrigation management and insufficient drainage. It is often associated with shallow water tables and sodic soils. Pumping groundwater for protection of farmland from rising groundwater combined with drainage from irrigated agriculture produces a large volume of water, too saline for reuse on pastures or crops. This waste water has traditionally been disposed of in evaporation basins, but there are opportunities for its reuse for growing tree plantations (see later sections). Between about 1999 and 2009, the dry conditions prevailing throughout the MDB resulted in substantial decreases in water table depths and soil salinity (Department of Environment and Climate Change NSW 2009; Gill et al. 2012), but there has been a limited rising trend since then in some catchments with the onset of wetter conditions (Gill et al. 2012).

Southwest Western Australia

Land salinization in Western Australia has become a major problem since land clearing for agricultural and pastoral pursuits, much of which occurred since the 1950s (Harper et al. 2012). Most river systems now have high salt concentrations. Rehabilitation of salinized waterways is a high priority for water supply catchments. Secondary (dryland) salinity caused by rising groundwater (85 % of which is in local groundwater systems) currently affects about 10 % (about 1.1 million ha) of the 450–750 mm rainfall zone, with up to one third at risk by 2050 (McFarlane et al. 2004). The area affected by salinity in the more steeply incised, higher rainfall zone (550–750 mm) where land was cleared more than 50 years ago is almost at equilibrium. Further water table rise in these areas is likely to result in increased rates of saline discharge rather than expansion of the area affected (George et al. 2012).

More than 50 % of farms are affected by salinity across Western Australia (Kington and Pannell 2003), and the proportion affected by salinity increases in the drier, eastern parts of the wheat belt because of flat topography, slow groundwater movement, and extensive salt storages. Salinity is still developing downslope of land cleared more recently. Hydrologic equilibrium is not likely to be achieved for a further 50–100 years in lower rainfall (450–550 mm) landscapes and flatter areas to the east and may result in 20–30 % of the landscape being affected by shallow water tables (State Salinity Council 2000). Loss of biodiversity and surface soil fertility, including soil organic matter, is also a major issue.

Southeast South Australia and Southwest Victoria

The area currently affected by salinity is about 250,000 ha in southeast SA, almost exclusively in regional groundwater flow systems. Groundwater quality is generally good in the Lower South East and in high demand for domestic and agricultural purposes due to unreliable surface water supplies. Scattered occurrences of land salinization in the Glenelg and Hopkins catchments in Victoria representing more than 27,000 ha are primarily in local groundwater flow systems and, therefore, more amenable to recovery through revegetation strategies. In-stream salinity levels are rather high.

Nature

Soil salinity is usually described in terms of the electrical conductivity (EC), measured in a 1:5 soil: water suspension ($EC_{1:5}$). This value can be converted to EC_e (EC of the extract from a saturated soil paste) by applying conversion factors based on soil texture (e.g., Slavich and Petterson 1993). Soils affected by salinity are usually grouped into classes for the purpose of classifying plant response. Marcar and Crawford (2004) used the following classification: nonsaline ($EC_e<2$ dS m^{-1}), slight (EC_e 2–4 dS m^{-1}), moderate (EC_e 4–8 dS m^{-1}), high (EC_e 8–16 dS m^{-1}), and extreme ($EC_e>16$ dS m^{-1}) saline. These classes reflect average salinity of plant root zones (typically 0.5 to several meters depth). Electrical conductivity gives an approximation of salt concentration but does not provide information on salt composition, which determines specific ion toxicities. A broad range of dissolved salts is known to occur in saline soil and water, principally sodium chloride (NaCl), but other ions such as calcium (Ca), magnesium (Mg), sulfate (SO_4), and bicarbonate (HCO_3) may also be found in large amounts, depending on geology and soil types (Fitzpatrick et al. 2000).

Occurrence of soil salinity varies markedly across catchments, farms, and paddocks. Figure 2 shows the variation in soil electrical conductivity (EM) across a small sub-catchment in central west New South Wales, which includes the paddock described in Fig. 3. These data have not been calibrated to represent soil salinity. However, it can be seen that high (>100 mS m^{-1} = 10 dS m^{-1}) EM readings are confined to the paddock, which is only a small proportion of the sub-catchment. Soil salinity distribution within a paddock on a farm in central west New South Wales is shown in Fig. 3.

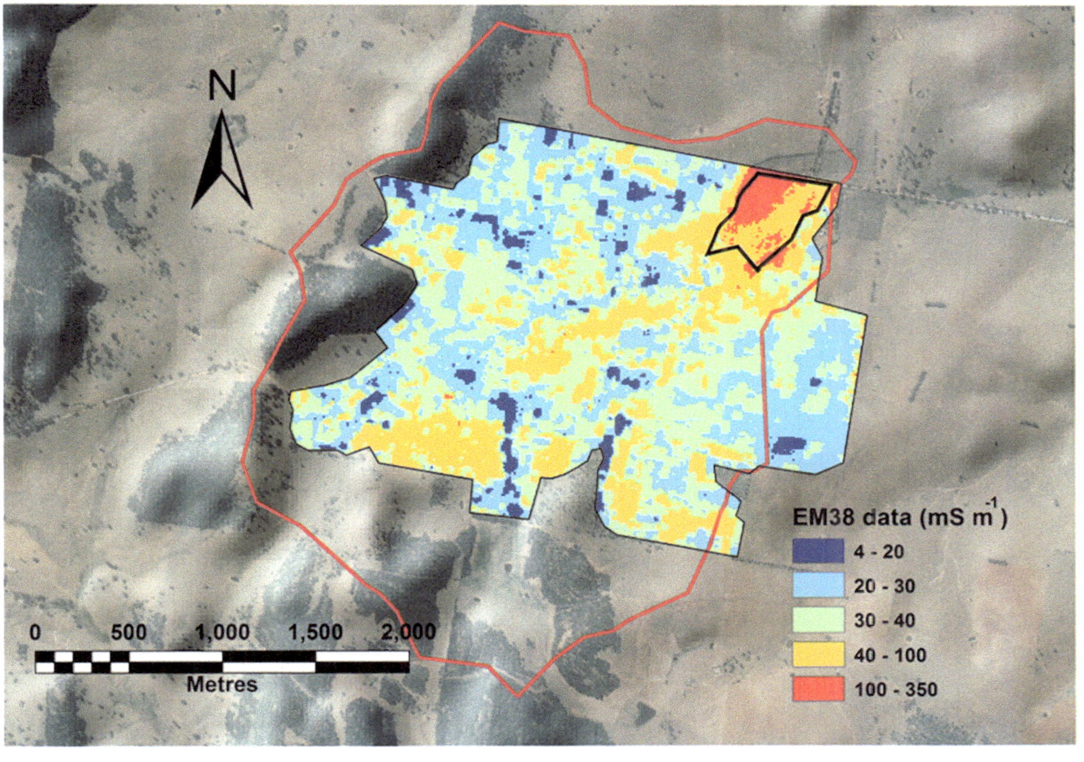

Fig. 2 Variation in apparent electrical conductivity (EC_a), measured by electromagnetic induction (EM–38), across a small sub-catchment in central west New South Wales (Source: Marcar and Morris (2005)

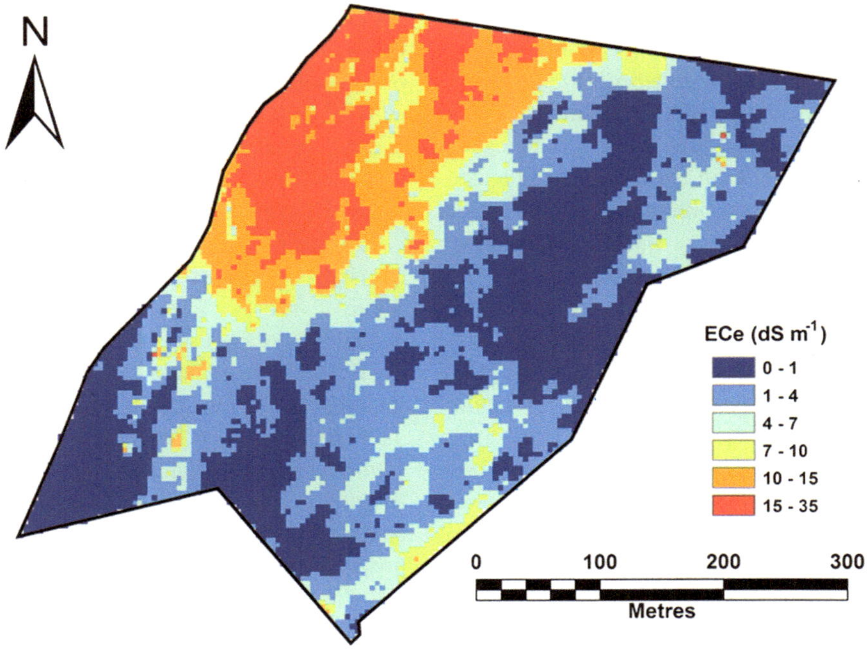

Fig. 3 Variation in soil salinity (EC_e) calculated from an electromagnetic induction (EM–38) survey, across a paddock of a farm in central west New South Wales (Source: Marcar and Morris (2005)

This map shows how soil salinity, calculated as average EC_e over 1.5 m depth from EM–38 (electromagnetic induction meter) measurements, varies over short distances. Such information is critical for determining which tree species or other species are suitable for planting in different parts of a farm.

Seasonal fluctuations in salinity within the soil profile can also be expected, within the salinity classes shown in Fig. 3. In dryland discharge locations, surface soil salinities are often much higher than those in the subsoil in summer months due to high evaporation and low rainfall, while salinity will be lower in winter due to lower evaporation and to dilution and leaching of salt from higher rainfall. The situation can be very different on irrigated land, where subsoils are often highly saline, even though surface soils are not, due to leaching of salts. Surface soil salinities may range from areas with EC_e of 2–6 dS m^{-1}, with corresponding changes from salt-sensitive clovers to salt-tolerant grasses (such as sea barley grass), rushes, and sedges, to EC_e of 6–15 dS m^{-1}, characterized by bare soil and scalding. At EC_e levels above 15 dS m^{-1}, salt crusting can often be seen on the soil surface (Semple and Williams 2002).

Saline soils may become seasonally waterlogged during winter and spring due to shallow water tables, seepage from perched water tables, or increased sodicity (especially on heavy-textured soils where infiltration at the soil surface is reduced by low permeability). Soil sodicity is often described in terms of the concentration of exchangeable sodium (Na) relative to the total quantity of exchangeable cations (cation exchange capacity) or the exchangeable sodium percentage (ESP). Three sodicity classes are commonly recognized: non-sodic (ESP < 6), sodic (ESP 6–14), and strongly sodic (ESP > 14) (Marcar and Crawford 2004).

Tree Planting Options for Managing Land and Water Salinity in Recharge and Discharge Locations

Where recharge and discharge areas in dryland catchments are well defined and both groundwater and surface water flow pathways are well understood and believed to play an important role, revegetation of key recharge areas (nonsaline or low salinity) for reducing deep drainage and lateral flow has been the key focus for land and catchment managers. Revegetation options include using trees and shrubs and establishment of perennial pasture such as deep-rooted lucerne (*Medicago sativa*) and native grasses. Tree establishment includes commercial tree plantations or agroforestry for multiple benefits, including mitigating salinity and land degradation, shelter and shade for stock, commercial wood products, increased carbon storage, habitat for wildlife, and improved biodiversity (Ferguson and Kirby 1992; Alexandra 2002; Bartle et al. 2007). With the advent of increasing efforts to curb increasing human-induced atmospheric CO_2 concentrations, tree planting for salinity management can also be claimed for the purpose of gaining carbon credits in Australia, as long as the location to be planted and the methods used can be shown to provide salinity benefits (Department of Climate Change and Energy Efficiency 2011, Mitchell et al. 2012). Several studies have included estimation of carbon storage of tree plantings on saline land (Harper et al. 2011, 2012; Sochacki et al. 2012).

In addition, productive use and repair (stabilization) of salt-affected land including the use of saltbush and strategically placed trees has also been emphasized (Marcar and Crawford 2004; Barrett-Lennard et al. 2005; George et al. 2012). Engineering options such as drainage and pumping of saline groundwater may be appropriate in certain circumstances including in irrigation areas (NDSP 2001).

Managing Recharge Areas for Salinity Mitigation

Catchment planning places emphasis on reducing recharge and salt mobilization from recharge areas. Planting trees in appropriate locations in landscapes, within an integrated land-use planning framework at farm and catchment scales, can make a significant contribution to limiting the rise of water tables and thus lower the risk of salinization.

Over the last two decades, considerable progress has been made in predicting impacts of tree planting and other land-use changes on stream and land salinization at regional (Polglase et al. 2006; Cheng et al. 2014), catchment (van Dijk et al. 2004; Beverly et al. 2006; Polglase et al. 2006), and sub-catchment scales (Beverly et al. 2004; Marcar et al. 2010) by using hydrological and combined hydrological and forest growth/land-use modeling and also interactions with farm and regional economics to suggest least cost options for maximizing outcomes (Bathgate et al. 2009; Finlayson et al. 2010; Nordblom et al. 2010).

Marcar et al. (2010) report on the use of the forest growth model 3PG in concert with the Catchment Analysis Tool (CAT) to predict growth, carbon sequestration, and both stream and land salinization impacts from various reforestation scenarios in the Corangamite Catchment, southern Victoria. Effects of forestry scenarios on stream flow and salt load varied with species (e.g., eucalypt species or *Pinus radiata*), scenario (e.g., long vs. short rotation), and sub-catchment. This means that there will be trade-offs between the reduction in stream flow and the salinity of these streams for different parts of the catchments depending on which species is planted and whether the system is a long (e.g., sawlog production) or short (e.g., pulpwood production) rotation. Generally, by planting trees in those parts of the catchment where water moves more freely and salts are more prevalent, there will be a tendency for greater reduction in movement of salts and water to streams. However, predicted stream salinity varied with the relative impact of stream flow and salt load. Modeling results would also be expected to differ if only restricted parts of landscape within the catchment were targeted for forestry. Figure 4 shows estimated plantation growth and change in stream salinity for the northern part of the Corangamite Catchment.

Further calibration and testing of hydrological models and improved links to tree growth models (e.g., Feikema et al. 2010) are ongoing

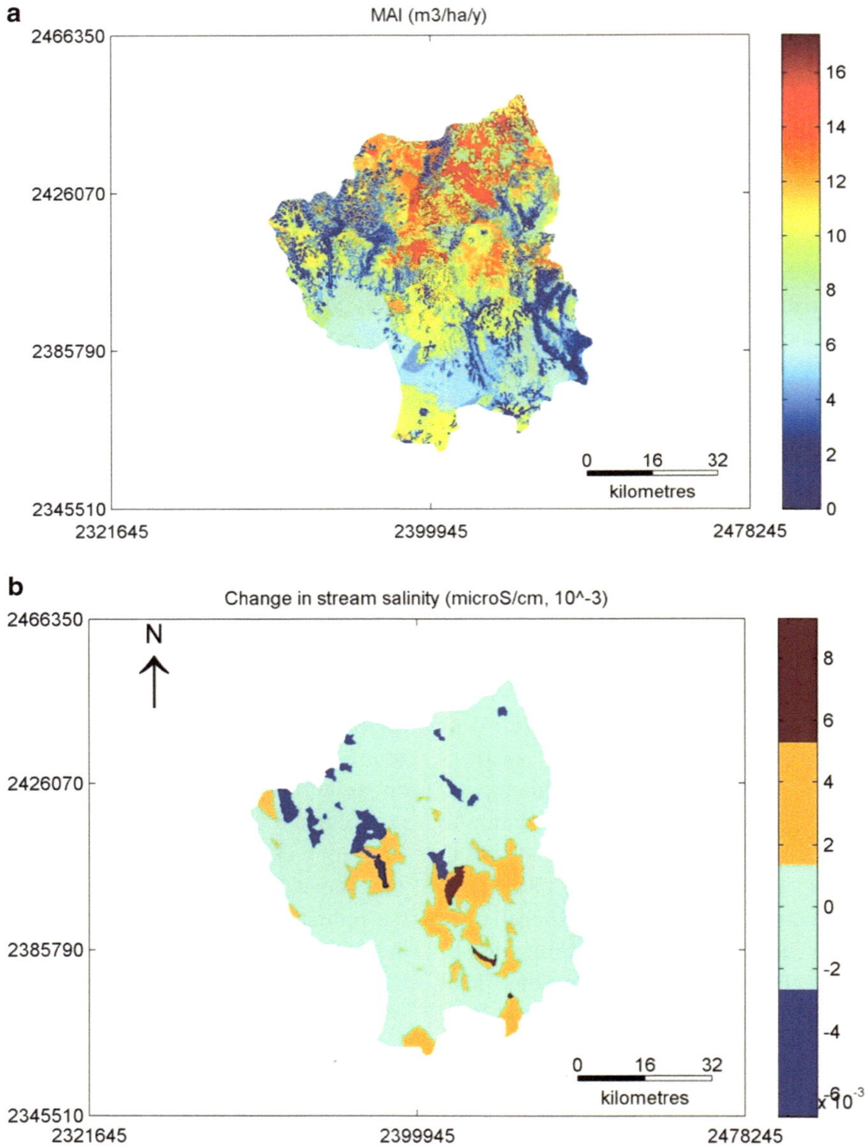

Fig. 4 Spatial output layer for the northern part of the Corangamite Catchment (Victoria) from CAT_3-PG+ modeling for (**a**) mean annual increment in stem volume (MAI, $m^3\ ha^{-1}\ year^{-1}$) and (**b**) change (positive number means an increase and negative number means a decrease) in stream salinity ($\mu S\ cm^{-1} \times 10^{-3}$) for *P. radiata* (30 year sawlog rotation, initial stocking of 1500 stems per ha, final stocking of 250 stems per ha) (Source: Marcar et al. (2010))

to better predict off-site impacts at farm scales and design appropriate plantation size, location, and management. Revegetation options include a spectrum from environmental plantings to farm forestry and industrial-scale plantations.

Managing Saline Areas for Rehabilitation and Productive Use

Trees can be established on or adjacent to discharge areas (seeps and scalds) in a variety of configurations: in woodlots and blocks, as shelter belts and alley plantings, and as belts around saline seeps and scalds, or scattered. Major goals of revegetation and native vegetation management in discharge areas are to lower locally high groundwater levels, to reduce salt transport to streams, and stabilize these areas (Marcar and Crawford 2004). A considerable proportion of salt entering streams can come from soil surface run-off during high rainfall events (Hume and Cawley 2004) and from subsurface flow of water. Salt transport to streams through these processes could be reduced if planting trees and other vegetation can reduce run-off and lower water tables in order to keep salts at depth. However, some of the salt that accumulates in subsoils under trees may be transported to streams through run-off if tree growth and water use were reduced as salinity increased and capillary rise brought salts to the surface (Fawcett 2004).

Water tables will be lowered only if the following conditions are met: trees are planted over large enough areas, the trees can reduce recharge and/or use groundwater, and lateral flows from surrounding areas to the planted areas are relatively small. Several studies have attempted to unravel the contributions of recharge reduction and groundwater use in lowering water tables. Reduction in recharge is believed to be operating near Boundain (Stolte et al. 1997) and near Merredin (George and Frantom 1990) in Western Australia, on sites where groundwater salinities were very high (George et al. 1999) and following initial drawdown of water tables by direct use of groundwater.

George et al. (1999) reviewed the effects on groundwater of tree planting on 47 discharge sites in Western Australia. They concluded that:

- Changes in water table ranged from increases of 1 m to decreases of 2.5 m, with the majority of sites showing a decrease.
- Tree planting is more effective at lowering groundwater if water salinity is low and where local flow systems, including perched water tables, predominate. This is because trees are less able to take up water with salinity above EC 5–10 dS m^{-1} and tree growth is impaired at such salinity levels.
- Trees had little or no impact on the water tables more than 10–30 m away from the area planted.

Opportunities to grow trees at a rate viable for commercial return are greater where groundwater is less than 4–5 m deep and its salinity is less than EC about 5 dS m^{-1}. Planting trees as shelterbelts, on the borders of paddocks, and along water channels in irrigation areas can help to maintain or reduce groundwater levels. In addition, trees can be irrigated with saline drainage water or groundwater pumped to plantations with proper irrigation scheduling.

Effect of Salinity on Plantation Growth and Water Use

The amount and concentration of salt to which tree roots may be exposed in saline soils varies with location of trees in the landscape, salt load in the soil regolith, site and stand management practices, and the extent to which saline soil water and groundwater are used by trees. Characteristics of the main sites and plantations referred to in this chapter are summarized in Table 1.

Salinity and Growth

Salinity in soil and/or water reduces tree growth. The degree of growth reduction at a given salinity level varies with genotypes. Key physiological reasons for reduced growth include reduced water uptake and photosynthesis resulting from decreased stomatal conductance, increased respiration associated with the processes of salt exclusion from the root and containment of salt within leaf cells, reduced turgor in growing tissues, and interference with the activity of some enzymes (Munns and Tester 2008). Apart from halophytic species, salt tolerance is usually a result of the ability to exclude salt from the root and restrict transport to the shoot. Waterlogging in the presence of salinity usually reduces the capacity of plant roots to exclude salt. This will reduce leaf area and hence photosynthetic assimilation and growth.

Tree plantations growing in saline soils commonly show a progressive decline in leaf area and productivity as salinity increases, often with enhanced mortality compared to nonsaline sites. Occasionally plantations surviving on highly

Table 1 Biophysical characteristics of research sites referred to in this review

Site	Soil description	Rainfall (mm $year^{-1}$)	Potential evapotranspiration (mm $year^{-1}$)	Irrigation		Groundwater	
				Volume (ML ha^{-1} $year^{-1}$)	EC (dS m^{-1})	Depth (m)	EC (dS m^{-1})
Boundain, WA	Sand over clay duplex soils; EC_e 0–21 dS m^{-1}	505	1600	0	0	1–2	0.5–41
Deniliquin, NSW	Deep sandy loam; EC_e 2–12 dS m^{-1}	410	1733	0	0	1–3	2–11
Girgarre, NSW	Loam over medium to heavy clay; sodic; EC_e 2–8 dS m^{-1}	450	1540	0	–	1–3	5–10
Kyabram, Vic.	Loam over medium clay	450	1550	0	–	2–10	4–20
Mt Scobie, Vic.	Medium clay	450	1550	3–9	8–10	1.5–2	5–12
Shepparton, Vic.	Silty clay loam over medium to heavy clay	490	1475	5.2–9	1.5	1–3	1–22
Timmering, Victoria	Clay loam over heavy clay	450	1550	4–6	10	1.5–3	1–10
Undera, Victoria	Clay loam over medium clay	463	1520	9	10	1.5–2.5	1–22
Wagga Wagga, NSW	Sandy loam to 0.4 m over sandy to medium clay	526	1712	4.9–8.6	<0.5 and 2.2	12–14	n/a
Wakool, NSW	Medium clay; EC_e 3–14 dS m^{-1}	410	1733	–	–	1–2	n/a
Wellington, NSW	Fine sandy loam over light to medium clay; EC_e 0–20 dS m^{-1}	656	1780	0	0	0.5–5	2–20

saline sites have shown a sudden and widespread scorching of the foliage in hot dry weather, from which only the more salt-tolerant trees have recovered (Morris et al. 1994). This stress is often associated with high Na and Cl concentrations (Marcar and Termaat 1990; Thomson 1988) in leaves and reduced K concentration (Morris 1981). The ion imbalance induced by salt uptake may lead to visual symptoms of stress including leaf chlorosis, necrosis, and premature leaf senescence.

Information on tolerance to salinity and waterlogging at the species level has been compiled by Marcar and Crawford (2004), Marcar et al. (2003), House et al. (1998), Niknam and McComb (2000), and others. Table 2 (modified from Marcar and Crawford 2004) is a summary of a broad classification of salt tolerance for species of *Acacia, Casuarina, Eucalypt, Melaleuca,* and *Pinus* potentially suitable for planting in southern Australia. Species are grouped into categories of tolerance from slight to extreme. This ranking of species is based on a general, conservative assessment of their performance for a given category. Wherever possible, summary information has been provided on the basis of results of field studies where response to salinity and other factors has been determined. It would be expected that each group of species would achieve good to very good survival but would grow up to 25 % slower compared with their performance on nonsaline soil. Therefore, it might be worth testing the performance of some species and provenances at a level of salinity higher than the one attributed to them in Table 2. However, a number of local environmental conditions, such as seasonal waterlogging, would interact to reduce growth at a site. Species highlighted in bold type are moderately or highly tolerant of waterlogging. Other factors, such as stocking density can affect the productivity response to salinity; for example, Sochacki et al. (2012) showed that biomass production for *E. occidentalis* was more sensitive to increasing salinity at 2000 stems than at 500 stems per ha at EC_e about 10 dS m^{-1} after 8 years.

Commercially grown eucalypts, such as *E. globulus* and *E. grandis*, are slightly salt-tolerant, with growth reduction expected at salinity levels above about EC_e 2 dS m^{-1} (e.g., Feikema and Baker 2011). In contrast, some species with less commercial potential—such as *Acacia stenophylla*, *E. occidentalis*, *E. sargentii*, and *E. spathulata*—are highly salt-tolerant, with little growth reduction expected even at EC_e about 10 dS m^{-1} (Marcar et al. 2003; Zohar et al. 2010). *E. spathulata* and *E. occidentalis* can survive and grow well when irrigated with saline drainage water of EC about 10 dS m^{-1} where soil drainage is satisfactory (Greenslade et al. 1999). *E. camaldulensis*, widely grown around the world on drier, saltier sites, is moderately salt tolerant (e.g., Marcar and Crawford 2004; Zohar et al. 2010; Feikema et al. 2010). Benyon et al. (1999) observed that the growth of *E. camaldulensis* was reduced by 10 % (stem diameter and height) at EC_e (mean over 0–60 cm soil depth) about 2 dS m^{-1}, whereas similar growth reduction occurred at about 10 dS m^{-1} in *E. occidentalis*. Responses of *E. globulus*, *E. grandis*, and *E. camaldulensis* on a saline water-irrigated site near Timmering, Victoria, (Feikema and Baker 2011) are shown in Fig. 5.

Slow growth rates of trees on saline sites are usually observed, with rates dependent on genotype (species and provenance), soil and groundwater conditions, climate, and site management (including stocking rate). Bennett and George (1996) found mean stem volume increment for provenances of *E. camaldulensis* from Lake Albacutya (northwestern Victoria) of 10 m^3 ha^{-1} $year^{-1}$ over 14 years on moderately saline and waterlogged soil, with little growth reduction until EC_e reached 3–4 dS m^{-1}. At age 5 years, *E. camaldulensis* provenances grown in small plots at Wellington (NSW) grew at 2–3 m^3 ha^{-1} $year^{-1}$ (Marcar, unpublished data). By comparison, the growth rate for *E. occidentalis* at the same site was 5–6 m^3 ha^{-1} $year^{-1}$. At Deniliquin (NSW), the growth rate for a wide range of *E. camaldulensis* provenances at age 3 years varied from 3 to 9 m^3 ha^{-1} $year^{-1}$. At Wakool (NSW), the mean stem diameters of sixteen *E. occidentalis* provenances ranged from 5.2 to 7.1 cm at age 6 years (Marcar, unpublished); differences in survival among provenances were not as large as those for

Table 2 Tolerance of selected trees and shrub species to salinity and waterlogging and potential suitability for planting in southern Australia

	Root-zone soil salinity ($EC_e dS\ m^{-1}$)			
Size	Slight (2–4)	Moderate (4–8)	High (8–16)	Extreme (>16)
Tree	*Acacia mearnsii*	*A. pendula*	*A. salicina*[b]	*A. stenophylla*
	A. melanoxylon[a]	*A. luehmannii*[a,d]	*Casuarina glauca*	*C. obesa*
	C. citriodora subsp. *variegata*[c]	*A. verticillata*	*Eucalyptus occidentalis*	
	C. maculata	*C. cristata*[b]	*E. platypus* subsp. *platypus*	
	E. aggregate	*C. cunninghamiana* subsp. *cunninghamiana*[b]	*E. sargentii*	
	E. botryoides	*E. astringens* subsp. *astringens*	*E. spathulata*	
	E. brockwayi[a]	*E. camaldulensis*		
	E. camphora subsp. *humeana*	*E. campaspe*[b]		
	E. cinerea subsp. *cinerea*	*E. largiflorens*		
	E. cladocalyx[a]	*E. leucoxylon* subsp. *leucoxylon*		
	E. coolabah[a]	*E. melliodora*[a]		
	E. cornuta[a]	*E. moluccana*		
	E. crenulata	*E. polybractea*		
	E. globulus subsp. *bicostata*	*E. raveretiana*		
	E. globulus subsp. *globulus*	*E. robusta*		
	E. grandis[a]	*E. salicola*		
	E. loxophleba subsp. *lissophloia*[a]	*E. tereticornis* subsp. *tereticornis*		
	E. microcarpa	*M. styphelioides*		
	E. ovate var. *ovata*	*P. pinaster*		
	E. saligna	*P. radiata*		
	E. sideroxylon[a]			
	E. tricarpa[a]			
	E. viminalis subsp. *viminalis*			
	P. brutia			
Shrub	*A. implexa*	*A. acuminata*	*M. cuticularis*	*M. halmaturorum*
	A. iteaphylla	*A. retinodes*	*M. lanceolata*	
	A. longifolia[a]	*A. saligna*		
		A. victoriae		
		M. acuminata		
		M. armillaris subsp.*armillaris*		
		M. bracteata		
		M. decussata		
		M. ericifolia		
		M. lateriflora		
		M. linariifolia		
		M. squarrosa		
		M. uncinata		

Source: Marcar and Crawford (2004)
Tree (>5 m); shrub (<5 m)
[a]Species which overlap slight to moderate salinity category
[b]Species which overlap moderate to high salinity category
[c]*C* = *Corymbia*
[d]*A* = *Allocasuarina*

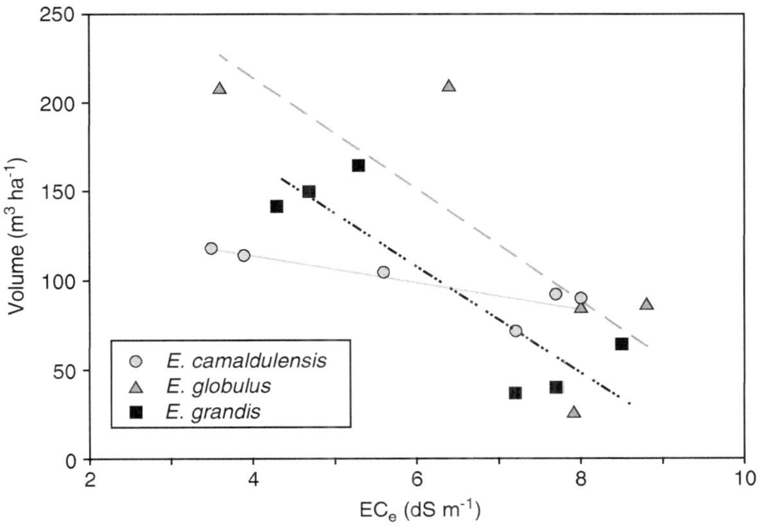

Fig. 5 Relationships between live stem volume at 10 years of age and soil salinity as EC_e(0–150 cm depth). Slopes and elevation of linear regressions of volume and EC_e were the same ($p > 0.05$) for *E. globulus* and *E. grandis* and lower ($p < 0.05$) for. *E. camaldulensis* (Source: Feikema and Baker (2011))

diameter, but there were marked differences between families for both survival and growth.

Although several studies have shown significant variation in mean tree growth on saline sites among provenances, differences in response to salinity among provenances (and families) are small compared to the effects of salinity on differences in actual growth among provenances (e.g., Marcar et al. 2002; Bush et al. 2013). However, in some instances, significant variation in responses to salinity has been observed for specific provenances and clones, including for *A. stenophylla* and *E. camaldulensis* (Marcar et al. 2003; Marcar and Crawford 2004).

Process-based growth models provide a tool to integrate factors likely to influence tree growth and to predict productivity over the life of a plantation. In addition to climate, site, and management-related factors, these include stresses in addition to salinity, for example, waterlogging, water stress, and sodicity; and spatial (with depth and position in a plantation) and temporal variation in root-zone salinity. However, such models need information on key parameters and the salt tolerance of a particular species. Morris and Collopy (2001) describe how soil salinity has been incorporated as a growth-modifying factor into the 3-PG forest model and validated against observed growth and water use of *E. globulus* and *E. grandis* plantations in northern Victoria. Parameters required by the model included the maximum soil water salinity tolerated by each species, which was estimated from observations of growth in saline conditions as 10 dS m^{-1} for *E. globulus* and 20 dS m^{-1} for *E. grandis*. Soil water EC will be higher than soil EC_e since salt concentration increases in the soil solution as soil dries. The study showed that the modified 3-PG model is capable of satisfactorily predicting water use of plantations using saline water. From this information, the potential accumulation of salt in the root zone can be calculated.

Genotypes selected in glasshouse screening trials that simulate both salt and waterlogging should perform better on salt-affected sites than those selected for salt tolerance alone, because soil salinity is often associated with waterlogging that can be prevalent particularly in winter and early spring in southern Australia. Selection of better-performing trees growing on saline sites has produced clones that grow better than clones from unselected trees when planted on saline sites (e.g., for *E. camaldulensis* and *C. obesa* (Bell et al. 1994). Genetic improvement through cloning will be more rapid when the variation between performance of individual trees and genotypes attributable to root-zone salinity (i.e., environmental effects) can be separated from genetic effects.

Research on clonal eucalypt hybrids has focused on combining the salt tolerance of *E. camaldulensis* with the growth rate, wood quality, and form of *E. grandis* and *E. globulus* (Odie and McComb 1996; Meddings et al. 2001; Dale and Dieters 2007). However, performance of hybrid clones has not been consistent in field trials on saline sites in both the MDB and Western Australia. The genetic background of clonal parents and the degree of soil salinity and waterlogging at the test site influence performance. Trials with *E. camaldulensis* x *globulus* hybrids (Odie and McComb 1996) indicate substantial variation in height growth and survival among clones, with responses to salinity being intermediate between parent species (J McComb, pers. comm.). However, at a saline site irrigated with groundwater (EC 5–10 dS m^{-1}) near Mount Scobie, Victoria, mean stem volume of *E. camaldulensis* x *grandis* and *E. camaldulensis* x *globulus* hybrid clones exceeded the best pure species parent by 140 and 252 % at age 7 years with gains in stem volume based on the mean of a commercial selection of hybrid clones ranging from 287 to 481 % (Dale and Dieters 2007). Hardner et al. (2010) genetically analyzed results of 21 trials and concluded that genotypes should be deployed on the basis of broad-scale adaptation.

Salinity and Tree Water Use

Several studies have suggested the need to revegetate some proportion of discharge locations within salinized catchments if rise of saline groundwater is to be slowed or prevented. Increasing soil and/or groundwater salinity reduces water uptake by trees, usually in response to reduced growth, transpiration rate, and leaf area production. Several studies have been undertaken to measure tree water use mostly directly using the heat pulse (sap flow) technique (Akilan et al. 1997; Marshall et al. 1997; Benyon et al. 1999; Crosbie et al. 2008) or using stable isotope analysis (Thorburn and Walker 1994) or both (Cramer et al. 1999). *E. camaldulensis* has been one of the most studied species, sometimes in comparison to other species. Some studies have also aimed to determine the relative proportion of water use derived from soil and groundwater as a function of soil and/or groundwater salinity and depth and species preference.

Akilan et al. (1997) showed that two clones of *E. camaldulensis* differed in the reduction of their water use per day when planted in upslope and downslope (more saline) positions. Benyon et al. (1999) found that salinity reduced stem growth and leaf area development rather than water use per unit of leaf area or sapwood area (sap flux density) in 6-year-old *E. camaldulensis* trees under slightly to moderately saline conditions (EC_e less than 8 dS m^{-1}) at Wellington (NSW). Similar results have been reported by Mahmood et al. (2001) for *E. camaldulensis* plantations in Pakistan, where sap flux density varied with season from 2000 to 12,000 L m^{-2} d^{-1} but was similar in plots with high (i.e., EC of a 1:1 soil water solution ($EC_{1:1}$) 5.7–8.5 dS m^{-1}) or low ($EC_{1:1}$ 3.2–4.2 dS m^{-1}) soil salinity. The finding that salinity does not reduce sap flux density implies that for trees of the same size, water use would be similar on both saline and nonsaline sites. The reduction of tree water use by salinity is therefore a result of reduction in the growth of leaves, stems, and roots. Salt-tolerant species probably maintain their sap flux density under increasing soil salinity by adaptive processes including regulation of stomatal function. Recovery of stomatal function after initial closure following exposure to salinity has been observed in tree seedlings (Thomson 1988; van der Moezel et al. 1989).

Species that maintain a relatively high growth rate under saline conditions are likely to have greater water use than those with slower growth. However, the difference in water use between fast-growing and slow-growing trees is not necessarily proportional to the difference in their growth rates. For example, annual water use by *E. camaldulensis*, 5–8 years old, drawing on saline groundwater near Girgarre, Victoria (water table depth varying between 0.7 and 1.5 m (in winter) and 3.0 m (in summer) with an EC of 5–10 dS m^{-1}), was 339 mm $year^{-1}$ (mean over 2 years), while that for *C. cunninghamiana*, which produced more than twice the basal area growth of *E. camaldulensis* in that period, was 359 mm year^{-1} (Morris and Collopy 1999). Sap flux density of the *C. cunninghamiana* trees decreased as their sapwood area increased over the 2 years of measurement, possibly as a result of limited soil water availability. Best plantation growth and water use will occur when a species can tolerate salinity higher than the equilibrium soil salinity likely to develop as a result of saline groundwater use by trees.

Uptake of Shallow, Saline Groundwater by Trees

One purpose for planting trees in saline areas is to lower shallow water tables using trees to increase evapotranspiration. The key factors influencing groundwater use by trees are depth to groundwater and its salinity, root growth characteristics, salt tolerance of tree species, and soil properties that influence water availability. Evidence for use of saline groundwater by trees has usually been obtained or inferred from reductions in groundwater levels, higher evapotranspiration than rainfall by plantations, and more direct measurements such as stable isotope discrimination (Thorburn and Walker 1994) and chloride mass balance. Feikema and Baker (2011) suggest that although tree plantations may not provide much scope for increasing discharge and, therefore, lowering water tables on heavy-textured soils, they may provide the means of utilizing land that may otherwise be unsuitable for agricultural crops. Crosbie et al. (2008) showed that a mixed *Casuarina-Acacia* tree belt on a sloping, moderately saline discharge site, with a low tendency to accumulate root-zone salts, used more water than on a nearby recharge site and proposed that such a planting would be more useful for managing groundwater rise and have lower impact on farm production.

As soils dry, tree roots tend to grow deeper to maintain access to water. If the water table is within a few meters of the surface, they are more likely to reach the near-saturated capillary zone above the water table and take up groundwater. Groundwater uptake by trees tends to decrease as the water table falls, as a result of lower water potential, greater resistance through longer root–xylem transport, and limits on the capacity of the tree to supply respiratory substrate to deeper roots. Some tree species which grow in seasonally or perennially wet soil adapt to hypoxic (low oxygen) conditions (e.g., by developing aerenchymatous roots as in *E. camaldulensis*); this is essential for survival in waterlogged soils, but is not required for the uptake of groundwater from the capillary zone.

Soil texture has a significant impact on groundwater use by trees, although the confounding effects of site and stand management make interpretations from studies difficult (Barrett-Lennard et al. 1999). The rate of water uptake per unit of root length will be lower from clay than from loamy and sandy soils, mainly because of reduced rate of delivery of water to tree roots. If clay subsoils with low hydraulic conductivity are present, the water table will fall as the water used by trees is replaced more slowly by restricted lateral and upward groundwater flow. Further extension of the root system is then required to follow the source of water to greater depth (Vertessy et al. 2000).

Based on evaluation of ten sites planted with trees, Thorburn (1996) concluded that uptake of saline groundwater by plants is often no more than would be expected by discharge from bare soil alone. Nevertheless, trees can use groundwater for growth in the absence of sufficient soil water, under favorable conditions, such as when depth is less than 5 m and salinity (EC) less than 5–10 dS m^{-1} (George et al. 1999). Trees can thereby help to maintain or lower shallow, saline water tables in discharge areas. This requires trees to access and use groundwater and planting on a scale appropriate to match the rate of groundwater recharge or inflow to the affected area.

The effectiveness of trees in using soil water and groundwater will depend on the maximum leaf area index (LAI, leaf area per unit ground area) attained, the time taken to reach that maximum, the length of time high leaf area can be maintained, root growth, and the availability and salinity of soil water and groundwater. Most tree species can use groundwater of good quality (EC less than 2 dS m^{-1}) to grow. Those able to continue using groundwater of increasing salinity require progressively higher salt tolerance and growth potential (Marcar and Crawford 2004). Although many tree species can take up water with an EC of 5–10 dS m^{-1}, only highly salt-tolerant species can use highly saline groundwater (EC greater than 15–20 dS m^{-1}).

Mensforth and Walker (1996) showed that natural stands of *Melaleuca halmaturorum* (salt paperbark) can use highly saline groundwater. However, groundwater use declined as salinity increased. For example, at one site, groundwater use was 1–4 mm d^{-1} (up to 590 mm $year^{-1}$) over groundwater of moderate salinity (EC about 8–14 dS m^{-1}), whereas over groundwater of high salinity (about 60 dS m^{-1}) uptake was 1–2 mm d^{-1} (up to 290 mm $year^{-1}$). A greater proportion of water transpired was from groundwater as distinct from soil water in late summer/autumn, and this corresponded with higher surface soil salinities and deeper water tables. Costelloe et al. (2008) showed that *E. coolabah* in arid zone central Australia can use highly saline water but that they preferentially grow in zones most frequently flushed by infiltrating stream flow.

Fraser et al. (1996) estimated that *C. glauca* trees in an agroforestry planting overlying a water table at 1.6 m with an average salinity of EC about 11 dS m^{-1} in southern Queensland were obtaining about 23 % of their water from the water table. Up to 35 % of transpiration of 8-year-old *E. camaldulensis* trees was derived from saline groundwater with EC up to 20 dS m^{-1} at Wellington (NSW) when root systems were isolated from surrounding soil and rainfall was excluded from roots by undercanopy shelters (Marcar and Benyon 2000).

Saline groundwater use by trees is likely to result in some degree of salt accumulation in the zone of water uptake, because plants exclude most of the salt in the soil solution (Stirzaker 2002). Evidence for this is presented in Fig. 6 for a groundwater-dependent *E. grandis* plantation near Kyabram (from Vertessy et al. 2000). Fine roots were most abundant in usually-dry soil near the surface, but high root density was also observed lower in the profile above the water table, associated with a zone of high salt concentration at 600–800 cm soil depth. However, over a 2-year monitoring period, the EC in this zone fluctuated by 10–20 dS m^{-1} (not shown in this Figure) as salts moved downward or upward with seasonal fluctuations in water table depth (Vertessy et al. 2000). Such large seasonal changes in root-zone salinity indicate that salt accumulation in the soil as a result of groundwater uptake is being balanced to some extent by salt removal.

The long-term sustainability of plantations in discharge locations (saline soils and/or over shallow, saline water tables) has been challenged on the basis that salts accumulate in the soil and groundwater as trees use water but exclude almost all dissolved salts (Thorburn 1996; Stolte et al. 1997). Salt accumulated in root zones is more likely to be leached and therefore not pose a significant risk to tree growth and water use if rainfall

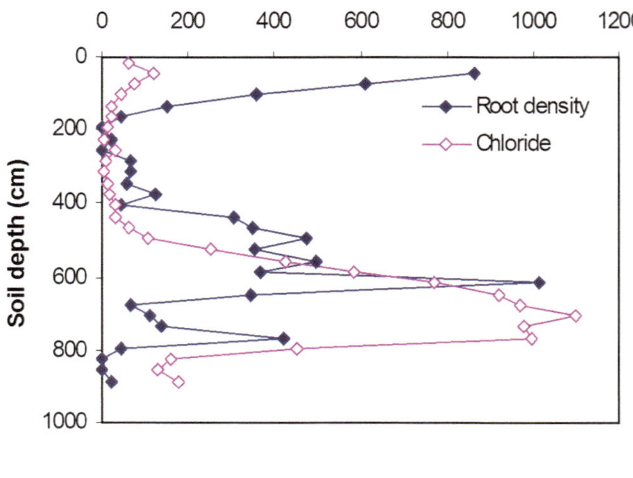

Fig. 6 Distribution of root density and soil chloride concentration with soil depth beneath an 18-year-old *E. grandis* plantation on heavy clay soil, drawing on shallow saline groundwater near Kyabram (Source: Marcar and Morris (2005))

is high. The long-term growth of groundwater-dependent plantations may be at risk when rainfall is lower than 500–600 mm $year^{-1}$ and where soils are heavier-textured and relief is low, and root-zone salinity continues to increase.

However, both the rate of accumulation and the maximum salt concentration that can develop are restrained by a reduction in groundwater use that tends to occur as water tables move deeper and become more saline (Morris and Collopy 1999). In addition, as the salt concentration in soil increases, the rate of salt diffusion from the near-saturated capillary zone back to groundwater will increase and will eventually equal to the rate of salt accumulation. When the rate of downward salt diffusion is equal to the rate of upward salt movement in groundwater moving into the capillary zone, an equilibrium condition may be reached. Morris (1999) showed that the equilibrium soil solution salinity may be around EC of 15–25 dS m^{-1}, similar to that observed by Vertessy et al. (2000) for the *E. grandis* plantation near Kyabram. Other observations of salt accumulation in subsoils under trees include 5–8-year-old *E. camaldulensis* and *C. glauca* plots in southern Queensland (Fraser et al. 1996); an agroforestry planting of several eucalypt species near Boundain, Western Australia (Stolte et al. 1997; Archibald et al. 2006); tree species trial plots near Hamilton, Victoria (Fawcett 2004); and 10-year-old *E. globulus*, *E. grandis*, and *E. camaldulensis* at an irrigated trial site near Timmering, Victoria (Feikema and Baker 2011).

As a result of the effects of salt accumulation and redistribution on the productivity and water use of plantations, their long-term effectiveness for lowering shallow groundwater may be less than that observed in younger plantations established on farmland. However, plantations growing over shallow, saline water tables may continue to grow provided that equilibrium salt concentrations do not exceed the maximum salinity tolerated by tree species, i.e., by correct selection of species. The following examples of plantations growing over shallow water tables support this view:

- *E. grandis* (slightly salt-tolerant) plantations have continued growth over 20 years over a shallow water table site near Kyabram (Victoria) at about 20 m^3 ha^{-1} $year^{-1}$ for stem volume (Vertessy et al. 2000). Further analysis of data from this site for *E. grandis* and *E. camaldulensis* indicates that any deleterious effect of increasing salinity in parts of the root zone mainly occur in summer months when trees are more reliant on groundwater rather than rainfall-derived soil water (Feikema and Baker 2011).
- *E. occidentalis* (highly salt-tolerant) showed continued healthy growth at age 14 on a moderately to highly saline site (Bandara et al. 2002).
- *Melaleuca halmaturorum* (extremely salt-tolerant) in natural stands draws on saline groundwater in extremely saline swamps of the lower southeast of South Australia (Mensforth and Walker 1996).
- *E. spathulata*, *E. occidentalis*, and *E. sargentii* have maintained high survival and good growth rates (about 1 m^3 ha^{-1} $year^{-1}$ stem volume for 160 trees ha^{-1}) at a highly saline site (EC_e mean 6.3 dS m^{-1} and maximum 21.5 dS m^{-1}) near Boundain, Western Australia, over 20+ years even though subsoil salinity has increased significantly during this period (Archibald et al. 2006).
- *E. camaldulensis* maintained stem volume growth around 10 m^3 ha^{-1} $year^{-1}$ over 10 years at soil EC_e up to 8 dS m^{-1} near Timmering, Victoria (Feikema and Baker 2011).

Irrigation with Saline Water

Access to good-quality irrigation water for plantation forestry or agroforestry is limited. However, considerable volumes of saline groundwater and drainage water are available within irrigation areas of the MDB (Murray–Darling Basin Commission 1999). Successful irrigation practice requires sufficient water to be applied to leach salts that accumulate over successive irrigation events in soil or groundwater below the main part of the root system. Marcar and Morris (2005) reported data from Bandara et al. (2002) on the changes in soil Cl concentrations with depth at five plantation sites, with varying soil texture conditions, and irrigated with water of different salinities. The trend of peak soil Cl concentration observed at each site generally followed the level of applied irrigation-water salinity, with no or little salt accumulation at those sites irrigated with low or slightly saline water, but with clear evidence of elevated Cl concentrations in the upper 3 m of soil at those sites irrigated with water of EC 5 and 10 dS m^{-1} respectively, with soils at these latter sites being heavier-textured than the others.

Impacts on plantation growth should be more pronounced where saline water is applied to clay soils rather than to sandy soils. Feikema and Baker (2011) further report on mean soil EC_e (0–150 cm soil depth) changes over time at one of these sites (Timmering, Victoria) irrigated with either good-quality water (EC 0.1–0.5 dS m^{-1}) or highly saline pumped groundwater (EC 2.9–8.8 dS m^{-1}). Figure 7 shows a clear difference in trends over time between the low and high salinity water treatments. Both the low and high salinity water treatment soil profiles had peaks in salinity around 90–150 cm depth at about 10 years. Average soil EC_e in the low salinity irrigation treatments was 4.3–5.0 dS m^{-1} and 7.6–8.2 dS m^{-1} in the high salinity irrigation treatments.

With adequate leaching, on soils with good hydraulic conductivity, salts can be moved out of the root zone. Myers et al. (1998) showed that two 200 mm irrigation events with low salinity (EC 0.5 dS m^{-1}) effluent were sufficient to leach the accumulated salt out of the root zone of *E. grandis* and *Pinus radiata* plantations near Wagga Wagga, NSW. Sufficient leaching to avoid salt accumulation may be restricted when:

- The root zone is deep—trees, depending on their growth rates, can exploit much of the soil-available water, and hence large volumes of water may be required to leach salts.
- A low infiltration rate in heavy-textured soils limits the volume of water that can be applied in an irrigation event.
- Leaching of salt below the root zone may raise the water table and increase the likelihood of upward capillary rise of salt.
- Subsurface drainage is required to facilitate leaching of salt.

Modeling by Thorburn et al. (1995) suggested that if salinity in tree root zones does increase for extended periods above levels tolerated by particular species, roots are likely to become confined to an increasingly shallow part of the upper profile above the zone of salt

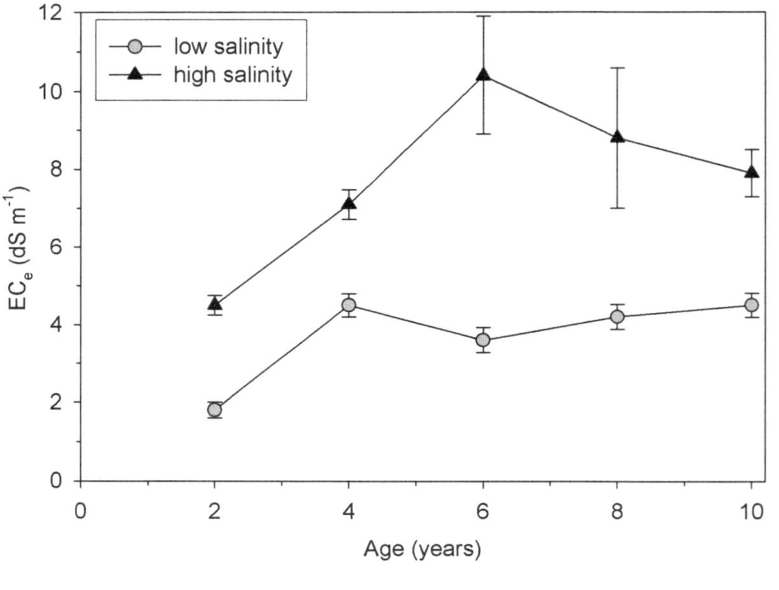

Fig. 7 Average soil salinity as EC_e (0–150 cm depth) with age since planting in the low and high salinity water irrigation treatments at Timmering, Victoria. Error bars indicate one standard deviation (Source: Feikema and Baker (2011))

accumulation, thereby reducing plantation productivity. More salt-tolerant trees will maintain a deeper functional root zone, but long-term survival and growth will only be possible where salt can be leached. Marcar and Morris (2005) suggest that as salt accumulation in the zone of maximum water uptake reaches a high enough concentration to limit root growth and function, water will infiltrate more deeply and carry salt with it. In that case, root system recovery in the now-leached uptake zone may follow, and it may be possible for a plantation to survive indefinitely with a cycle of processes: root-zone salt accumulation–reduced water uptake–root-zone leaching. However, survival in these cyclically salinized conditions will come at a cost to productivity. Genotypes with adequate salt tolerance and high growth potential are likely to be capable of both acceptable survival and productivity under saline irrigation.

A plantation irrigated with saline water and operating as described above would allow reuse of saline water with relatively low environmental impact, assuming that salt accumulation is confined to the subsoil below the rooting depth. Most studies on saline water-irrigated plantations in the MDB have only been established and monitored since 1990. Their long-term productivity needs to be systematically studied. Theiveyanathan et al. (2003) have developed a spreadsheet model for determining irrigation scheduling for plantations using saline water by determining leaching fraction requirements to minimize salt accumulation in the root zone based on salinities tolerated by different species. Further validation and application of this model may provide ways to better manage saline water–irrigated plantations.

Serial biological concentration (SBC), which incorporates adequate drainage capacity, has been tested at Undera (Victoria) for irrigating plantations with saline water without unduly increasing soil salinity. This system represents an alternative to evaporation basins for saline wastewater disposal (Heath and Heuperman 1996). Ground water with approximate EC of 8.5 dS m^{-1} was pumped and applied to large blocks of *E. camaldulensis* and *Atriplex nummularia*, planted on saline, clay soil. Tile drains at 1.9 m below the surface at a spacing of 28 m collected drainage into mariculture ponds and a basin for harvesting salt. Results show that this system has been able to stabilize root-zone EC_e levels at about 15 dS m^{-1} (Heuperman et al. 2002). However, SBC plantations are relatively expensive to establish and manage. Hence, they are likely to be attractive to farmers only where there is a substantial reduction in the land area required for managing excess saline drainage water compared to the alternative of a simple evaporation basin. SBC plantations may be an option for managing saline effluent from industries.

Conclusions

Salinity is extensive in southern Australia with outbreaks typically patchy and of varying intensity at catchment, farm, and paddock scales in the east and more widespread in the west. Tree plantations targeted to appropriate landscape positions can make an important contribution to the mitigation and management of salinity, both by reducing recharge to groundwater and by lowering water tables in discharge areas. The likelihood of plantation establishment on the scale required for significant regional impact will be increased if plantations offer commercial yields of tree products in addition to environmental benefits.

Saline soil and groundwater conditions, and associated stresses of waterlogging and sodicity, reduce tree growth. The extent of growth reduction varies between genotypes and sites. Shallow water table sites that are sandy are likely to supply water for tree growth more readily than those that are clay-dominated. At high soil and groundwater salinities, growth and water use of plantations will be considerably less than at low salinities. Maintenance of adequate long-term productivity requires the use of salt-tolerant species, provenances, and/or clones coupled with good site and stand management practices.

Accumulation of salt in the root zone of trees using saline groundwater or irrigation water is likely to reduce long-term growth rates where effective leaching is limited by heavy soils, deep root systems, or insufficient water supply. In other circumstances, salt accumulation may be reduced by managing the leaching of salts. Although tree water use is also reduced under saline conditions, plantation uptake of saline groundwater and consequent lowering of shallow water tables has been demonstrated on a range of sites. Knowledge of the equilibrium root-zone salinities likely to develop under the rainfall and/ or irrigation schedules used and plant genotypes that can grow at acceptable rates under these conditions would help better matching of site, species, and silviculture. Research has been important for identifying and developing salt-tolerant genotypes and testing hydrological models to indicate the location and extent, and hence economic viability, of tree plantings on farmland needed for slowing and possible reversal of salinization processes.

Acknowledgements The author acknowledges use of material from his previous works, with due citations, whilst an employee of CSIRO and also engaged in CRC Salinity projects.

References

Akilan K, Farrell RCC, Bell DT, Marshall JK (1997) Responses of clonal river red gum (*Eucalyptus camaldulensis*) to waterlogging by fresh and salt water. Anim Prod Sci 37:243–248

Alexandra J (2002) Landscape change in the Goulburn–Broken Catchment. Final report for the Goulburn–Broken Catchment Management Authority (Alexandra & Associates) Eltham, Victoria. November 2002

Archibald RD, Harper RJ, Fox JE, Silberstein RP (2006) Tree performance and root-zone salt accumulation in three dryland Australian plantations. Agrofor Syst 66:191–204

Australian Bureau of Statistics (2002) Australian year book: environment, salinity and land management. Australian Bureau of Statistics, Canberra

Bandara G, Morris J, Stackpole D, Collopy J (2002) Soil profile monitoring at irrigated plantation sites in northern Victoria: progress report 2001. CFTT Report No. 2002/025. Forest Science Centre, Melbourne, Australia

Bann GR, Field JB (2006). Dryland salinity in south-east Australia: which scenario makes more sense? In: Proceedings of the Australian Earth Sciences Convention, July, Melbourne, 9 p. Available at: www.earth2006.org.au

Bann G, Field J (2010) Dryland salinity on the uplands of southern Australia: a top-down soil degradation process, or a bottom-up deep hydrology (groundwater) process? In: Gilkes RJ, Prakongkep N (eds) World congress of soil science 2010. Conference Organising Committee, Brisbane, pp 17–20

Barrett-Lennard EG, Speijers J, Morris J, Marcar N (1999) Transpiration by trees on land with shallow water-tables: a survey of the literature suggests that transpiration is affected by soil texture. In: Redesign of Australian plant production systems- deep drainage and nitrate losses in the Mediterranean climate region. Land and Water Resources Research and Development Corporation, Canberra

Barrett-lennard EG, George RJ, Hamilton G, Norman HC, Masters DG (2005) Multi-disciplinary approaches suggest profitable and sustainable farming systems for valley floors at risk of salinity. Aust J Exp Agric 45:1415–1424

Bartle J, Olsen G, Cooper D, Hobbs T (2007) Scale of biomass production from new woody crops for salinity control in dryland agriculture in Australia. Int J Global Energy Issues 27:115–137

Bathgate A, Seddon J, Finalyson J, Hacker R (2009) Managing catchments for multiple objectives: the implications of land-use change for salinity, biodiversity and economics. Anim Prod Sci 49:852–859

Bell DT, McComb JA, van der Moezel PG, Bennett IJ, Kabay ED (1994) Comparisons of selected and cloned plantlets against seedlings for rehabilitation of saline and waterlogged discharge zones in Australian agricultural catchments. Aust For 57:69–75

Bennett D, George RJ (1996) River red gums (*E. camaldulensis*) for farm forestry on saline and waterlogged land in south-western catchments of Western Australia? In: Proceedings of the 4th NBeational conference and workshop on the productive use and rehabilitation of saline lands, Albany, WA, Australia (Promaco Conventions: Perth), 25–30 March 1996, pp 345–358

Benyon RG, Marcar NE, Crawford DF, Nicholson AT (1999) Growth and water use of *Eucalyptus camaldulensis* and *E. occidentalis* on a saline discharge site near Wellington, NSW, Australia. Agric Water Manage 39:229–244

Beverly C, Bari M, Christie B, Hocking M, Smettem K (2004) Understanding catchment dynamics from land use change; comparison between a rapid assessment approach and a detailed modelling framework. In: Ridley A, Feikema P, Bennet S, Rogers MJ, Wilkinson R, Hirth J (eds) Salinity solutions, proceedings of the salinity solutions conference 'Working with Science and Society' Bendigo, Victoria, Australia, 2–5 August 2004. CRC for Plant-Based Management of Dryland Salinity, Perth. CD ROM

Beverly C, Christy B, Weeks A (2006) Application of the 2CSalt-model to the Bet Bet, Wild Duck, Gardiner and Sugarloaf catchments in Victoria. Department of Primary Industries, Victoria

Blackmore DJ, Goss KF, Newman RJ, Powell J (1999) Salinity management in the Murray–Darling Basin. In: The spirit of the snowies—fifty years on. Australian Academy of Technological Sciences and Engineering, Canberra. http://www.atse.org.au/index.php?sectionid=284

Bush D, Marcar N, Arnold R, Crawford D (2013) Assessing genetic variation within *Eucalyptus camaldulensis* for survival and growth on two spatially variable saline sites in southern Australia. For Ecol Manage 306:68–78

Cheng X, Benke KK, Beverly C, Christy B, Weeks A, Barlow K, Reid M (2014) Balancing trade-off issues in land use change and the impact on stream flow and salinity management. Hydrol Process 28:1641–1662

Coram J, Dyson P, Houlder P, Evans R (2000) Australian groundwater flow systems contributing to dryland salinity. Report by the Bureau of Rural Sciences for the National Land and Water Resources Audit, Canberra

Costelloe JF, Payne E, Woodrow IE, Irvine EC, Western A, Leaney FW (2008) Water sources accessed by arid zone riparian trees in highly saline environments, Australia. Oecologia 156:43–52

Cramer VA, Thorburn PJ, Fraser GW (1999) Transpiration and groundwater uptake from farm forest plots of *Casuarina glauca* and *Eucalyptus camaldulensis* in saline areas of southeast Queensland, Australia. Agric Water Manage 39:187–204

Crosbie RS, Wilson B, Hughes JD, McCulloch C, King WM (2008) A comparison of the water use of tree belts and pasture in recharge and discharge zones in a saline catchment in the Central West of NSW, Australia. Agric Water Manage 95:211–223

Dahlhaus PG, Cox JW, Simmons CT, Smitt CM (2008) Beyond hydrogeologic evidence: challenging the current assumptions about salinity processes in the Corangamite region, Australia. Hydrogeol J 16: 1283–1298

Dale G, Dieters M (2007) Economic returns from environmental problems: breeding salt- and drought-tolerant eucalypts for salinity abatement and commercial forestry. Ecol Eng 31:175–182

Department of Climate Change and Energy Efficiency (2011) Carbon farming initiative. Preliminary Estimates of Abatement, p 29. www.climatechange.gov.au/

Department of Environment and Climate Change NSW (2009) Salinity audit upland catchments of the New South Wales Murray–Darling. http://www.environment.nsw.gov.au/resources/salinity/09153SalinityAudit.pdf

Fawcett J (2004) Assessment of revegetated discharge sites on the Dundas Tablelands. In: Ridley A, Feikema P, Bennet S, Rogers MJ, Wilkinson R, Hirth J (eds) Salinity solutions, proceedings of the salinity solutions conference 'Working with Science and Society', 2–5 August 2004, Bendigo, Victoria, Australia; CRC for Plant-Based Management of Dryland Salinity, Perth. CD ROM

Feikema PM, Baker TG (2011) Effect of soil salinity on growth of irrigated plantation Eucalyptus in south-eastern Australia. Agric Water Manage 98:1180–1188

Feikema PM, Morris JD, Beverly CR, Collopy JJ, Baker TG, Lane PNJ (2010) Validation of plantation transpiration in south-eastern Australia estimated using the 3PG+ forest growth model. For Ecol Manage 260:663–678

Ferguson IS, Kirby G (1992) Economic analyses of tree-growing options in the irrigation zone of north-central Victoria. In: Proceedings of the seminar on commercial tree-growing for land and water care in some irrigation and associated dryland areas in the Murray–Darling Basin, July 1992, University of Melbourne, Parkville, pp 14–24

Finlayson JD, Bathgate A, Nordblom TT, Farquharson B, Crosbie R, Mitchell D, Hoque Z (2010) Balancing land use to manage river volume and salinity: economic and hydrological consequences for the Little River catchment in Central West, New South Wales, Australia. Agric Syst 103:161–170

Fitzpatrick R, Merry R, Cox J (2000) What are saline soils? What happens when they are drained? Nat Resour Manage (special issue June 2000):26–30

Fraser GW, Thorburn PJ, Taylor DW (1996) Growth and water use of trees in saline land in south-east Queensland? In: Proceedings of the fourth national conference and workshop on the productive use and rehabilitation of saline lands, 25–30 March 1996, Albany, WA, Australia, Promaco Conventions, Perth, pp 219–224

George RJ, Frantom PWC (1990) Preliminary groundwater and salinity investigations in the eastern wheatbelt. 2. Merredin Catchment. Division of Resource Management technical report 89. Government of Western Australia, 75 pp

George RJ, Nulsen RA, Ferdowsian R, Raper GP (1999) Interactions between trees and groundwaters in recharge and discharge areas—a survey of western Australian sites. Agric Water Manage 39:91–113

George SJ, Harper RJ, Hobbs RJ, Tibbett M (2012) A sustainable agricultural landscape for Australia: a review of interlacing carbon sequestration, biodiversity and salinity management in agroforestry systems. Agric Ecosyst Environ 163:28–36

Gill B, Zydor H, Hekmeijer P (2012) Victorian dryland salinity (2012) Post-drought groundwater response. Technical report, Department of Primary Industries, Epsom, Victoria

Greenslade R, Heupermann AF, Marcar N, McKinnon L (1999) Saline groundwater management at Undera, Victoria. In: Marcar NE, Hossain AKMA (eds) Managing saltland into the twenty-first century: dollars and sense from salt. Proceedings of the fifth national conference on the productive use and rehabilitation of saline land, Tamworth, NSW, Australia; National Committee for the Productive Use and Rehabilitation of Saline Land, March 1998, pp 123–126

Hardner CM, Dieter M, Dale G, DeLacy L, Basford KE (2010) Patterns of genotype-by-environment interaction in diameter at breast height at age 3 for eucalypt hybrid clones grown for reafforestation of lands affected by salinity. Tree Genet Genomes 6:833–851

Harper RJ, Smettem K, Sochacki S, Nakagima Y, Honda S, Takahashi BJ (2011) Using carbon reforestation for water and environmental restoration. J Arid Land Stud 21:57–61

Harper RJ, Okom A, Stilwell A, Tibbett M, Dean C, George S, Sochacki S, Mitchell C, Mann S, Dods K (2012) Reforesting degraded agricultural landscapes with Eucalypts: effects on carbon storage and soil fertility after 26 years. Agric Ecosyst Environ 163:3–13

Heath J, Heuperman A (1996) Serial biological concentration of salts—a pilot project. Aust J Soil Water Conserv 9:27–31

Heuperman A, Mann L, Greenslade R, Heath J (2002) Piloting farm scale serial biological concentration. Final report for MDBC project D115. Victorian Department of Natural Resources, Melbourne

House S, Nester M, Taylor D, King J, Hinchley D (1998) Selecting trees for the rehabilitation of saline sites. Technical paper no. 52. Queensland Department of Primary Industries, Brisbane

Hume I, Cawley R (2004) Murrumbidgee river salinity: a new perspective. In: Ridley A, Feikema P, Bennet S, Rogers MJ, Wilkinson R, Hirth J (eds) Salinity solutions, proceedings of the salinity solutions conference working with science and society (2–5 August 2004), Bendigo, Victoria, Australia (CRC for Plant-Based Management of Dryland Salinity, Perth) CD ROM

Jolly I, Walker G, Stace P, van der Wel B, Leaney R (2000) Assessing the impacts of dryland salinity on South Australia's water resources. A sub-project of the National Land and Water Resources Audit – Dryland Salinity Theme Project 1. CSIRO Land and Water, Canberra, ACT. Technical Report 9/00, 85 pp

Jones CE (2000) The great Australian salinity debate: Part II. Why the recharge-discharge model is fundamentally flawed. Stipa News Lett 14:6–11

Kington EA, Pannell DJ (2003) Dryland salinity in the Upper Kent River catchment of Western Australia: farmer perceptions and practices. Aust J Exp Agric 43:19–28

Mahmood K, Morris J, Collopy J, Slavich P (2001) Groundwater uptake and sustainability of farm plantations on saline sites in Punjab province, Pakistan. Agric Water Manage 48:1–20

Marcar NE, Benyon RG (2000) Managing trees in saline environments: interactions between trees and salt. In: Nambiar EKS, Cromer RN, Brown A (eds) Restoring tree cover to the Murray–Darling Basin. CSIRO Forestry and Forest Products, Canberra, pp 21–26

Marcar NE, Crawford DF (2004) Trees for saline landscapes. Rural Industries Research and Development Corporation, Canberra

Marcar NE, Morris J (2005) Plantation productivity in saline landscapes. In: Nambiar EKS, Ferguson I (eds) New forests: wood production and environmental services CSIRO. Collingwood, Australia, pp 51–74

Marcar NE, Termaat A (1990) Effects of root-zone solutes on *Eucalyptus camaldulensis* and *Eucalyptus bicostata* seedlings: responses to Na^{+}, Mg^{2+} and Cl^{-}. Plant Soil 125:245–254

Marcar NE, Crawford DF, Saunders A, Matheson C, Arnold RJ (2002) Genetic variation among provenances and families of *Eucalyptus grandis* W. Hill and *E. globulus* Labill. subsp. *globulus* seedlings in response to salinity and waterlogging. For Ecol Manage 162:231–249

Marcar NE, Crawford DF, Hossain AKMA, Nicholson AT (2003) Survival and growth of tree species and provenances on a saline discharge site. Aust J Exp Agric 43:1293–1302

Marcar N, Theiveyanathan T, Crawford D, Hawkins C, Jovanovic T, Polglase P, Siggins A (2010) Predicting growth, carbon sequestration and salinity impacts of forestry plantations. In: Plant adaptation and phytoremediation. Springer, Dordrecht, pp 143–149

Marshall JK, Morgan AL, Akilan K, Farrell RC, Bell DT (1997) Water uptake by two river red gum (*Eucalyptus camaldulensis*) clones in a discharge site plantation in the Western Australian wheatbelt. J Hydrol 200:136–148

McFarlane DJ, George RJ, Caccetta PA (2004) The extent and potential area of salt-affected land in Western Australia estimated using remote sensing and digital terrain models. In: 1st national salinity engineering conference, Perth, Western Australia, 9–12 November 2004

Meddings RLA, McComb JA, Bell DT (2001) The salt-waterlogging tolerance of *Eucalyptus camaldulensis* x *E. globules* hybrids. Aust J Exp Agric 41:787–792

Mensforth LJ, Walker GR (1996) Root dynamics of *Melaleuca halmaturorum* in response to fluctuating saline groundwater. Plant and Soil 184:75–84

Mitchell CD, Harper RJ, Keenan RJ (2012) Status and prospects of carbon forestry in Australia. Aust For 75:201–213

Morris JD (1981) Factors affecting the salt tolerance of eucalypts. In: Old K, Kile G, Ohmart C (eds) Eucalypt die-back in forests and woodlands. CSIRO Division of Forest Research, Canberra, pp 190–204

Morris JD (1999) Salt accumulation beneath plantations using saline groundwater: lessons from the Kyabram plantation study. In: Thorburn P (ed) Agroforestry over shallow water tables. Water and salinity issues in agroforestry no. 4. Publication No. 99/36, Rural Industries Research and Development Corporation, Canberra, pp 27–29

Morris JD, Collopy J (1999) Water use and salt accumulation by *Eucalyptus camaldulensis* and *Casuarina cunninghamiana* on a site with shallow saline groundwater. Agric Water Manage 39:205–227

Morris JD, Collopy JJ (2001) Validating plantation water use predictions from the 3PG forest growth model. In: Ghasemi F, Post D, Sivapalan M, Vertessy R (eds) Proceedings MODSIM 2001 international congress on modelling and simulation, vol I. MSSANZ, Canberra, pp 437–442

Morris JD, Bickford RG, Collopy JJ (1994) Tree and shrub performance and soil conditions in a plantation irrigated with saline groundwater. Research report 357, Forest Research and Development Branch. Department of Conservation and Natural Resources, Victoria

Munns R, Tester M (2008) Mechanisms of salinity tolerance. Annu Rev Plant Biol 59:651–681

Murray–Darling Basin Commission (1999) The salinity audit of the Murray–Darling Basin: a hundred-year perspective. Salt: our challenge for the twenty-first century. Murray–Darling Basin Commission, Canberra

Murray-Darling Basin Commission (2004) 2003–2004 basin quality management strategy annual implementation report–summary. Murray-Darling Basin Commission, Canberra

Myers BJ, Benyon RG, Theiveyanathan S, Criddle RS, Smith CJ, Falkiner RA (1998) Response of effluent-irrigated *Eucalyptus grandis* and *Pinus radiata* to salinity and vapor pressure deficits. Tree Physiol 18:565–573

National Land and Water Resources Audit (2001) Australian dryland salinity assessment 2000: extent, impacts, processes, monitoring and management options. National Land & Water Resources Audit. Land Water Australia. Canberra, ACT, 129 pp

NDSP (2001) Assessment of the efficacy of engineering options for the management of dryland salinity. National Dryland Salinity Program research report to land and water resources research and development corporation

Niknam SR, McComb J (2000) Salt tolerance screening of selected Australian woody species-a review. For Ecol Manage 139:1–19

Nordblom T, Christy B, Finlayson J, Roberts A, Kelly J (2010) Least cost land use changes for targeted salt-load and water-yield impacts in south-eastern Australia. Agric Water Manage 97:811–823

Odie RLA, McComb JA (1996) *Eucalyptus camaldulensis* x *globulus* hybrids for saline land. In: Proceedings of the fourth national conference and workshop on the productive use and rehabilitation of saline lands, Albany, WA, Australia (Promaco Conventions, Perth, pp 367–371), 25–30 March 1996

Oliver M, Wilson S, Gomboso J, Muller T (1996) Costs of salinity to government agencies and public utilities in the Murray–Darling Basin. ABARE research report 96.2. Australian Bureau of Agriculture and Resource Economics, Canberra

Polglase P, Booth T, England J, Falkiner R, Hawkins C, Jovanovic T, Marcar N, Paul K, Theiveyanathan T, van Dijk A, Freudenberger D, Cawsey M, Barrett G (2006) Commercial environmental forestry: targeting new forests for multiple benefits. In: Sustainable forestry – everybody benefits. Proceedings of the Australian forest growers international biennial conference, Launceston, Australia, 22–25 October 2006, pp 168–175

Rengasamy P (2002) Transient salinity and subsoil constraints to dryland farming in Australian sodic soils: an overview. Aust J Exp Agric 42:351–361

Semple W, Williams B (2002) Saline seepage scalds—can they be usefully classified? Nat Resour Manage 5:13–21

Slavich P, Petterson GH (1993) Estimating the electrical conductivity of saturated paste extracts from 1:5 soil: water suspensions and texture. Aust J Soil Res 31:73–81

Sochacki SJ, Harper RJ, Smettem KRJ (2012) Bio-mitigation of carbon following afforestation of abandoned salinized farmland. GCB Bioenergy 4:193–201

State Salinity Council (2000) Natural resource management in Western Australia: the salinity strategy. Government of Western Australia, Perth

Stirzaker R (2002) Planting trees over shallow, saline water tables. In: Vertessy R, Silberstein R, Sarre A (eds) Trees, water and salt: an Australian guide to using trees for healthy catchments and productive farms, Chapter 7. RIRDC/LWRRDC/FWPRDC Joint Venture Agroforestry Program, Canberra, pp 94–109

Stolte WJ, McFarlane DJ, George RJ (1997) Flow systems, tree plantations and salinization in a West Australian catchment. Aust J Soil Res 35: 1213–1229

Summerell GK, Miller M, Beale G, Emery K, Lucas S (2005) Current and predicted minimum and maximum extents of land salinisation for the upland NSW portion of the Murray Darling Basin. In: Zerger A, Argent RM (eds) MODSIM 2005 international congress on modelling and simulation. Modelling and Simulation Society of Australia and New Zealand, Melbourne

Theiveyanathan S, Benyon RG, Myers BJ, Polglase PJ, Marcar NE, Falkiner RA (2003) An irrigation-scheduling model for sustainable application of saline water to tree plantations. For Ecol Manage 193:97–112

Thomson L (1988) Salt tolerance in *Eucalyptus camaldulensis* and related species. PhD thesis, University of Melbourne, Melbourne

Thorburn PJ (1996) Can shallow water tables be controlled by the revegetation of saline lands? Aust J Soil Water Conserv 9:45–49

Thorburn PJ, Walker GR (1994) Variations in stream water uptake by *Eucalyptus camaldulensis* with differing access to stream water. Oecologia 100:293–301

Thorburn PJ, Walker GR, Jolly ID (1995) Uptake of saline groundwater by plants: an analytical model for semi-arid and arid areas. Plant Soil 175:1–11

van der Moezel PG, Watson LE, Bell DT (1989) Gas exchange responses of two *Eucalyptus* species to salinity and waterlogging. Tree Physiol 5:251–257

van Dijk AIJM, Cheng, X, Austin J, Gilfedder M, Hairsine BH (2004) Predicted stream flow and salinity changes after afforestation in the southwest Goulburn. Report for the Commercial Environmental Forestry. CSIRO Land and Water, Canberra

Vertessy R, Connell L, Morris J, Silberstein R, Heuperman A, Feikema P, Mann L, Komarzynski M, Collopy J, Stackpole D (2000) Sustainable hardwood production in shallow water table areas. RIRDC Publication No. 00/163. Rural Industries Research and Development Corporation, Canberra

Zohar Y, Di Stefano J, Bartle J (2010) Strategy for screening eucalypts for saline lands. New For 78:127–137

Models for Estimating Evapotranspiration of Irrigated Eucalypt Plantations

S. Theiveyanathan, R.G. Benyon, V. Koul, R.K. Yadav, and R.I.S. Gill

Abstract

Evapotranspiration, a major component of water balance and net primary productivity in plant-based terrestrial production systems at local and regional scale, is difficult to measure. In order to better understand tree growth and water-use relationships, and to design plantations and optimize their irrigation schedules, it is important to estimate the climatically induced evapotranspiration demand of tree crops. This demand, considered as the maximum evapotranspiration (ETm), is regulated by the resistances imposed by canopy surfaces during the process of evapotranspiration. This chapter describes several simple methods that have been proposed previously to estimate ETm and compares various process-based estimates of ETm with water-use rates determined from a water balance study. The observations from the study conducted at Forest Hill near Wagga Wagga, NSW, Australia, show that ETm can be estimated from standard meteorological parameters as a one-step approach using the Penman-Monteith equation. In the absence of required climatic data, ETm can be estimated from the radiation using Priestley-Taylor technique. For irrigation scheduling, however, ETm may be estimated from pan evaporation data using an estimated pan factor. This factor is site specific and varies with the season and the age of the plantations. For purposes of design and scheduling of irrigation, monthly pan factors can also be determined from climatic data using the Penman-Monteith equation.

S. Theiveyanathan (✉)
Formerly, Commonwealth Scientific & Industrial Research Organisation (CSIRO), Sustainable Ecosystems, GPO Box 284, Canberra, ACT 2601, Australia
e-mail: tivi@internode.on.net

R.G. Benyon
Forest Hydrology, Department of Forest and Ecosystem Science, University of Melbourne, 221, Bouverie Street, Parkville, VIC 3010, Australia
e-mail: rbenyon@unimelb.edu.au

V. Koul
CSIRO Ecosystem Sciences, GPO Box 284, Canberra, ACT 2601, Australia
e-mail: Vijay.Koul@csiro.au

R.K. Yadav
Central Soil Salinity Research Institute, Karnal 132001, Haryana, India
e-mail: rk.yadav@icar.gov.in

R.I.S. Gill
Department of Forestry and Natural Resources, Punjab Agricultural University, Ludhiana, India
e-mail: rishigill@pau.edu

J.C. Dagar, P.S. Minhas (eds.), *Agroforestry for the Management of Waterlogged Saline Soils and Poor-quality Waters*, Advances in Agroforestry 13, DOI 10.1007/978-81-322-2659-8_5

Introduction

Annual average terrestrial rainfall in the world is estimated at 750 mm, and about two-thirds are being returned back to the atmosphere as evapotranspiration (ET), which makes ET the largest single component of the terrestrial hydrologic cycle. Vegetation ET and CO_2 exchange maintain a dynamic but continuous exchange between land surface, plantation, and atmosphere (Savabi and Stockle 2001). Greenhouse effect-induced warming under changing climate has further accelerated the need for understanding the disturbances in hydrologic cycle (Kaczmarek et al. 1996). As such, forest plantations have strong influence on the hydrologic and carbon cycles and salt balance of the site (Musselman and Fox 1991). Actual measurement of evapotranspiration on the trees is very difficult, but certain methods have been developed for regular measurement of humidity and wind velocities using the instrumentation mounted above the plantation and thus estimate (model) the water fluxes occurring out of the canopy at the plantation stand scale. Since the majority of the precipitation returns to the atmosphere as evapotranspiration, the most difficult process to physically measure in hydrologic cycle, an effective estimation of the evapotranspiration is important to understand the terrestrial ecosystem water balance.

In addition to a strong influence on site hydrologic and carbon cycles, plantations also have the potential that can be used for a variety of nontraditional roles such as recharge interception, maintaining favorable water, and salt balance in soil profile and for the productive and ecologically sound reuse of municipal sewage and industrial effluents or agricultural drainage water. An accurate estimate of the rate of water use of plantations is required for evaluating the effectiveness of plantations for achieving the same. The actual evapotranspiration (ETa) of a plantation is a complex interaction and aggregation of transpiration by individual tree in the overstory plantation, transpiration by the understory vegetation and evaporation from the soil surface. A number of methods exist for direct measurement of water use by individual components of the plantation (e.g., tree sap flow measurements for transpiration) or by integrated portions of the plantation (e.g., Bowen ratio energy balance, eddy correlation methods, ventilated chambers, and weighing lysimeters). Integrated plantation ETa is most commonly estimated by solving the water balance equation in which all input and output components are measured. However, all of these methods are cumbersome, labor intensive, and require complex set of instrumentations, which is expensive and not easily transportable and applicable at every site. Therefore, development and use of models in describing water fluxes out of the stand canopy is a necessary step in understanding the effect of plantations on sustainability and optimal utilization of available water resources (Kite 1998). For these reasons, many empirical and physical models have been developed to estimate evapotranspiration based on climatic data (Penman 1948; Monteith 1965; Priestley and Taylor 1972; Perry 1987). Vorösmarty et al. (1998) used water balance model and compared nine models on the watersheds of the continental USA. ETm models have also been compared on sparsely vegetated rangeland (Stannard 1993), wild land vegetation in semiarid rangeland (Dye 1993), partial canopy/residue-covered fields (Farahani and Ahuja 1996), maize with bare soil (Farahani and Bausch 1995), and barley (Tourula and Heikinheimo 1998). Federer et al. (1996) compared ETm models at seven locations, but did not compare the ETm estimates with actual measurements. These models estimate ET from calculated potential evaporation using a range of environmental, physical, and physiological factors, including temperature, humidity, radiation, wind speed, canopy height and configuration, and stomatal conductance. However, ET estimated using different models varies widely because of the differences in the parameters used for estimating the evapotranspiration among the models (VEMAP Members 1995; Ford et al. 2007). Only a few studies (Joshua et al. 2005; Domec et al. 2012) have analyzed evapotranspiration dynamics in forest ecosystems not only because of the general focus has been on agriculture, but also due to the difficulty of obtaining evapotranspiration measurements in forests. Therefore, to compare the plantations, ET estimated using different models with that of actual ETa, the non-limiting growth *Eucalyptus* (*Eucalyptus grandis*) and pine (*Pinus radiata*) plantations established at Wagga Wagga, Australia, and irrigated with effluents to maintain soil water at field capacity were monitored regularly at 15 days interval continuously for 4 years for their water balance. The observations were then used to model the pathways of water and nutrient use and develop the guidelines for optimizing their design and management (Myers et al. 1999). The evapotranspiration of a crop freely supplied with water is determined by the energy supply and resistances to vapor transport across the leaf and soil surfaces and out of the canopy (Monteith 1986). It was, therefore, assumed that ETa measured in this study was the maximum possible rate of evapotranspiration (ETm) by these plantations under the prevailing meteorological conditions.

Rationale to Develop and Use ET Models

Evapotranspiration has been the focus of quantitative agronomic studies because its quantitative link or relationship to growth has been well established by many research workers in the past (Passioura 1977; Doorenbos and Kassam 1979; Fischer 1979; Perry 1987; Wallace 1994; Myers et al. 1999). Since the capacity of plantations to use irrigation water may be limited by the evaporative demand of the environment or by nutrient availability, the maximum water use of plantation needs to be estimated to determine actual application rates. ET is the largest component of the water balance in the irrigated areas, but it is the most difficult process to estimate because it involves the integrated effects of soil, plant, and climate. The actual measurement of ETm is a site-specific, cumbersome, expensive, and time-consuming process because it requires a complex set of lysimetric and instrumentation installations with continuous precise monitoring of all input and output components of water balance like precipitation, irrigation, interception losses, changes in soil profile moisture, and deep drainage to arrive at actual water balance. For this reason, only limited published information is available on measurements of ETm of young irrigated plantations (Dunin and Mackay 1982; Myers et al. 1999). There are many empirical and physical models of measuring evapotranspiration based on climatic data, which are being used for designing and scheduling irrigation and for predicting crop growth. The water, carbon, and energy fluxes as well as meteorological variables above the forest ecosystem canopy are measured using FLUXNET network of towers across the world (Goldstein et al. 2000). Although the first modeling and analysis of forest evapotranspiration was done in the 1970s (Spittlehouse and Black 1979) as the novel way of acquisition of flux data from this tower, this facility, as part of AmeriFlux network, has been used by (Joshua et al. 2005) for comparing five potential evapotranspiration (PET) models on ponderosa pine forest ecosystem in Northern California at larger scale.

Prior to 1948, two theoretical approaches, viz., the aerodynamic ability of removal of moisture from any surface in relation to the turbulent transport of vapor by the process of eddy diffusion and the partition of incoming net radiation between sensible and latent heat transfer were used to estimate the crop evaporative demand. Penman combined these two approaches and termed the new term "the combination equation" to estimate open water evaporation (Penman 1948). Later, a seasonal factor "*f*" was introduced to derive evapotranspiration of the crops using the evaporation from an open water surface and coined the term potential evapotranspiration (PET). This was applied to areas of actively growing short green crops such as alfalfa or grass of uniform height of 15–30 cm, completely covering the soil surface, well supplied with water and about 100 m from the upwind edge of the crop (Penman 1948, 1956). Penman's combined equation method has been widely used to determine the maximum water requirement of crops because growth of most crops is highest when the water supply is non-limiting.

Since Penman's studies, several modifications have been incorporated in the combination equation in order to make it more applicable under variable conditions. A number of simple empirical forms of the aerodynamic wind speed functions have been derived by researchers for making it suitable under particular sets of conditions. In Australia, Dilley and Shepherd (1972) derived an empirical wind speed function for a potato crop at Aspendale, Victoria. Thom and Oliver (1977) incorporated a generalized ventilation term into the combination equation to estimate evapotranspiration of crops ranging from short grass to tall pine trees. Monteith (1965) revised the combination equation from first principles to include resistances to water transport from the soil to atmosphere. The evapotranspiration of a crop freely supplied with water (wet soil) is governed by the energy supply and the resistances to vapor transport across the leaf and soil surfaces (canopy resistance, or more precisely, surface resistance, e.g., Monteith 1986) and out of the canopy (aerodynamic resistance). This rate of water use, maximum evapotranspiration (ETm), postulates that the freely evaporating crop does not behave as the completely wet system to which crops were initially compared in Penman's definition of potential evapotranspiration (ETp). ETm is, however, the important upper boundary of actual evapotranspiration for any given crop (Monteith 1965; Tanner 1967; Ritchie and Burnett 1971; Jenson 1973; Connor 1975; Doorenbos and Pruitt 1977; Passioura 1977; Stewart et al. 1977a, b). These models estimate actual evapotranspiration from calculated potential evaporation using a range of factors or constants. In the present chapter, the accuracy and reliability of following six different models for predicting ETm of irrigated *Eucalyptus* plantations from time of planting to the stage of canopy closure have been compared by using data sets of Wagga Wagga experiments on sewage water use. These included (i) Penman universal combination equation (Penman 1948, 1956), (ii) A Penman-Monteith equation (Monteith 1965) using wind function of Thom and Oliver (1977), (iii) radiation equation (Priestley and Taylor 1972), (iv) two saturation deficit equations (Dilley and Shepherd 1972), and (v) class A pan evaporation measured on site. These models were chosen because they are commonly used in water balance studies (Arnell and Reynard 1996) and hydroinformatics (Naoum and Tsanis 2003).

Site Description

For comparing the ET of plantations estimated using different models with that of actual ETa, plantation was established adjacent to the sewage treatment works of Forest Hill town Wagga Wagga, NSW, Australia (35°10'S, 147°28'E). The site receives 570-mm mean annual rainfall with relatively more winter-dominant distribution. Annual pan evaporation is 1860 mm and strongly seasonal, varying from a monthly low of 35 mm in June and July to a peak of 320 mm in January. The mean minimum temperature of 3 °C is observed in the coldest month of June, while the mean maximum temperature of the hottest month of January is 31 °C. The site experiences on an average of about 13 frost days per year. Soils of the site are having a well-drained sandy loam or sandy clay loam A horizon (20–45-cm deep) overlaying a sandy-clay to medium-clay B horizon. These are classified as Red Chromosol and Red Kandosols and Red Podsolic and red earth, respectively, in the Great Soil Group Classification. The land was previously used for wheat cropping and sheep grazing. Meteorological data (rainfall, air temperature, humidity, solar radiation, pan evaporation, wind speed, wind run, and direction) were recorded at an hourly interval using an automatic weather station (Starlog, UNIDATA Australia, Perth, Western Australia) established at the site. While the supplementary data required for the analyses (e.g., 3-hourly wet and dry bulb temperatures and air pressure) were obtained from the Bureau of Meteorology weather station located 3 km from the site, daytime positive net radiation above the crop canopy (*Rn*) was estimated from daytime global radiation (*Rs*) using the following equations of models of Linacre (1968) and Leuning et al. (1991a):

$$Rn = Rs(1-\alpha) - Rl \tag{1}$$

$$Rl = \left[0.1 + (1-0.1)\left(\frac{n}{N}\right)\right].\left[(\varepsilon_a - \varepsilon_c).\sigma.(T+273)^4\right] \tag{2}$$

where Rl is the net upward long wave radiant flux density, α is the canopy albedo (0.15 for eucalypts and 0.25 for grass), n is daily duration of clear sunshine (hrs), N is the time from sunrise to sunset in hours, ε_a is the clear sky emissivity, ε_c is the canopy emissivity (0.96 for eucalypts), σ is the Stefan-Boltzmann constant, and T is the near-ground air temperature (°C). Daily mean saturation deficit (es-ea) was calculated from 3-hourly observations of wet and dry bulb temperature taken between dawn and dusk (Lowe 1977).

Plantation Establishment and Management

Six-month-old *Eucalyptus grandis* seedlings were planted at 2 m×3 m spacing (1667 trees ha^{-1}). Irrigation treatments, applied in duplicate on 0.2 ha plots of 300 trees each, were based on the water-use rates of the plantations and varied seasonally in response to the climate and canopy development. Under-tree micro-sprinklers were used to apply secondary-treated municipal sewage effluent at a discharge rate of 4.6 mmh^{-1}. The medium (M) irrigation treatment consisted of application of effluents at the estimated water-use rate of the plantation less rainfall. The aim of applying irrigation at this rate in M treatment was to maintain the deep drainage near to naturally occurring rate, while other two irrigation treatment plots received nominally twice and half as much effluent, respectively. Two plots of only pasture without *Eucalyptus* plantation were also irrigated at their rate of water use and less rainfall. Complete details of the experimental design used and treatments applied are as per that of Myers et al. (1999). Irrigation scheduling was controlled, and the applied volumes were logged, by a PC-based irrigation program (IRRICOM, Peter Cornish & Associates Pty Ltd, Canberra, Australia). Irrigation was applied at night to minimize seasonal variation in irrigation interception loss. For the purpose of irrigation scheduling, plantation water use was estimated using the water balance equation over 2-weeks intervals. Inputs used were rainfall, canopy interception, irrigation volume, and changes in soil water storage. Soil water content of the plantation and the irrigated pasture was measured with a neutron probe (503 Hydroprobe, Campbell Pacific, Pacheco, CA, USA) every 2 weeks in three access tubes per plot at nine depths to 2 m. Plots were irrigated weekly during the first 3 years and twice weekly subsequently. The irrigation aimed to fill the top meter of soil to 90 % of the drained upper limit (DUL) of soil water holding capacity. DUL was taken as the wettest drained profile recorded, 2 days after substantial rain in spring. A refill level of 90 % was used to reduce the risk of drainage occurring if rain fell shortly after irrigation, and complete details of the irrigation strategy and scheduling were followed as suggested by Myers et al. (1999).

Weather During Study Period

Mean daily climatic data are presented for each month of the 4-year period in Table 1, and a comparison of seasonal daily variation of solar radiation (Rs), wind speed (U), average temperature (T ave), relative humidity (RH), and day length (DL) is shown in Fig. 2. These data illustrate that the trees were exposed to variation in the annual and seasonal climatic conditions. As the season progressed from July (midwinter) to January (midsummer), there was increase in radiation (195–488 W m^{-2}), temperature (8.1–22.9 °C), vapor pressure deficit (relative humidity decreased from 80 to 48 %), and hence evaporative demand (pan evaporation increased from 1.2 to 8.6 mm). During hot summer periods, higher radiation and higher vapor pressure deficit produced a greater evaporative demand (Epan of 7.7–8.6 mm d^{-1}). The wind speed was higher in December–January (2.3–2.4 m s^{-1}) which also contributed to increased evaporative demand.

Estimation and Measurements of Actual Evapotranspiration (ETa)

Surface runoff, subsurface lateral flow, deep drainage, ET (using crop factor of that stage), and interception losses are the components that need to be recorded for arriving at the periodic water balance at any given site. However, under our site (Wagga Wagga, NSW, Australia) conditions, neither surface runoff nor subsurface lateral flow (as monitored by logging piezometers) was observed during the experimental period. It implied that soil has

Table 1 Monthly climatic data compared with pan evaporation from class A pan for Wagga Wagga, NSW, during 1991–1995

	Jul	Aug	Sep	Oct	Nov	Dec	Jan	Feb	Mar	Apr	May	Jun
Total rainfall (mm)	67	46	69	70	55	63	40	57	66	35	56	47
Average daily temp (°C)	8.1	8.4	10.6	14.9	17.5	21.5	22.9	22.8	19.0	15.2	11.5	8.6
Average daily relative humidity (%)	80	73	70	62	54	50	48	52	55	58	73	80
Average daily global radiation (W m^{-2})	198	269	331	422	463	435	488	437	411	341	271	195
Average daily wind speed (m s^{-1})	1.2	1.6	1.9	2.0	2.2	2.3	2.4	2.2	2.0	1.6	1.2	1.3
Average daily sunshine (h)	4.3	6.0	6.8	9.1	9.5	8.9	10.2	9.6	9.0	8.4	5.7	4.2
Average daily Epan (mm)	1.2	1.8	2.6	4.5	6.4	7.7	8.6	7.7	6.0	3.7	1.9	1.1

high hydraulic conductivity and the drainage was the main component of water loss other than evapotranspiration and interception. However, since the irrigations were designed to leave a soil water deficit of 25 mm or more in the top meter of soil, it was assumed that with < 25-mm rain in a cycle, drainage below 1 m would be negligible. The water balance equation could be solved for desired period of interval; in our case it was for every 2-week cycle. For the dry cycles, ETa was calculated as:

$$\Delta WU = \frac{WU}{(t2 - t1)} = SWt1 - SWt2 + \sum_{t2}^{t1} IRn + \sum_{t2}^{t1} Pn \qquad (3)$$

where ETa is the mean daily evapotranspiration between time t1 and t2, SWt1 and SWt2 are the soil water storage at time t1 and t2, and $\sum Pn$ and $\sum IRn$ are the cumulative net precipitation and net irrigation between time t1 and t2. A crop (plantation) factor was calculated for each of these dry 2-week cycles as the ratio between ETa and measured pan evaporation (Ep). When there was more than 25 mm of rainfall in a cycle, a crop factor for that wet cycle was calculated as the mean of the crop factors of the preceding and subsequent dry cycle. This was applied to the measured Ep to estimate the water use for the cycle. In this way, the water use of every 2-week cycle was either measured (dry cycles) or estimated (wet cycles). Only 31 % of all cycles were wet, and these were predominantly in winter when up to 50 % of cycles were wet. The proportion of wet cycles during the summer irrigation seasons was zero in the second year and 15 % in the third year. Net precipitation and irrigation (i.e., total minus interception loss) were measured during a number of rainfall and irrigation events using interception troughs and time domain reflectometry (TDR). Scaled from these measurements and those reported by Myers and Talsma (1992) and Myers et al. (1996), interception loss was calculated as a fixed rate per rain or irrigation event within an irrigation season. We calculated ETa from age 6 months to the time when the canopy was closed (age 2 years) with the foliage biomass of ~5 t ha^{-1} and the leaf area index (LAI) > 4.0. To partition the total plantation water use into tree and understory evapotranspiration prior to canopy closure, two pasture plots were also irrigated at their estimated water-use rates determined in the same manner as the *Eucalyptus* plots. Eighty-six fortnightly measurements of water-use data sets covering a range of LAI and atmospheric conditions were compared with the model predictions.

The biweekly measurements of ETa made from 6 months of tree growth to the stage of canopy closure, i.e., 4 years are presented in Fig. 3 and

Table 2 Measurements[a] of monthly mean daily actual evapotranspiration (ETa, mm d^{-1}) of an irrigated eucalypt plantation for 4 years from planting

Season	Jul	Aug	Sep	Oct	Nov	Dec	Jan	Feb	Mar	Apr	May	Jun
1991–1992								3.8	3.0	2.1	1.1	0.8
1992–1993	0.9	1.5	1.8	2.8	3.3	4.4	6.1	6.0	4.7	4.3	2.5	1.7
1993–1994	1.6	1.9	2.3	3.5	5.9	6.3	7.3	6.2	4.6	3.3	2.3	1.1
1994–1995	1.1	1.1	2.5	3.6	5.3	6.5	6.1	6.8	6.1	3.5	1.9	1.5

[a]The measurements were taken at 14-day intervals

Table 2. Because these measurements were made under non-soil water-limited conditions, the magnitude of ETa and ETm in this study was found to be the same. These data were used in the comparisons presented in the following section. As the season progressed, ETa ranged from 1.1 to 7.3 mm d^{-1}. During the nonirrigation period of winter season, mean monthly ETa was around 1.1–2.3 mm d^{-1} increasing to 7.3 mm d^{-1} in the summer irrigation season. The results show that there was a sharp variability between the seasons reflecting a similar variation in the driving environmental variables of radiation, temperature, and wind. The seasonal variation in evapotranspiration was recorded to be most prominent after the canopy closure stage. Between the months of February and June, for example, ETa varied from 3.8 to 0.8 mm d^{-1} in 1992 compared with 6.2 to 1.1 mm d^{-1} in 1994. The results also show that evapotranspiration in the 1993–1994 season was higher than in 1994–1995 (e.g., in December, ETa of the eucalypt plantation was 7.3 mm d^{-1} in 1993 compared with only 6.1 mm d^{-1} in 1994) which again reflected the effect of environmental variation dominating the evaporative demand. The water balance measurements were made with every possible accuracy and care, and because ET was the only unmeasured component of the water balance during fortnights with low rainfall (dry cycles), the water use calculated by water balance was considered accurate enough to be used in comparing other indirect techniques (models) of estimating ET.

Models for Prediction of Maximum Evapotranspiration (ETm)

The six models of increasing complexity were assessed based on the precision and accuracy of their estimates of ETm compared to ETa for 86 2-week cycles using regression techniques. The coefficient of efficiency (E), coefficient of determination (R^2), the parameters to assess the closeness of fit of the regressions to the 1:1 line, and the precision of the estimates when forced through the origin, respectively, were determined statistically using the procedures of Aitken (1973) in following equation:

$$E = \frac{\sum (Y_i - \bar{Y})^2 - \sum (Y_i - X_i)^2}{\sum (Y_i - \bar{Y})^2} \quad (4)$$

Prior to canopy closure, plantation water use can be estimated using the Penman-Monteith method by the sum of the tree water use and grass water use based on the proportion of the trees covering the ground within the plots. The ground cover estimates could be derived by fitting a logistic curve for the observed data taken from photographs (Fig. 1). In our studies, we assumed that evaporation from understory was negligible after the canopy closure stage, i.e., when the LAI>4.0. Details of functions and utility conditions of each model used are seen below.

(i) Penman Universal Combination Equation

This model assumes that for a crop/plantation freely supplied with water, potential evapotranspiration (ETp) can be calculated by combining the energy balance and aerodynamic equations developed using the meteorological data from nearby observatory. The combination equation used in this model involves the following functions:

$$\mathrm{ETp} = c(Er + Ea) \quad (5)$$

$$Er = \frac{\Delta R_n}{(\Delta + \gamma)} \quad (6)$$

$$Ea = \frac{gf(u)(e_s - e_a)}{(\Delta + g)} \quad (7)$$

$$f(u)=0.35(0.5+0.01*u) \tag{8}$$

Here, in equations from 5 to 8, the two terms *Er* and *Ea* are the energy and aerodynamic components, respectively, and *c* is the dimensionless correction factor which has been set equal to 1 in our study, Δ is the slope of the saturation vapor pressure versus temperature curve at mean air temperature, γ is the psychrometric constant, and *es* and *ea* are the saturated and actual vapor pressures, respectively. $f(u)$ is a wind speed function and *u* is wind speed measured in miles d^{-1}, and *f* is obtained from regression analysis of ETm and ETp. ETm was then estimated from ETp using the equation

$$\text{ETm} = f\ \text{ETp} \tag{9}$$

(ii) Penman-Monteith Combination Equation

The Penman combination equation was further generalized to a significant extent with incorporation of canopy responses to evaporative demand of the environment by Monteith (1965). In this one-step approach, the evapotranspiration (ETpm) can be calculated based on radiation and resistances to evaporation imposed by the atmosphere and the canopy. Under irrigated conditions, in soil moisture maintained at field capacity, evapotranspiration is assumed to be equal to maximum evapotranspiration (ETm). It involves the following function:

$$\text{ETm} = \text{ETpm} = N\left(\frac{1}{\lambda}\right)\left(\frac{\varepsilon(cRn) + \dfrac{\rho\lambda D}{ra}}{\varepsilon + 1 + \dfrac{rs}{ra}}\right) \tag{10}$$

where *N* stands for the period from sunrise to sunset in seconds, *c* is a dimensionless parameters assumed to be equal to 0.1 (Raupach 1995), ε is a dimensionless slope of the saturated specific humidity curve, respectively. λ is the latent heat of vaporization (J kg^{-1}), ρ is the density of air (kg m^{-3}), and *D* is the saturation deficit. *rs* and *ra* represent the canopy and aerodynamic resistances to water vapor transport from the soil and soil surface through the plant to the canopy surfaces and from there into the atmosphere.

Though the aerodynamic resistance, *ra,* has many formulations, some are based entirely on empirical evidence, and the others are related to the mixing length theory, but Thom and Oliver (1977) developed following general wind speed function by including crop geometry and height, which were used in present comparison of models.

$$r_a = \frac{\left(\ln\dfrac{Z-d}{Zo}\right)\left(\ln\dfrac{Z-d}{ZoHE}\right)}{k^2U} \tag{11}$$

Here, *U* stands for the average wind speed in m s-1 measured at *Z* m above the plantation canopy, *d* the zero plane displacement, *k* the von Karman constant, and *Zo* and Z_{oHE} are the roughness lengths (m) of the surface for momentum transfer and heat and water vapor transfer. These were calculated as explained in Raupach (1995).

Canopy resistance, the parallel sum of the stomatal resistances of the photosynthetically active leaves and dependent on LAI, varies significantly with the time of the day and therefore requires complex models to estimate (Dolman et al. 1991; Wallace 1994). In our studies at the Wagga Wagga, LAI was measured at 6-monthly intervals using both destructive and nondestructive techniques. These data were then used to derive a logistic function to obtain estimates of LAI (Fig. 1). Trees also control the environmental effect on evapotranspiration through opening and closing of stomata, and thus canopy resistance is an additional function of radiation, saturation deficit, and soil moisture stress as expressed in the following relationship of r_s to these important variables in an empirical equation (Leuning et al. 1991a; Raupach 1995; Shuttleworth 1989).

$$r_s = \frac{r_{s(\min)}}{f_R(Rs)\times f_D(D)\times f_W(W)\times f_L(L)} \tag{12}$$

In the above equation, $r_{s(\min)}$ represents the minimum value of the canopy resistance (27 s m^{-1} for *Eucalyptus* trees and 60 s m^{-1} for grass) under optimal growth conditions, and f_R, f_D, f_W, and f_L are dimensionless environmental functions, which range between 0 and 1. The evaporative stresses on the trees are caused by lower incoming solar radiation (*Rs*), higher saturation deficit (*D*), higher soil

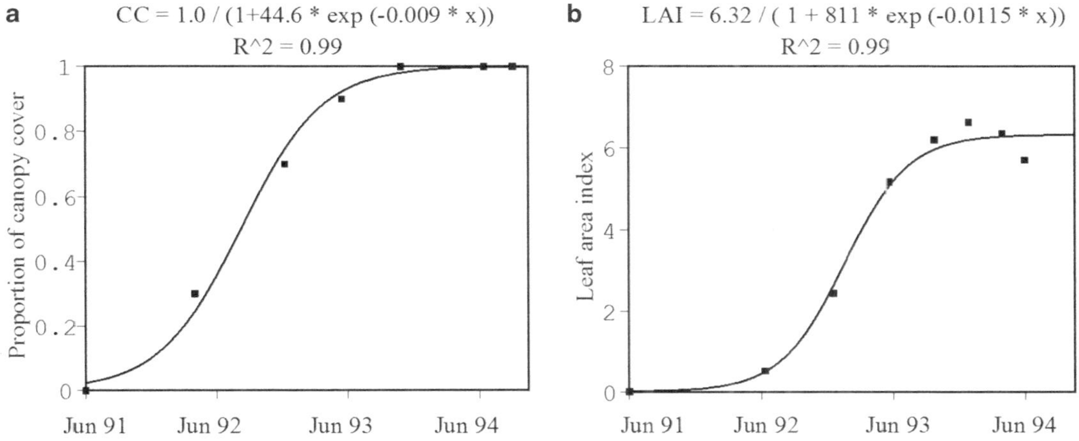

Fig. 1 Measurements showing (**a**) the proportion of tree canopy cover, CC, and (**b**) changes in the leaf area index, LAI, in relation to the age of the eucalypt plantation. The seedlings were planted in June 1991. The lines were fitted through the observed data using logistic functions

water deficit (*W*), and lower LAI. Threshold values for solar radiation (350 W m^{-2}; Leuning et al. 1991b), saturation deficit (35 g Kg^{-1}; Hookey et al. 1987), and LAI (3.0–3.5; Dunin and Aston 1984; Persson 1995) were used in this study. Since the irrigation strategy is based on bringing one-meter depth of soil twice every week to 90 % of the DUL of the soil moisture holding capacity, the average *fW* during the experimental period was assumed to be equal to 1.0. This model has also been used earlier to estimate ETm of Eucalyptus trees and grass by setting Zo to one tenth of plant height (Szeicz et al. 1969; Watts and Hancock 1984) (Fig. 2).

(iii) Radiation Equation (Priestley-Taylor Model)

Radiation methods assume that the ultimate source of energy required for evaporation is the sun. Priestley and Taylor (1972) used the radiation term of Penman's combination equation (Er) for estimate ETm as under and established values of α, the adjustment factor or widely known as Priestley-Taylor parameter, from comparisons with lysimetric data (e.g., McNaughton and Black 1973).

$$\mathrm{ETm} = \alpha Er = \alpha \frac{\Delta \mathrm{Rn}}{\Delta + \mathrm{g}} \qquad (13)$$

Using selected days of measurements, they found α to vary between 1.08 and 1.34 with an overall mean of 1.26. They derived from the Penman combination equation the limits of variation of α under potential conditions to be $1 < \alpha < (\Delta + \gamma)/\Delta$). However, Pereira and Villa-Nova (1992) suggested that the fluctuations for α are governed primarily by the sensible heat flux and Stannard (1993) using multiple linear regression analysis established the dependence of α on LAI and the time of rainfall under sparse vegetation in San Luis Valley in southern Colorado. For a Douglas fir forest, McNaughton and Black (1973) found $\alpha = 1.05$. Davies and Allen (1973) reported α values between 1.01 and 1.34. Pereira and Villa-Nova (1992) showed that the fluctuations of α, either on an hourly or on a daily basis, are governed primarily by the sensible heat flux variations. Viswanadham et al. (1991) obtained a mean value of 1.16 for the Amazon forest. Shuttleworth and Calder (1979) showed that the Priestley-Taylor model is of limited use for tall vegetation, where there are large atmospheric exchange coefficients. However, Viswanadham et al. (1991) found good agreement between the model and independent measurements made with an eddy correlation technique. In our studies, we estimated ETm using the Priestley-Taylor model by obtaining a value for α through -regression analysis of the measured data at the site.

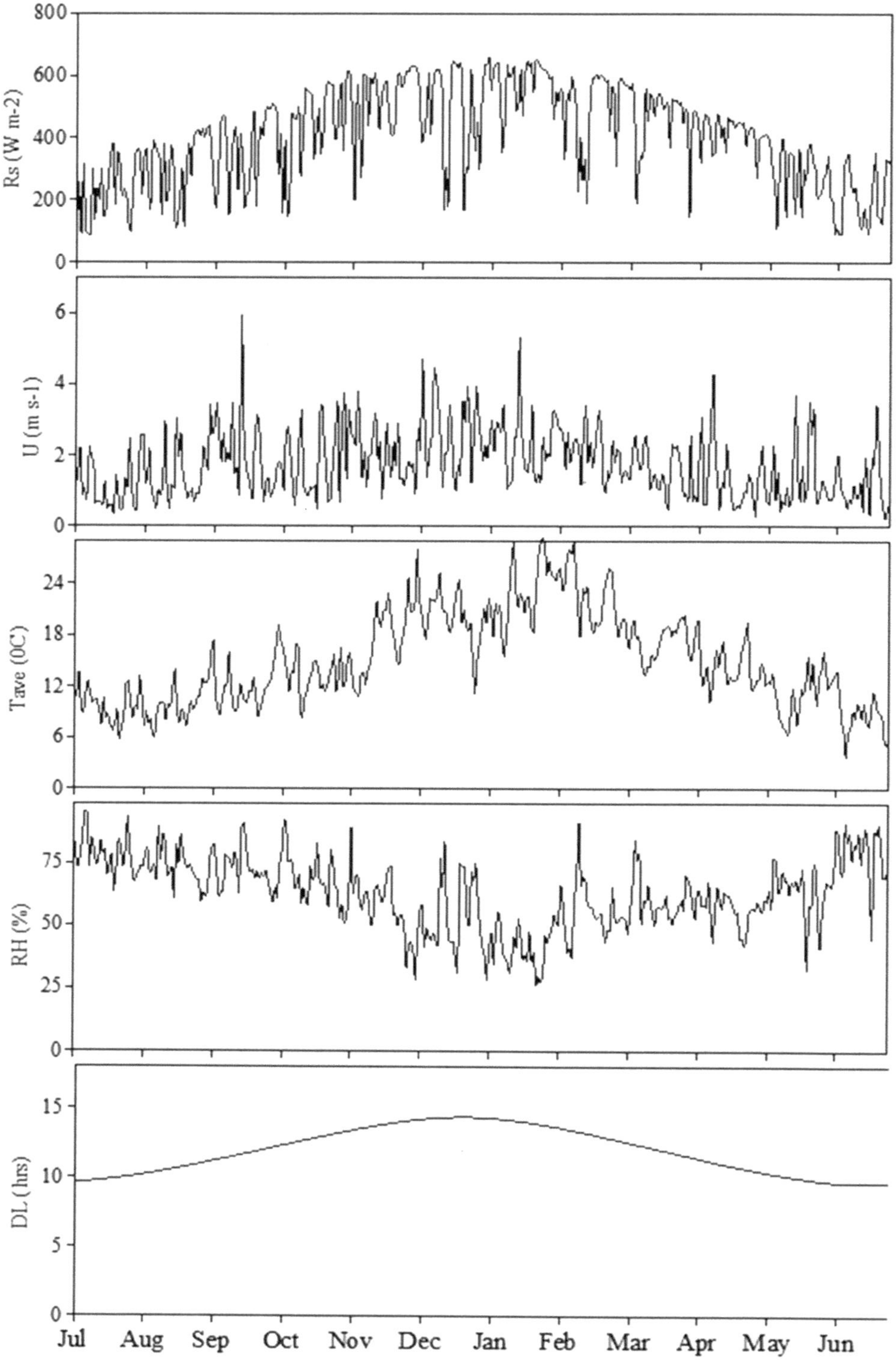

Fig. 2 Variation in daily values of daytime solar radiation (Rs), wind speed (U), average temperature (Tave), relative humidity (RH), and daylength (DL) for Wagga Wagga, New South Wales, during a year (1993–1994) of the experimental period

(iv) Two Saturation Deficit Equations

Correlation between ETp and saturation deficit was first pioneered by Dalton (1802) and later on used to estimate ETm of various crops (Tanner and Sinclair 1983) and also for trees (Perry 1987). Responses of trees to atmospheric humidity tend to limit water use during high atmospheric demand. However, Australian studies of water use indicate that although *Eucalyptus* species tend to have effective stomatal response mechanism to soil moisture deficits, but certain species may not have a well-developed response to atmospheric humidity (Carbon et al. 1981; Colquhoun et al. 1984; Greenwood et al. 1985). However, Dye (1993) observed that the stomatal response to atmospheric humidity in *Eucalyptus grandis* varied seasonally and at high saturation deficit, the response was less. This model assumes that the evaporative demand of the atmosphere, ETm, can be adequately represented by the vapor pressure gradient above the tree canopy. Two forms of this model were used in our study. The first and complex form (VPD1 method, Eq. 13) included both the canopy and aerodynamic resistances to estimate ETm and referred as Esd1 as under.

$$\mathrm{ETm} = \beta \mathrm{Esd1} = \frac{\rho \times Cp}{\gamma} \frac{(e_s - e_a)}{(r_s + r_a)} \quad (14)$$

In this, β is a constant of proportionality, ρ is density of air (kg m^{-3}), Cp is specific heat of air (J kg^{-1} °C^{-1}), and γ is the psychrometric constant (mb °C^{-1}). The second form (VPD2 method, Eq. 14) used a simpler relationship between ETm, and vapor pressure deficit referred as Esd2 as depicted below.

$$\mathrm{ETm} = \beta 1 \mathrm{Esd2} = \frac{\gamma}{\Delta + \gamma}(e_s - e_a) \quad (15)$$

Here, $\beta 1$ stands for a constant of proportionality (mm mb^{-1}), and Δ is slope of the vapor pressure curve at the daily average air temperature point (mb °C^{-1}).

(v) Class A Pan Evaporation (Epan)

This model, a simple method of estimating ET, requires only evaporation data from an open container such as an evaporation pan, which are easily available from Class A pans. Many researchers have used this method to estimate daily ETm of various crops including trees. Many irrigation models use pan evaporation data for estimating ET and to schedule irrigation (Jones and Bauder 1987; Smittle and Dickens 1992; Smittle et al. 1992; Theiveyanathan et al. 2004). It is always easier to collect data from an evaporation pan than from vegetation (Doorenbos and Pruitt 1977; Myers and Talsma 1992). The pan evaporation method is a two-step process involving conversion of pan evaporation to ETp using pan coefficient followed by calculation of ETa by using crop coefficient (Pereira et al. 1995). In the present study, both these coefficients have been combined as a single pan factor (Kp) and used to estimate ETm from the pan evaporation data as seen below.

$$\mathrm{ETm} = \mathrm{Kp}\,\mathrm{Epan} \quad (16)$$

In the past, pan factors have been derived and used for monthly intervals (Doorenbos and Pruitt 1977), for daily intervals (Smittle and Dickens 1992), and for 5-day intervals (Chiew et al. 1995). In this study, 2 years of weekly totals of pan evaporation and ETa data were used to derive monthly pan factors for the irrigated *Eucalyptus* plantations. ETa is obtained from the Penman-Monteith ET estimation. The weekly totals were used because in many cases the daily pan evaporation data suffer from measurement errors especially when evaporation was low during winter months. The weekly total pan evaporation helped to mitigate the errors of daily measurements.

Comparison of Model Estimates and Actual Evapotranspiration

The 2-weekly actual evapotranspiration and the calculated estimates from the two forms of the combination equation (i.e., ETp and ETpm), from radiation (Er), from two forms of saturation deficit (Esd1 and Esd2), and from pan evaporation (Epan) were plotted against time for the plantation growth period of 6 months to canopy closure stage of 4 years (Fig. 3). The monthly mean values of the estimates of Er, Esd1, Esd2, ETp, ETpm, and Epan for

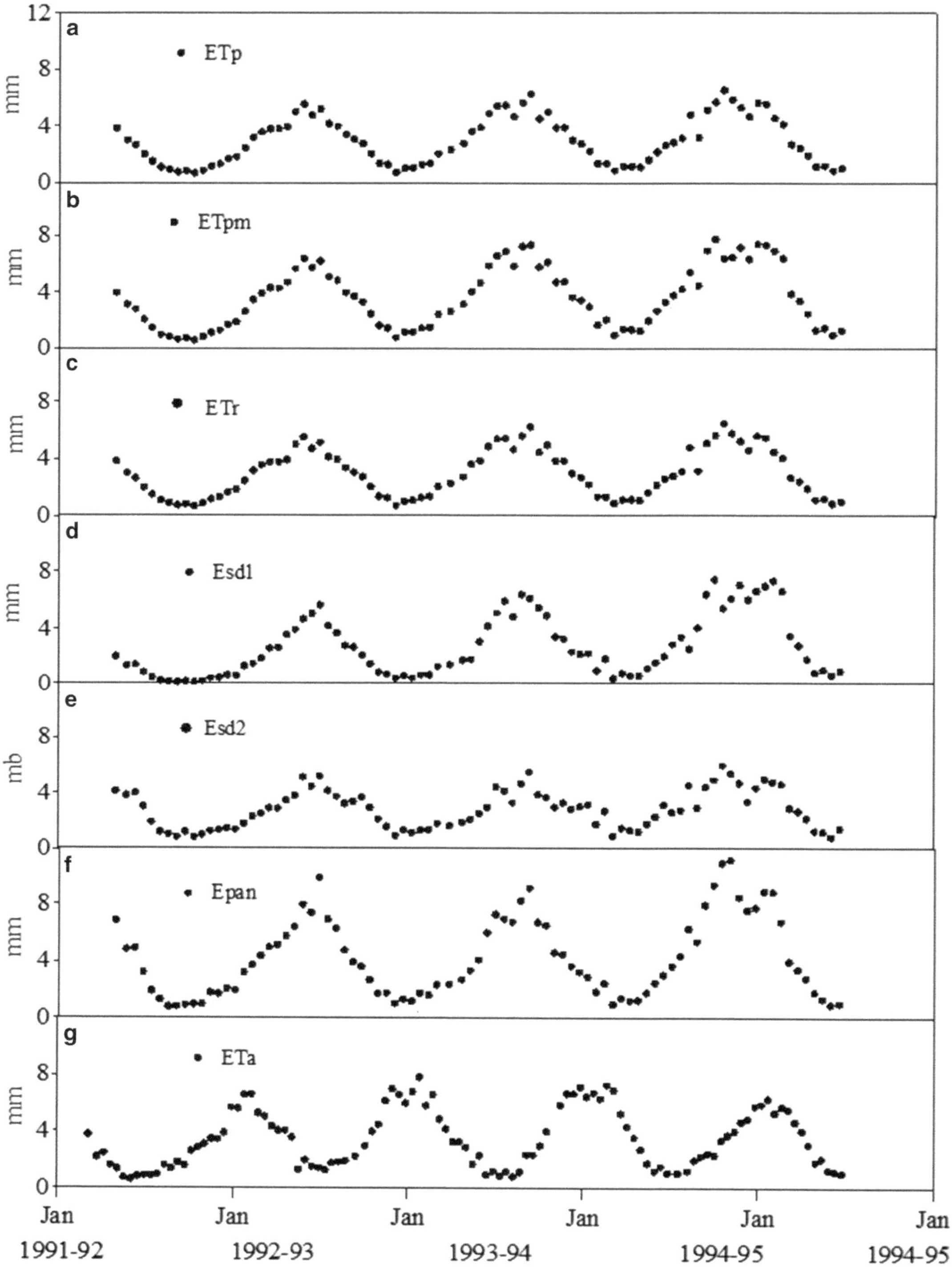

Fig. 3 Fourteen-day mean daily estimates of maximum evapotranspiration (ETm) using the following methods: (**a**) Penman combination (ETp), (**b**) Penman-Monteith (ETpm), (**c**) Priestley and Taylor (Er), (**d**) and (**e**) saturation deficit method 1 and 2 (Esd1 and Esd2), (**f**) pan evaporation (Epan), and (**g**) mean daily measurements of actual evapotranspiration (ETa)

the study period years (1991–1995) are also shown in Table 3.

During the 4-year period, all the estimates of ETm showed marked variation between seasons, fluctuating from a minimum value in June–July to a maximum value in December–January (Table 3). Pan evaporation was consistently higher than the other estimates throughout the period especially during summer months when it was the highest. Joshua et al. (2005) observed that, for all potential evapotranspiration models, simulated ETm compared reasonably well with measured evapotranspiration at the beginning of the summer season (April–May). However, as the soil moisture decreased through the summer, all models tended to overpredict evapotranspiration because these were designed for well-watered soil conditions rather than natural summertime Mediterranean drought conditions.

In our studies, various estimates of ETm were also plotted against measured ETa (Fig. 4), and a statistical analysis of these relationships was also carried out. The coefficient of efficiencies for all the data sets was estimated from the regression analysis. With the exception of the Esd1 estimate, all relationships were forced through the origin to result in zero measured water use implying zero estimate of ETm and valid comparisons of the relationships made in terms of their slope and coefficient of determination. Further, a power relationship was used to relate the response of Esd1 estimate to Eta. During hot summer months of December and January when the vapor pressure deficit was high, the ETpm method consistently overestimated the actual water use by 0.5–1 mm. By varying the threshold vapor pressure deficit in the Penman-Monteith combination equation from 0.010 to 0.045 kg kg^{-1}, a better prediction of actual water-use rate was obtained for *Eucalyptus* trees at this site compared to use of a constant threshold vapor pressure deficit of 0.023 kg kg^{-1} (Fig. 4).

Comparisons Across Seasons

Eighty-six measurements of fortnightly mean daily water use recorded from 6 months of tree growth to the stage of canopy closure were compared against the estimates of ETm to produce the relationships between the estimated and measured data (Figs. 4 and 6). The models, which were used to describe the regression analysis of the two forms of the vapor pressure deficit methods, ES1 and Es2, were different: Esd1 was fitted with a power function and Esd2 was fitted with a linear function. The most important parameter observed in these analyses was the coefficient of determination, which varied from 0.72 for Esd2 to 0.92 for ETpm as shown in Table 4. This parameter was used to define the consistency of the relationship and hence models' predictive capacity over the range of measured values.

When we compared these models for their consistency, the order of decreasing consistency was ETpm>Er>ETp>Epan>Esd1>Esd2. The most noteworthy features of the comparison are the poor performance of Esd2 and the ability of Er to match the high performance of ETpm. The second important parameter is the coefficient of efficiency (E), which varied from 0.65 for Esd2 to 0.90 for ETpm (Table 4). This parameter determined the closeness of agreement of the relationships between the observed and the estimated values. High E values in the Penman and radiation methods indicated the high quality of estimation by these models. However, it should be noted that ETm is the only one-step approach of calculating ETa rather than estimating it (i.e., coefficient of efficiency is 0.90 and the slope of the regression analysis was almost equal to 1). In the case of Er, the slope of the relationship with ETa, i.e., the value of α was 1.17, which was consistent with that of 1.1–1.3 depending upon the surface conditions as observed by Priestley and Taylor (1972). With the exception of the Esd1 estimate, all other estimates showed a linear response to ETa (Fig. 4). Esd1 showed a power relationship with the ETa, indicating that at higher vapor pressure deficits, the response to evapotranspiration rate was lower.

Comparisons Between Seasons

As the season progressed to summer, the evaporative demand increased (Fig. 3), and, importantly, the relative contribution of radiant and aerodynamic energy to total evaporation changed. During winter when the evaporative demand was less,

Table 3 Monthly mean daily estimates of evapotranspiration (mm d^{-1}) using six methods[a] for the years 1991–1995

Methods	Jul	Aug	Sep	Oct	Nov	Dec	Jan	Feb	Mar	Apr	May	Jun
1991–1992												
Penman (ETp)	0.8	1.3	2.1	3.5	4.4	4.2	5.0	3.9	3.4	2.2	1.3	0.8
P-M (ETpm)	0.7	1.2	2.0	3.7	4.7	4.5	5.1	4.0	3.5	2.3	1.2	0.7
P & T (Er)	0.8	1.2	2.1	3.5	4.4	4.2	5.0	3.8	3.4	2.2	1.2	0.8
VPD1 (Esd1)	0.2	0.4	0.6	1.3	2.5	2.6	2.8	2.1	1.6	1.1	0.3	0.2
VPD2 (Esd2)	0.9	1.4	1.8	3.4	5.0	5.1	5.3	4.4	3.9	3.3	1.5	1.0
Pan eva (Epan)	1.0	1.7	2.7	5.3	7.9	8.1	9.4	7.3	5.7	3.9	1.5	0.8
1992–1993												
Penman (ETp)	0.8	1.2	1.8	3.0	3.8	4.1	5.2	4.8	3.7	2.9	1.6	1.0
P-M (ETpm)	0.7	1.2	1.8	3.3	4.3	4.8	6.1	5.8	4.4	3.5	1.8	1.1
P & T (Er)	0.8	1.2	1.8	3.0	3.8	4.1	5.2	4.7	3.6	2.8	1.6	1.0
VPD1 (Esd1)	0.2	0.4	0.7	1.5	2.4	3.3	4.8	5.0	3.2	2.4	1.0	0.5
VPD2 (Esd2)	1.0	1.3	1.5	2.2	2.9	3.3	4.8	4.8	3.5	3.6	2.3	1.2
Pan eva (Epan)	1.0	1.6	2.1	3.7	4.9	5.6	7.7	8.6	5.4	3.8	2.0	1.2
1993–1994												
Penman (ETp)	1.1	1.7	2.4	3.4	4.8	4.9	6	4.8	3.9	2.8	1.6	1.0
P-M (ETpm)	1.2	2.0	2.8	3.9	5.7	6.1	7.4	6.0	4.7	3.5	2.2	1.2
P & T (Er)	1.1	1.7	2.4	3.4	4.8	4.9	5.9	4.7	3.8	2.7	1.6	1.0
VPD1 (Esd1)	0.5	1.0	1.5	2.0	3.9	5.3	6.4	5.2	3.3	2.3	1.6	0.7
VPD2 (Esd2)	1.2	1.6	1.8	2.1	3.2	3.7	5.1	3.8	3.2	3.1	2.4	1.4
Pan eva (Epan)	1.5	2.0	2.4	3.4	5.8	6.7	8.8	6.7	4.5	3.5	2.4	1.3
1994–1995												
Penman (ETp)	1.2	1.9	2.7	4.2	4.5	6.2	5.1	5.5	4.3	2.5	1.3	1.0
P-M (ETpm)	1.4	2.3	3.5	5.2	6.2	6.6	7.0	7.2	6.6	3.4	1.4	1.2
P & T (Er)	1.2	1.9	2.7	4.2	4.5	6.1	5.1	5.4	4.2	2.5	1.3	1.0
VPD1 (Esd1)	0.6	1.3	2.5	3.4	5.9	6.0	6.7	6.7	6.9	2.9	0.8	0.8
VPD2 (Esd2)	1.4	2.1	2.8	3.9	4.1	5.6	4.2	4.7	4.6	2.8	1.3	1.2
Pan eva (Epan)	1.3	2.1	3.3	5.9	7.5	10.5	8.4	8.2	7.6	3.5	1.9	1.1

[a]The six methods are *Penman* = Penman combination equation, ETp; *P-M* = Penman-Monteith combination equation, ETpm; *P & T* = Priestley and Taylor method, Er; *VPD1* = vapor pressure deficit method 1, Esd1; *VPD2* = vapor pressure deficit method 2, Esd2; and *pan eva* = pan evaporation method, Epan

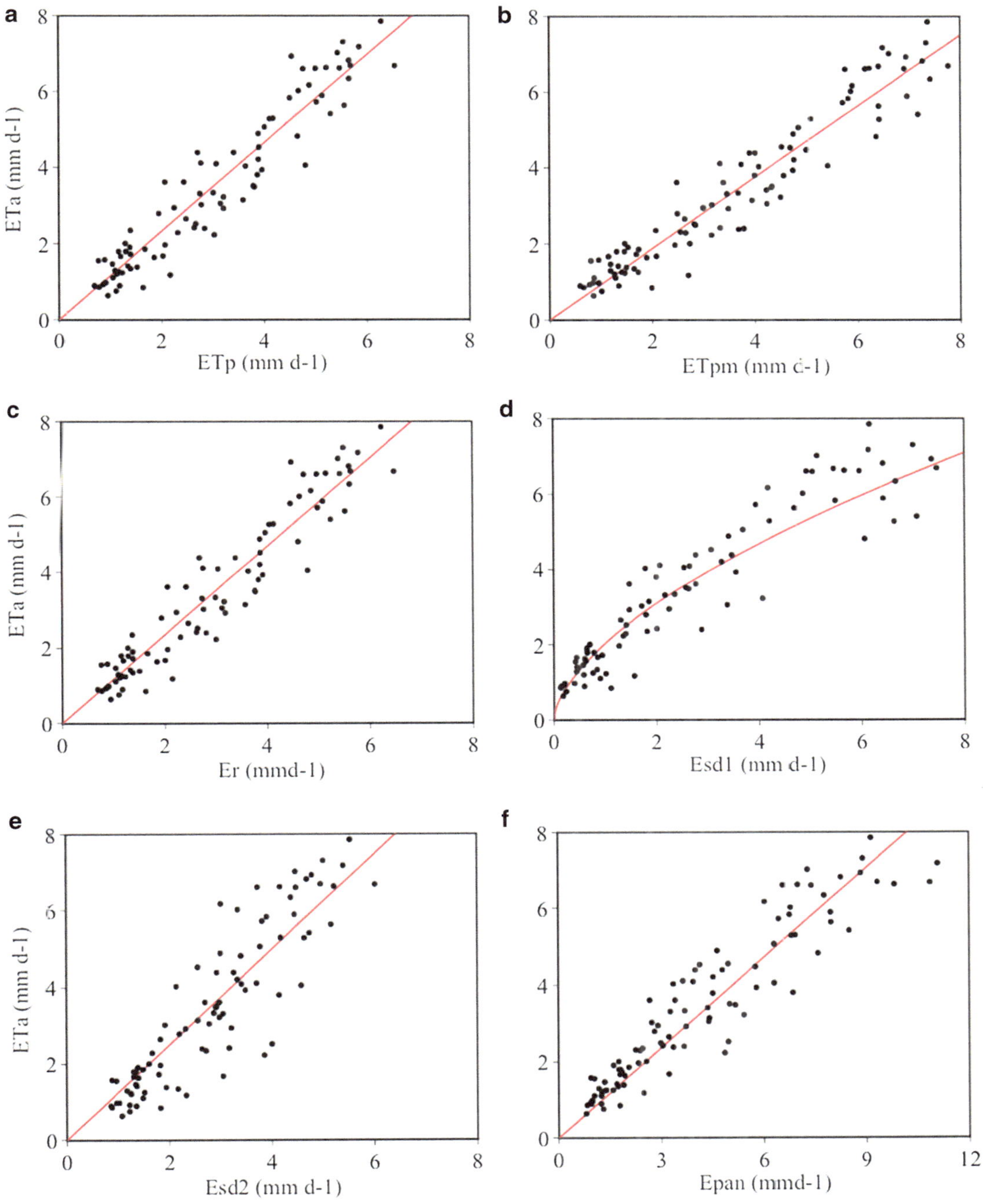

Fig. 4 The correlation between 14-day mean daily measurements of actual evapotranspiration (ETa) and the estimates of maximum evapotranspiration (ETm) during the years 1991–1995 using the following methods: (**a**) Penman combination (ETp), (**b**) Penman-Monteith (ETpm), (**c**) Priestley and Taylor (Er), (**d**) vapor pressure deficit 1 (Esd1), (**e**) vapor pressure deficit 2 (Esd2), and (**f**) pan evaporation (Epan)

Table 4 Overall regression analysis of the estimated and measured ET data for the irrigated eucalypt plantations during the 4 years of 1991–1995

Estimating method	Coefficient of efficiency (E)	Regression equation	1991–1995 Parameter values	CD (R^2)
Penman (ETp)	0.84	y=a * x	a=1.16	0.88
P-M (ETpm)	0.90	y=a * x	a=0.95	0.92
P & T (Er)	0.83	y=a * x	a=1.17	0.91
VPD1 (Esd1)	0.71	y=a * x^b	a=2.07, b=0.63	0.81
VPD2 (Esd2)	0.65	y=a * x	a=1.25	0.72
Pan eva (Epan)	0.57	y=a * x	a=0.78	0.82

CD=coefficient of determination. The six methods are *Penman*=Penman combination equation, *P-M*=Penman-Monteith combination equation, *P & T*=Priestley and Taylor method, *VPD1*=vapor pressure deficit method 1, *VPD2*=vapor pressure deficit method 2, and *pan eva*=pan evaporation method

Table 5 Regression analysis between the estimated and measured ET data for irrigated (November–April) and nonirrigated (May–October) seasons during the 4 years of 1991–1995

Estimating method	Regression equation	Irrigated season Parameter values	Irrigated season CD (R^2)	Nonirrigated season Parameter values	Nonirrigated season CD (R^2)
Penman (ETp)	y=a * x	a=1.18	0.83	a=1.07	0.69
P-M (ETpm)	y=a * x	a=0.96	0.87	a=0.91	0.79
P & T (Er)	y=a * x	a=1.19	0.83	a=1.08	0.69
VPD1 (Esd1)	y=a * x^b	a=2.07	0.75	a=1.85	0.69
		b=0.63		b=0.45	
VPD2 (Esd2)	y=a * x	a=1.30	0.50	a=1.04	0.52
Pan eva (Epan)	y=a * x	a=0.78	0.62	a=0.90	0.70

CD=coefficient of determination. The six methods are *Penman*=Penman combination equation, *P-M*=Penman-Monteith combination equation, *P & T*=Priestley and Taylor method, *VPD1*=vapor pressure deficit method 1, *VPD2*=vapor pressure deficit method 2, and *pan eva*=pan evaporation method

wind speed was the predominant factor contributing toward evaporation (Table 1), but during summer, solar radiation was noticed as the major determinant of evaporation. Fortnightly, mean daily water-use data of 40 biweekly measurements of the nonirrigation season during winter and 46 measurements of the irrigation season during summer were compared against the estimates of ETm to produce relationships between the estimated and measured data for these two seasons, and the results of the analyses are given in Table 5.

The ETpm method showed consistently higher R^2 than the other methods for the irrigated and nonirrigated seasons, 0.87 and 0.79, respectively (Table 5). The other important features of the comparison are the overall poor performance of Esd1 and Esd2 and the ability of Epan to predict better during the summer months (R^2=0.70) than in winter months (R^2=0.62). The ETpm method also gave a slope of the relationship with ETa of 0.96 and 0.91 for both seasons indicating that the method clearly takes care of the effect of environment on evaporation of trees (Fig. 5a).

Er, although overall slightly inferior to ETpm, gave the next best analysis of evaporation during the summer period (R^2=0.83). The slope of the relationship with ETa also varied from 1.19 to 1.08 between the two seasons which was within the limits of Priestley and Tailor's estimates. Joshua et al. (2005) observed that modified Priestley-Taylor model was found to perform well given its relative simplicity; however, they suggested that a soil moisture function should be integrated in all the ET models for improving their accuracy in simulations of actual evapotranspiration under variable soil moisture conditions.

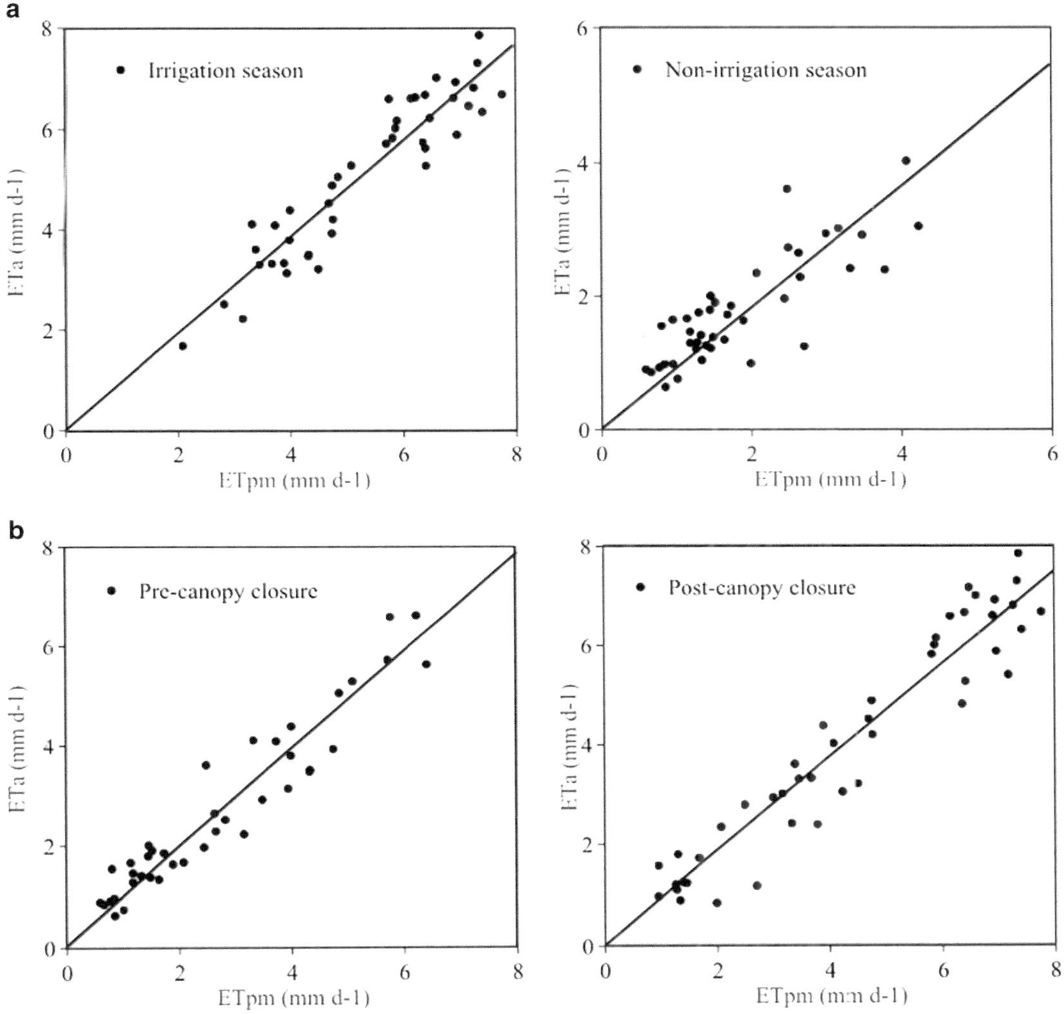

Fig. 5 The correlation between fortnightly measurements of mean daily evapotranspiration (ETa) and the Penman-Monteith estimates evapotranspiration (ETpm) during (**a**) irrigation and nonirrigation seasons and (**b**) pre-canopy closure (1991–1993) and post-canopy closure (1994–1995) periods

Similarly, Ge Sun et al. (2010) also observed that the growing season ET from wet forests was generally higher while those from woodlands in the arid and semiarid regions were lower than ETo. Esd methods did not show any difference between periods of lower and higher evaporative demand (i.e., during winter and summer), whereas ETp gave its best estimates during summer period (R^2=0.83), but Epan showed poor correlation during summer (R^2=0.62).

Comparisons during Pre- and Post-Canopy Closure

To investigate the predictive ability of the functions through this changing pattern of tree structure and environment, an analysis of the relationships between the five methods and ETa was made for the periods before and after canopy closure (Table 6). The ETpm model gave good estimates throughout the period of growth (R^2 of

Table 6 Regression analysis between the estimated and measured ET data for the pre- and the post-canopy closed eucalypt plantations from 1991 to 1995

Estimating method	Regression equation	Pre-canopy closure		Post-canopy closure	
		Parameter values	CD (R^2)	Parameter values	CD (R^2)
Penman (ETp)	y = a * x	$a=1.12$	0.87	$a=1.19$	0.93
P-M (ETpm)	y = a * x	$a=0.98$	0.91	$a=0.94$	0.94
P & T (Er)	y = a * x	$a=1.13$	0.87	$a=1.20$	0.93
VPD1 (Esd1)	$y=a*x^b$	$a=2.24$	0.92	$a=1.64$	0.87
		$b=0.58$		$b=0.72$	
VPD2 (Esd2)	y = a * x	$a=1.12$	0.79	$a=1.33$	0.81
Pan eva (Epan)	y = a * x	$a=0.75$	0.85	$a=0.80$	0.82

CD = coefficient of determination. The six methods are *Penman* = Penman combination equation, *P-M* = Penman-Monteith combination equation, *P & T* = Priestley and Taylor method, *VPD1* = vapor pressure deficit method 1, *VPD2* = vapor pressure deficit method 2, and *pan eva* = pan evaporation method

0.91 and 0.94 for pre-canopy closed and post-canopy closed conditions, respectively). This model also gave a slope of 0.98 and 0.94 for the two periods indicating its reliability of application at all stages of development of the plantation (Fig. 5b). Er recorded the best analysis of evaporation after canopy closure ($R^2=0.93$). The Esd methods gave better estimates when the data was analyzed separately for pre- and post-canopy closed conditions. The Esd1 method showed consistently higher reliability of prediction of ETa during pre-canopy closure stage of the plantation but the ETp produced best estimates after the canopy was closed. Epan showed similar correlation with ETa during pre- and post-canopy closed conditions.

Monthly Pan Factors

Since pan data is being used to estimate evapotranspiration and to schedule irrigation in many irrigated plantations, monthly pan factors were determined using analyses between weekly totals of pan evaporation and ETpm for mature plantations after canopy closure (Table 7). ETpm is used here as a surrogate for ETa because it has been conclusively identified from the above analyses that ETpm is an accurate one-step approach for estimating ETa. The measured daily pan evaporation and the estimated daily ETpm for the *Eucalyptus* plantation were used to obtain weekly mean Epan and ETpm for the study period. A statistical analysis of the relationships between estimated ETpm and Epan for each month was carried out. The slope of the regression line gives an estimate of the pan coefficient for that month. These relationships have also been forced through the origin and assessed in terms of slope and the coefficient of determination (Table 7).

Table 7 Estimated monthly pan factors (Kp) and coefficient of determination (CD) obtained using a regression analysis between the weekly totals of measured Epan and estimated ETpm for the irrigated eucalypt plantation at Wagga Wagga

Months	Pan factor (Kp)	CD (R^2)
January	0.70	0.75
February	0.70	0.67
March	0.77	0.76
April	0.81	0.42
May	0.80	0.67
June	0.85	0.49
July	0.84	0.51
August	0.89	0.79
September	0.94	0.70
October	0.84	0.76
November	0.76	0.68
December	0.65	0.75

Poor relationships were obtained during the winter months of April ($R^2=0.42$) and June ($R^2=0.49$). On average, there was greater consistency in the predictive ability of Epan during summer months than in winter months. Pan factors ranged from 0.65 in December when the evaporative demand is high to 0.94 in September when the demand is low. Lower pan factors during summer months from November to March (0.76–0.77) and higher pan factors from April to October (0.81–0.84) indicate that pan evaporation is effected by different processes to plantation evapotranspiration.

Conclusions

Reliable estimates of plantation ET are fundamental to improve the understanding of the relationships between soil moisture content and water fluxes from the soil and the vegetation and ultimately the ecosystem hydrology and environmental management. Direct measurements are usually too expensive, laborious, and time-consuming in most cases. Our study evaluated five methods commonly used to estimate ETm of tree plantations, by comparison with measurements made by the water balance technique.

The two methods, ETp and ETpm, based on the combination equation are technically the most satisfying because they represent the processes of evaporation from tree canopies more completely. At the Wagga Wagga site, these two methods consistently performed better than the other methods used to estimate ETm (particularly the ETpm method as shown in Fig. 5). The disadvantage of these methods, however, is that they require comprehensive measurements of climatic conditions and crop structure.

Methods based upon component processes, e.g., radiation (Er) and saturation deficit (Esd), require less input data, and for this reason have been widely used. For success, these techniques rely upon the existence of strong correlations between the individual processes of evaporation. In these data, Esd1 gave a hyperbolic response indicating that at higher deficits, the rate of evapotranspiration begins to decline. This observation highlights the importance of selecting the right species when designing irrigated tree plantations for climates where the vapor pressure deficits may rise above the threshold level of some species. The poor performance of the Esd2 method is of interest in the light of much recent work on crop water-use efficiency where saturation deficit has been used to standardize the performance of crops in different seasons and in different locations. The two humidity methods (Esd1 and Esd2) assume that the process of evapotranspiration affects the changes in the vapor pressure levels above the tree canopy, and hence if such changes could be monitored it would be possible to estimate ETm using an empirical relationship with saturation deficit. Although theoretically this procedure appears to be reasonably good, accurate measurements of the vapor pressure deficit above the tree canopy are often very difficult to obtain.

Er performed well with comparable consistency of prediction to ETp in all years. However, the value of α needs to be estimated for a particular location because it is sensitive to the fluctuations in sensible heat flux and advection. Epan was poorly related to ETa. Pan evaporation involves the same basic processes as evapotranspiration, and therefore it is possible to calculate an estimate of evapotranspiration from measured pan data. However, it is very difficult to make a general and a practical use of pan data except in special situations. The pan evaporation method also assumes that the pan behaves in the same way as a crop with zero resistance. Pans unlike crops are small in area and therefore experience advection differently. Measurements of pan evaporation are also well known to suffer from inappropriate siting and often from inadequate maintenance of the pans. The present data did not suffer in this way as these were collected in similar way as the other climatic data at the site and compared against the data collected by the Bureau of Meteorology, 2 km from the experimental site. The pan evaporation data collected from these two sites gave a 1:1 relationship, indicating the high level of accuracy of measurements collected at the site.

Pan evaporation has been found more sensitive to surrounding conditions than the tree vegetation. The albedo of trees is about 0.25 whereas the surface albedo of water varies from 0.02 to 1. Heat storage within the pan is large compared to that of soil, and the roughness to air movement also differs between vegetation and pan. Heat transfer through the sides of the pan and the turbulence, temperature, and humidity of the air immediately above its surface compared with that above plantation canopy changes the relationship between the measured Epan and ETm. Also, there may be more evaporation from the pan at night compared to plantation ET. It is reasonable therefore to assume that as a standard meteorological measurement, pan evaporation does not define ETa of plantations without calibration. Further, the low correlation ($R^2 < 0.4$) during winter months casts doubt on the general utility of pan evaporation measurements during these periods even with calibration. It seems that there is a good argument to replace such measurements with others that could be used to calculate ETa of tree crops. In general, this would require substantial extension of the radiation network.

Irrigation scheduling programs require input data which could be either measured or estimated by the users. Daily evapotranspiration is one of them. The accuracy of its measurement depends on the accuracy of the estimation of other inputs and also on the accuracy of forecasting the weather. Under these circumstances, a simple method to estimate evapotranspiration is sufficient in designing and scheduling irrigated plantations. The pan evaporation method, though not very accurate, showed greater reliability and predictability when used on annual or seasonal basis than on a monthly, weekly, or a daily basis. Regression on a monthly basis was, however, better than on a weekly or daily basis, which could be employed in irrigation scheduling programs in plantation crops. For a site under an irrigated *Eucalyptus* plantation, monthly pan factors could be better derived from many years of historical climatic data using the Penman-Monteith combination equation (Fig. 6). These pan factors may then be used in designing and scheduling irrigation for areas of similar climatic zones. The pan factors derived for *Eucalyptus* plantation under irrigated conditions at the Wagga Wagga site can be compared with pan factors derived by other workers in similar climatic regions and under similar set of conditions.

The analyses presented here show that it is possible to estimate ETa of *Eucalyptus* plantation from standard meteorological data. The best technique, based upon the Penman-Monteith combination equation, requires a complete climatic

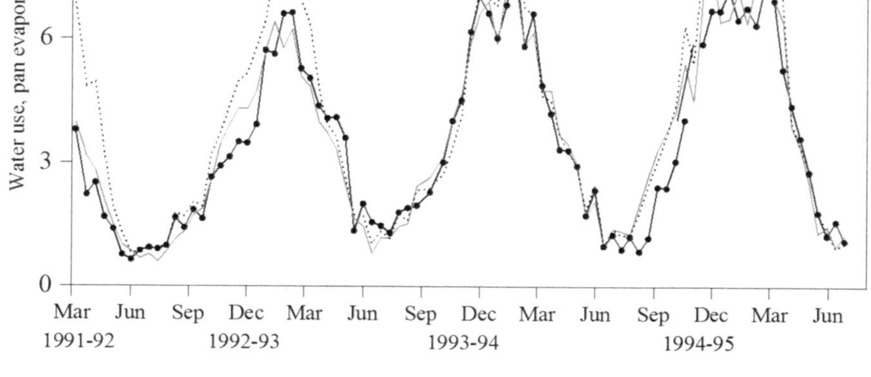

Fig. 6 Comparison of 14-day mean daily measurements of actual evapotranspiration (ETa) and pan evaporation (Epan) and the estimates of maximum evapotranspiration using Penman-Monteith equation (ETpm) of the eucalypt plantation from planting to 4 years. The comparison shows that ETpm was an accurate estimate of ETa during 4 years of growth

data set and additional information on tree height and cover. Almost comparable values can be calculated from radiation data alone (the Priestley-Taylor technique) which has the advantage of requiring considerably less input data. Estimates based on saturation deficit (Esd), although widely used in analyses of water-use efficiency, did not show strong linear correlation with ETa. Pan evaporation was shown to be poorly correlated with ETa if used for time steps less than the seasonal or monthly periods, adding weight to the growing concern about the utility of this measurement for daily predictions and the transportability of this relationship to different sites.

References

Aitken AP (1973) Assessing systematic errors in rainfall-runoff models. J Hydrol 20:121–136

Arnell NW, Reynard NS (1996) The effects of climate change due to global warming on river flows in Great Britain. J Hydrol 183:397–424

Carbon BA, Bartle GA, Murray AM (1981) Patterns of water stress and transpiration in jarrah (*Eucalyptus marginata* Donn ex Sm.) forests. Aust For Res 11:191–200

Chiew FHS, Kamaladasa NN, Malano HM, McMahon TA (1995) Penman-Monteith, FAO-24 reference crop evapotranspiration and class-A pan data in Australia. Agric Water Manage 28:9–21

Colquhoun IJ, Ridge RW, Bell DT, Loneragan WA, Kuo J (1984) Comparative studies in selected species of *Eucalyptus* used in rehabilitation of the northern Jarrah forest, Western Australia. 1. Patterns of xylem pressure potential and diffusive resistance of leaves. Aust J Bot 32:367–373

Connor DJ (1975) Growth, water relations and yield of wheat. Aust J Plant Physiol 2:353–366

Dalton J (1802) Experimental essays on the constitution of mixed gases; on the force of steam or vapour from water and other liquids in different temperatures, both in a Torricellian vacuum and in air; on evaporation and on the expansion of gases by heat. Mem Manch Lit Philos Soc 5:535–602

Davies JA, Allen CD (1973) Equilibrium, potential and actual evaporation from cropped surfaces in Southern Ontario. J Appl Meteorol 18:898–903

Dilley AC, Shepherd W (1972) Potential evaporation from pasture and potatoes at Aspendale. Agric Meteorol 10:283–300

Dolman AJ, Gash JHC, Roberts J, Shuttleworth WJ (1991) Stomatal and surface conductance of tropical rainforest. Agric For Meteorol 54:303–318

Domec J-C, Sun G, Noormets A, Michael J, Gavazzi-Emrys TA, Erika-Cohen JJ, Swenson S, McNulty G, John SK (2012) A comparison of three methods to estimate evapotranspiration in two contrasting Loblolly pine plantations: age-related changes in water use and drought sensitivity of evapotranspiration components. For Sci 58:497–512

Doorenbos J, Kassam AH (1979) Yield response to water. FAO Irrig Drainage Paper 33:193

Doorenbos J, Pruitt WO (1977) Guidelines for predicting crop water requirements. FAO Irrig Drainage Paper 24:207

Dunin FX, Aston AR (1984) The development and proving of models of large-scale evapotranspiration: an Australian study. Agric Water Manage 8:305–323

Dunin FX, Mackay SM (1982). Evaporation of eucalypt and coniferous forest communities. In: Proceedings of the first national symposium on forest hydrology, Melbourne, pp 18–35

Dye PJ (1993) Estimating water use by *Eucalptus grandis* with the Penman-Monteith equation. In: Swanson RH, Bernier PY, Woodward PD (eds) Forest hydrology and watershed management, proceedings of the Vancouver symposium, pp 329–337

Farahani HJ, Ahuja LR (1996) Evapotranspiration modeling of partial canopy/residue-covered fields. Trans ASAE 39:2051–2064

Farahani HJ, Bausch WC (1995) Performance of evapotranspiration models for maizedbare soil to closed canopy. Trans ASAE 38:1049–1059

Federer CA, Vorösmarty CJ, Fekete B (1996) Intercomparison of methods for calculating potential evaporation in regional and global water balance models. Water Resour Res 32:2315–2321

Fischer RA (1979) Growth and water limitations to dryland wheat yield in Australia; a physiological framework. J Aust Inst Agric Sci 45:83–95

Ford CR, Hubbard RM, Kloeppel BD, Vose J (2007) A comparison of sap flux-based evapotranspiration estimates with catchment-scale water balance. Agric For Meteorol 145:176–185

Ge Sun, Karrin A, Jiquan C, Shiping C, Chelcy RF, Guanghui L, Chenfeng L, Nan L, Steven GM, Haixia M, Asko N, James MV, Burkhard W, Melanie Z, Yan Z, Zhiqiang Z (2010) A general predictive model for estimating monthly ecosystem evapotranspiration. Ecohydrology 4:245–255

Goldstein AH, Hultman NE, Fracheboud JM, Bauer MR, Panek JA, Xu M, Qi Y, Guenther AB, Baugh W (2000) Effects of climate variability on the carbon dioxide, water, and sensible heat fluxes above a ponderosa pine plantation in the Sierra Nevada (CA). Agric For Meteorol 101:113–129

Greenwood EAN, Klein L, Beresford JD, Watson GD (1985) Differences in annual evaporation between grazed pasture and *Eucalyptus* species in plantations on a saline farm catchment. J Hydrol 77:237–252

Hookey GR, Bartle JR, Loh IC (1987) Water use of eucalypts above saline groundwater. Final report: Australian water resources. Council Research Project 84/166. Water Authority of Western Australia, Perth, p 40

Jenson ME (ed) (1973) Consumptive use of water and irrigation water requirements. American Society of Civil Engineers, New York

Jones AJ, Bauder JW (1987) Computer-assisted irrigation scheduling: an educational tool. Appl Agric Res 2:260–271

Joshua BF, Terry A, DeBiase YQ, Ming X, Goldstein AH (2005) Evapotranspiration models compared on a Sierra Nevada forest ecosystem. Environ Model Softw 20:783–796

Kaczmarek Z, Strzepek KM, Somlyo-dy L, Priazhinskaya V (eds) (1996) *Water resources* management in the face of climatic/hydrologic uncertainties. Kluwer, Hingham

Kite G (1998) Integration of forest ecosystem and climatic models with a hydrologic model. J Am Water Resour Assoc 34:743–753

Leuning R, Kriedemann PE, McMurtrie RE (1991a) Simulation of evapotranspiration by trees. Agric Water Manage 19:205–221

Leuning R, Wang YP, Cromer RN (1991b) Model simulations of spatial distributions and daily totals of photosynthesis in *Eucalyptus grandis* canopies. Oecologia 88:494–503

Linacre ET (1968) Estimating the net-radiation flux. Agric Meteorol 5:49–63

Lowe PR (1977) An approximating polynomial for the computation of saturation vapour pressure. J Appl Meteorol 16:100–103

McNaughton KG, Black TA (1973) A study of evapotranspiration for dryland fir forest using the energy balance application. Water Resour Res 9:1579–1590

Monteith JL (1965) Evaporation and environment. In: The state and movement of water in living organisms. Symposium of Society of Experimental Biology 19, pp 205–234

Monteith JL (1986) How do crops manipulate water supply and demand? Philos Trans Roy Soc Lond A 316:245–259

Musselman RC, Fox DG (1991) A review of the role of temperate forests in the global CO_2 balance. J Air Waste Manage Assoc 41:798–807

Myers BJ, Bond WJ, Benyon RG, Falkiner RA, Polglase PJ, Smith CJ, Snow VO, Theiveyanathan S (1999) Sustainable effluent-irrigated plantations: an Australian guideline. CSIRO Forestry and Forest Products, Canberra, p 286

Myers BJ, Talsma T (1992) Site water balance and tree water status in irrigated and fertilised stands of *Pinus radiata*. For Ecol Manage 52:17–42

Myers BJ, Theiveyanathan S, O'Brien ND, Bond WJ (1996) Growth and water use of effluent-irrigated *Eucalyptus grandis* and *Pinus radiata* plantations. Tree Physiol 16:211–219

Naoum S, Tsanis IK (2003) Hydroinformatics in evapotranspiration estimation. Environ Model Softw 18:261–271

Passioura JB (1977) Grain yield, harvest index and water use of wheat. J Aust Inst Agric Sci 43:117–120

Penman HL (1948) Natural evaporation for open water, bare soil and grass. Proc Roy Soc Lond A 193:120–146

Penman HL (1956) Estimating evaporation. Trans Am Geophys Union 37:43–50

Pereira AR, Villa-Nova NA (1992) Analysis of Priestley-Taylor parameter. Agric Forest Meteorol 61:1–9

Pereira AR, Villa-Nova NA, Pereira AS, Barbieri V (1995) A model for the class A pan coefficient. Agric Forest Meteorol 76:75–82

Perry MW (1987) Water use efficiency of non-irrigated field crops. In: Proceedings of the fourth Australian agronomy conference La Trobe University Melbourne, Victoria pp 83–99

Persson G (1995) Willow stand evapotranspiration simulated for Swedish soils. Agric Water Manage 28:271–293

Priestley CHB, Taylor RJ (1972) On the assessment of surface heat flux and evaporation using large-scale parameters. Mon Weather Rev 100:81–92

Raupach MR (1995) Vegetation-atmosphere interaction and surface conductance at leaf, canopy and regional scales. Agric For Meteorol 73:151–179

Ritchie JT, Burnett E (1971) Dryland evaporative flux in a subhumid climate: 1. Micrometeorological influences. Agron J 63:51–52

Savabi MR, Stockle CO (2001) Modeling the possible impact of increased CO_2 and temperature on soil water balance, crop yield and soil erosion. Environ Model Softw 16:631–640

Shuttleworth WJ (1989) Micrometeorology of temperate and tropical forest. Philos Trans Roy Soc Lond B 324:299–334

Shuttleworth WJ, Calder IR (1979) Has the Priestley-Taylor equation any relevance to forest evaporation? J Appl Meteorol 18:639–646

Smittle DA, Dickens WL (1992) Water budgets to schedule irrigation for vegetables. Hortic Technol 2:54–59

Smittle DA, Dickens WL, Hayes MJ (1992) An irrigation scheduling model for summer squash. J Am Soc Hortic Sci 117:717–720

Spittlehouse DL, Black TA (1979) Determination of forest evapotranspiration using Bowen ratio and eddy correlation measurements. J Appl Meteorol 18:647–653

Stannard ID (1993) Comparison of Penman-Monteith, Shuttleworth-Wallace, and modified Priestley-Taylor evapotranspiration models for wildland vegetation in semiarid rangeland. Water Resour Res 29:1379–1392

Stewart JI, Cuenca RH, Pruitt WO, Hagan RM, Tosso J (1977a) W-67 determination and utilization of crop water production functions for principal Californian crops. California Contributions Project Report, University of California, Davis, p 27

Stewart JI, Hanks RJ, Danielson RE, Jackson EB, Hagan RM, Pruitt WO, Franklin WT, Riley JP (1977b) Optimizing crop production through control of water and salinity levels in the soil. Utah Water Laboratory PRWG 151-1, Logan, p 191

Szeicz G, Endrodi G, Tajchman S (1969) Aerodynamic and surface factors in evaporation. Water Resour Res 5:380–394

Tanner CB (1967) Measurement of evapotranspiration. Agron J 11:534–574

Tanner CB, Sinclair TR (1983) Efficient water use in crop production. Research or re-*research*. In: Taylor MH, Jordan WR, Sinclair RT (eds) Limitations to efficient water use in crop production. American Society of Agronomy, Madison, pp 1–27

Thom AS, Oliver HR (1977) On Penman's equation for estimating regional evaporation. Quart J Roy Meteorol Soc 103:345–357

Theiveyanathan S, Benyon RG, Marcar NE, Myers BJ, Polglase PJ, Falkiner RA (2004) An irrigation-scheduling model for application of saline water to tree plantations. For Ecol Manage 193:97–112

Tourula T, Heikinheimo M (1998) Modeling evapotranspiration from a barley field over the growing season. Agric For Meteorol 91:237–250

VEMAP Members (1995) Vegetation/ecosystem modeling and analysis project: comparing biogeography and biogeochemistry models in a continental-scale study of terrestrial ecosystem responses to climate change and CO_2 doubling. Glob Biogeochem Cycles 9:407–437

Viswanadham Y, Silva-Filho VP, Andre RGB (1991) The Priestley-Taylor parameter α for the Amazon forest. For Ecol Manage 38:211–225

Vorösmarty CJ, Federer CA, Schloss AL (1998) Potential evaporation functions compared on US watersheds: possible implications for global-scale water balance and terrestrial ecosystem modeling. J Hydrol 207:147–169

Wallace JS (1994) Calculating evaporation: resistance to factors. Agric For Meteorol 73:353–366

Watts PJ, Hancock NH (1984) Evaporation and potential evaporation – a practical approach for agricultural engineers. In: Proceedings of conference on agriculture engineering Bundaberg, Queensland, Australia, pp 290–297

Perspectives for Bio-management of Salt-affected and Waterlogged Soils in Pakistan

A.S. Qureshi

Abstract

Waterlogging and salinity have afflicted about 4.5 million ha of irrigated lands in Pakistan which has reduced production potential of the Indus Basin by 25 %. Over the last 40 years, Pakistan has adopted engineering, reclamation, and biological measures to rehabilitate these soils. Engineering solutions involved implementation of large-scale Salinity Control and Reclamation Projects (SCARPs) covering 8 million ha of land with an estimated cost of US$ 2 billion. Reclamation of saline soils was mainly achieved by leaching excess salts and the use of chemicals such as gypsum and acids. Despite initial success, the success of these initiatives has been limited due to several factors. Operational and maintenance cost of drainage projects proved a financial burden for the government and was discontinued. Nonavailability of subsidized gypsum and chemicals restricted the capacity of farmers to reclaim saline soils. As an alternative to engineering and reclamation solutions, Pakistan has made considerable progress in bioremediation by identifying and establishing plant species that can effectively lower groundwater tables and reclaims saline soils. The bioremediation systems also help in sequestering carbon, diminishing the effects of wind erosion, providing shade and shelter, and in enhancing the biodiversity. In order to derive the benefits of such efforts, bioremediation plantations have to be extended over relatively large areas. Therefore, it may not be an attractive proposition to individual farmers unless there are cooperative efforts. Involving the farming communities in such bioremediation programs seems promising for their economic and social well-being in countries such as Pakistan where individual land holdings are small.

A.S. Qureshi (✉)
Research & Innovation Division, International Center for Biosaline Agriculture (ICBA), Dubai, United Arab Emirates (UAE)
e-mail: a.qureshi@biosaline.org.ae

J.C. Dagar, P.S. Minhas (eds.), *Agroforestry for the Management of Waterlogged Saline Soils and Poor-quality Waters*, Advances in Agroforestry 13, DOI 10.1007/978-81-322-2659-8_6

Introduction

Globally approximately one billion hectares of lands are salt-affected (Wicke et al. 2011). Despite large-scale contribution in securing food and fiber needs of the increasing population, irrigated agriculture is usually blamed for negative environmental impacts such as soil salinization. The salt-affected soils occupy more than 20 % of irrigated lands (Ghassemi et al. 1995), and this process continues to occur at an estimated rate of 0.25–0.50 million hectares each year (FAO 2000). The large extent of salt-affected soils is of special concern because they greatly reduce agricultural productivity. In South Asia, the continuous expansion of salt-affected soils is of even greater concern due to increasing demand for food for the fast-growing population. In Pakistan, salinity has affected about 4.5 Mha (Qureshi 2011). Poor irrigation management practices, lack of appropriate drainage systems, and the use of poor-quality irrigation water are the major reasons for waterlogging and salinity problems in India and Pakistan. In Bangladesh, seawater intrusion has caused soil salinity problems for about 1 Mha of coastal zone of the Ganges–Brahmaputra River Delta (Hossain 2010). This situation has threatened the food security of these countries and warrant immediate attention for inexpensive and environmentally acceptable solutions for reclaiming and managing these soils (Qadir and Oster 2002).

For ensuring future food security in arid and semiarid regions, the use of salt-affected soils and marginal-quality waters can play a key role to improve crop production and mitigate environmental effects. This, however, will require a comprehensive approach to soil, water, and crop management which necessarily should include rehabilitation of degraded lands, reclamation of new lands, and adoption of improved irrigation practices for increasing land and water productivity and decreasing environmental problems. In irrigated areas, irrigation management needs to be improved to reduce volumes of drainage water to avoid disposal and associated environmental problems (Wichelns 2009). As the increasing competition for freshwater resources has already reduced allocations of freshwater for agriculture in many regions (Tilman et al. 2002), the reuse of saline and sodic drainage water generated by irrigated agriculture and marginal-quality water generated by municipalities need to be explored (Rhoades 1999; Oster 2000; Qadir and Oster 2004).

Conventional agriculture is usually not considered economically viable on salt-affected soils due to low crop yields and low economic returns (Qadir and Oster 2004). Furthermore, physical remediation of salt-affected soils is complex, expensive, and time-consuming for many farmers due to lack of proper guidance and training. Therefore, for these soils, forestry and agroforestry systems could be an attractive alternative, because many tree species are tolerant to soil salinity and their cultivation can help in regeneration of these soils (Wicke et al. 2013). Agroforestrysystems have proved economically viable land use options in many areas of South Asia and the world (Kaur et al. 2002; Masters et al. 2007).

In Pakistan, about half of the 4.5 Mha of salt-affected lands are located in irrigated areas (Qureshi et al. 2008). It is estimated that salt-affected lands have reduced the production potential of irrigated lands by almost 50 % (Qureshi et al. 2008). Over the past four decades, numerous efforts have been made to reclaim and manage salt-affected and waterlogged soils through engineering, reclamation, and biological measures. However, the success has been limited due to several reasons. High operational and maintenance costs make drainage projects an economic burden for the governments, whereas nonavailability of subsidized gypsum and other amendments restricted farmers to reclaim their lands. As an alternative, bioremediation measures were adopted by farmers. This chapter discusses the causes and extent of soil salinity and waterlogging in Pakistan and reviews the impact of different interventions made for the management of salt-affected and waterlogged soils with special emphasis on bioremediation measures. Lessons learned and steps needed for future management of these troubled soils are also presented.

Extent of Salt-affected Waterlogged Soils in Pakistan

Out of the total geographical area of 79.61 Mha of Pakistan, more than 21 Mha is cultivated. About 59 % of the agricultural area is located in the Punjab province, which produce more than 90 % of the total grain production of the country. Next is the Sindh province with about 23 % of the total agricultural lands. Rest 18 % of the area is almost equally shared by Khyber Pakhtunkhwa (KP) and the Balochistan provinces. In arid and semiarid regions, drainage is considered as a complimentary activity with the irrigation to avoid groundwater table rise and consequent soil salinization. However, despite considerable progress in irrigation development, provision of drainage facilities was never given a priority in Pakistan. This negligence causes groundwater tables to rise in most of the canal command areas due to persistent seepage over the years from unlined earthen canals and from a large network of distributing channels and percolation losses from irrigated fields. As a result, the groundwater table rose rapidly in vast irrigated areas to within 1.5 m of the soil surface.

The groundwater table in the canal command areas of Pakistan exhibits an annual cycle of rise and fall. In most of the Indus plain, the groundwater table is at its lowest level prior to the monsoon (April/June) season, and as a result of the *kharif* (summer) canal supplies and the effects of the monsoon rains, it rises and comes close to the land surface in October, after which it begins to decline again. Based on their annual monitoring of groundwater levels all over Pakistan, Water and Power Development Authority (WAPDA) of Pakistan has estimated that for 4.7 million ha (30 % of the irrigated area), the groundwater table is within 1.5 m of the soil surface after the monsoon season. Prior to the monsoon season, this area is reduced to about 2 Mha or 13 % of the irrigated area (Fig. 1). The Punjab province has about 25 % of its irrigated area severely waterlogged, while in Sindh, this figure is about 60 %. Due to the presence of this shallow, saline groundwater, about 40,000 ha are annually abandoned within the Indus basin due to secondary salinization (WAPDA 2007).

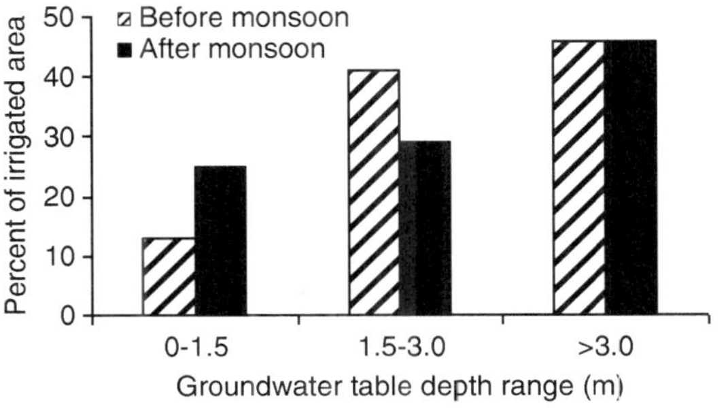

Fig. 1 Percentage of area under different water table depths before and after monsoon

The salt-affected soils associated with the use of poor-quality groundwater for irrigation have become an important ecological entity of the Indus basin. Latest soil salinity survey conducted by WAPDA shows that the extent of salt-affected lands has decreased to about 4.5 Mha from about 6 Mha in the 1980s (WAPDA 2007). The reduction in salinity (surface and profile) is primarily due to increased irrigation water supply from surface and groundwater sources, better water management, increased cropping intensity, and measures taken to reclaim the waterlogged and salt-affected lands. Surface salinity trends in four provinces of Pakistan are shown in Fig. 2 (WAPDA 2007). It is clear that still salinity problems in the Sindh are much more severe than in other provinces.

About 21 % of lands in Pakistan are saline, out of which about 7 % (1.5 Mha) is strongly saline. The problems of soil salinity are more serious in Sindh province (lower part of the Indus basin), where about 50 % of the area is saline (Qureshi 2011). This is mainly because of the presence of marine salts, poor natural drainage conditions, and the use of poor-quality groundwater for irrigation because surface water supplies in the Sindh province are far less than the actual crop water requirements. Furthermore, leaching opportunities are also very limited due to highly saline soils at shallow depths and highly saline groundwater at deeper depths (Bhutta and Smedema 2007). These problems have brought into question the sustainability of the system and the capacity of Pakistan to feed its growing population in the coming decade unless better management institutions are introduced.

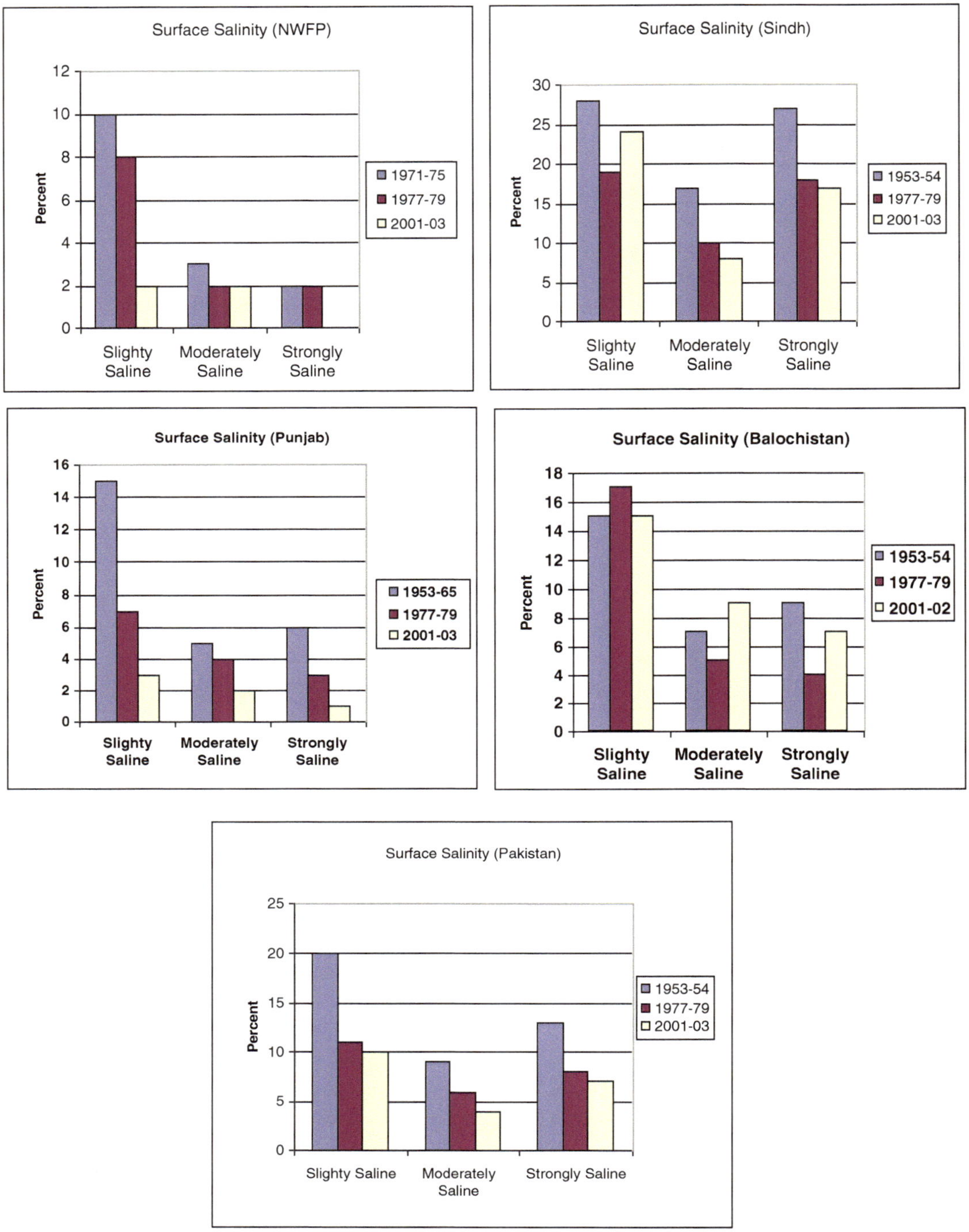

Fig. 2 Country- and province-wise distribution of salt-affected area (WAPDA 2007)

Table 1 Classification of salt-affected soils in Pakistan

Classification of salt-affected soils	Area affected (million ha)	Characteristics
Slightly saline sodic	0.6	Slight salinity-sodicity problem, occurring as patches (about 20 % of the area) in cultivated fields
Porous saline sodic	1.4	Saline sodic throughout the root zone, porous, and pervious to water
Severely saline sodic	0.8	Have high groundwater tables, dense, and nearly impervious to water
Soils with sodic tube well water	1.7	Severely sodic due to application of sodic tube well water, contain high concentrations of carbonates and bicarbonates, and almost impervious

Modified from Qureshi and Barrett-Lennard (1998)

The problems of soils in the Indus basin are not only of salinity but also of sodicity. About 70 % of the tube wells in the Indus basin pump sodic water, which contain high concentrations of carbonate and bicarbonate. Application of this quality of water for irrigation turns the soils to saline sodic affecting soil structure and infiltration rates, thereby restricting the growth of conventional crops. Salt-affected soils of the Indus basin are usually classified into four types. The area affected and the characteristics of these four soil types are given in Table 1.

The above facts indicate that the agricultural sector suffers deeply from both waterlogging and salinity. About 75 % of the population and about half of the gross national product (GNP) are directly or indirectly related to the agricultural sector. This shows that the problems of waterlogging and salinity are not just agricultural problems but that they do affect the country as a whole and ultimately the social fabric of Pakistani society. Waterlogging and salinity have very adverse social and economical effects on communities in Pakistan, causing poor living standards in affected areas and health problems for humans and animals. This situation has forced the local population to migrate to other areas.

Strategies Adopted in Pakistan

In Pakistan, engineering, reclamation, and bioremediation strategies have been used for the rehabilitation of salt-affected and waterlogged soils. Engineering approaches include installation of surface drainage, subsurface drainage, and vertical drainage systems to drain excess water from soils. Surface drainage systems were not only used to remove excess water from agricultural fields and also as conveyance channels for urban and industrial waste water and subsurface drainage systems. Due to mixing of industrial water with the agricultural drainage water, the quality of water in surface drains was deteriorated which restricted its reuse for agricultural purposes and increased the volume of disposal effluent. Reclamation of salt-affected soils was mainly done through leaching of salts by applying higher levels of irrigation, the use of gypsum or acids, physical removal of salts, and growing salt-tolerant plants. Bioremediation approaches involve using salt-tolerant crops on highly salt-affected soils with saline water. These approaches are discussed below in the context of Pakistan.

Engineering Strategies

After detailed survey of groundwater table depth and salinity in the 1950s, Pakistan decided to install 14,000 tube wells in fresh groundwater areas (covering 2.6 Mha of irrigated land) for lowering groundwater table to control waterlogging and to increase irrigation supplies at the farm gate by mixing pumped groundwater with the canal water. Under this Salinity Control and Reclamation Projects (SCARPs), 63 projects were completed during the last four decades to cover about 8 Mha with an estimated cost of US$ 2 billion (Qureshi 2011). The SCARPs project were successful in controlling or even reversing the waterlogging and salinity problems and providing additional water for irrigation. As a result, cropping intensities increased from 84 to 115 % in most SCARP areas. However, over time, increased operational and maintenance costs and increasing salinity of the pumped groundwater reduced the efficiency of SCARPs which resulted in rising water tables and lower crop yields.

In the mid-1970s, thinking shifted toward horizontal (pipe) drainage systems with the perception that over time drainage water quality will be improved which will increase the possibility of using drainage water for irrigation. Furthermore, the disposal problems will be reduced. Since then, about ten major horizontal drainage projects (12,600 km of pipe drains) have been completed in different parts of Pakistan (Qureshi et al. 2008). The major bottleneck in the successful operation of these drainage systems was the safe disposal of saline drainage effluent. To overcome this issue, Pakistan constructed a 2000-km long surface drain on the left side of the Indus River to transport drainage water of more than 500,000 ha of land to the sea (Qureshi et al. 2008). Initial results of this drain were very encouraging, but soon surrounding areas of this unlined drain started becoming waterlogged due to seepage. This escalated interprovincial dispute between Punjab and Sindh provinces, which resulted in impasse of drainage water from Punjab passing through Sindh province and finally into sea. This exacerbates waterlogging problems in Punjab.

Reclamation Strategies

The presence of excessive salts in the soil does not have adverse impacts on soil structure and its physical and hydraulic properties. In many cases, saline conditions may have favorable effects on soil structure (Quirk 2001). The soil structural problems mainly stems from the presence of sodium in the soil. Sodic soils exhibit structural problems created by physical process (slaking, swelling, and dispersion of clay) and specific conditions (surface crusting and hard setting) (Sumner 1993; Wicke et al. 2011). These soils can be reclaimed by providing a source of calcium (Ca^{2+}) to replace excessive sodium Na^{+} from the cation exchange sites (Qadir and Oster 2004). The replaced sodium is leached from the root zone through excessive irrigation.

In Pakistan, different strategies were applied for reclaiming sodic soils. Agricultural and industrial waste, e.g., farmyard manure and by-products of the sugar industry, have been used to improve sodic soils. A large range of chemicals (sulfur, sulfuric acid, and aluminum sulfate) have also been tested to provide calcium source to sodic soils. Gypsum was widely used as an amendment for sodic soils because it was highly subsidized by the government (US$ 0.25 per 50 kg bag) (Qureshi and Barrett-Lennard 1998). However, over time, the cost of these amendments has increased due to decrease in government subsidies to farmers. In addition, low quality of amendments and difficulties in their timely availability have discouraged reclamation through chemical means especially for the subsistence farmers. At the government level, there is no nationwide action program for promoting reclamation efforts. Efforts by local governments are mainly confined to supporting field-level research and providing subsidies to the farmers for the use of gypsum. Farmers in Pakistan have also used physical methods of surface scraping and deep plowing in a desperate attempt to reclaim their saline-(sodic) soils.

Planting of salt-tolerant crops has also been experimented as a reclamation intervention. Some crops are more salt tolerant at certain stages of development than others, e.g., sugar beet at germination, barley at the early seedling stage, and rice at the flowering stage. Although extremely salt sensitive, rice paddy has been found to be a better crop than others for reclamation of saline-sodic soils, as it provides sufficient water for leaching. Similarly, barley and cotton can withstand higher levels of salinity under good drainage conditions.

Bioremediation/Agroforestry Strategies

The limited success in controlling waterlogging and salinity problems despite huge investments prompted scientists and engineers to think about other options which are less expensive, more sustainable, and cost-effective. One of the promising solutions is lowering water table through biological means. The use of bioremediation in waterlogged areas is based on the concept of enhanced evapotranspiration (Jeet-Ram et al. 2011). The biological approach emphasizes the use of highly saline water and lands on a sustained basis through profitable and integrated use of the genetic resources embedded in plants, animals, fish, and insects and improved agricultural practices. This approach attempts to promote bio-reclamation techniques using salt-tolerant plants, bushes, trees, and fodder grasses. Plants, particularly trees, are commonly referred to as biological pumps and play an important role in the overall hydrological cycle in a given area. Bioremediation systems are considered beneficial compared to conventional subsurface drainage systems because in bioremediation systems we do not need to

- Induce water flow through the soil toward a tube well or pipe drain
- Install collector and main drains to transport water out of the draining area
- Operate pumps for evacuating drained water and then transport to disposal sites
- Create disposal sites (e.g., by evaporation ponds)

The long-term sustainability of bioremediation has been intensely debated. Smedema (1997, 2000) has suggested that bioremediation could be considered for waterlogged landscape depressions and canal seepage interception and could be applied in "parallel field drainage" arrangements as an alternative to conventional field drainage systems. In Australia, it is now widely accepted that enhanced evapotranspiration bioremediation sites will eventually succumb to salinity, unless conventional drainage systems are installed to control salt balance by removal of saline drainage effluent (Heuperman 2000).

Studies done in Pakistan (Qureshi and Barrett-Lennard 1998; Ghafoor et al. 2004; Shah et al. 2011) have shown that highly saline waters could be used to grow salt-tolerant fodder grasses to improve the quality and quantity of livestock. Management practices of these waters include the use of chemical amendments, organic matter, and mineral fertilizers and judicious selection of salt-tolerant forages and grasses. Trees and plants act as biological drainage agents, helping to lower water table depths—a very simple as well as energy-saving method. This is basically a "pro-poor" approach that enhances the income of poor farmers who otherwise might leave their lands barren.

Bioremediation approaches experimented in Pakistan mainly used salt-tolerant crops to get maximum benefit from saline land and water. Soils that are saline down to 0.45-m soil depth have been successfully reclaimed by growing *kallar* grass (*Leptochloa fusca*) over a period of 5 years, after which the field can be put to normal cropping. The growth of perennial forage grasses is also important in salt-affected areas. Rhoades grass (*Chloris gayana*), Tall wheatgrass (*Elytrigia elongata*), and many other species have been shown to have tolerance to salt stress. Saltbushes such as river saltbush (*Atriplex amnicola*) are highly salt-tolerant forage bushes (Qureshi and Akhtar 2002). *Kallar* grass is most commonly used as a forage plant. It does not retain most of the taken up salts; hence, it remains reasonably palatable for farm animals. As a result, large tracts of *kallar* grass are being used as a sole source of fodder for livestock (Shah et al. 2011). So far, no adverse effects of feeding *kallar* grass to animals have been reported. Therefore, its production can be further encouraged in salt-affected areas.

In Pakistan, bioremediation has also been used for the reclamation of waterlogged soils. Bioremediation involves growing certain categories of plants that habitually draw their main water demand directly from the groundwater. A number of trees are in this group, including poplar (*Populus deltoides*), *Eucalyptus*, *Tamarix*, mesquite (*Prosopis juliflora*), and *Acacia nilotica*. Similarly, nonwoody plants such as bushes, sedges, grasses, and herbs can develop deep-rooted systems that contact groundwater (Choudhry and Bhutta 2000). Any significant effect of such plantation on the water table would be expected only when the plants occupy a large enough portion of the catchments so that their total water use approaches the total recharge for the catchments. In Pakistan, the capacity of productive tree plantations to extract shallow groundwater is seen as a valuable tool for controlling rising water tables and salinity.

Bioremediation Species Used in Pakistan

Selection of plant species for bioremediation depends on the environmental conditions for which they are planned. Salt tolerance will be an important criterion for (potentially) saline discharge environments, and water use considerations will prevail in recharge control situations where salinity is of no concern and in channel seepage scenarios with low-salinity water supply. For agricultural crops, salt- tolerance information based on Maas and Hoffman (1977) is most commonly used. For nonagricultural tree and bush species, reliable information is more difficult to obtain. Marcar et al. (1995) provided detailed information on the use of 30 tree species for use on salt-affected lands and less detailed summary descriptions for an additional 30 species. Schulz (1994) provided comparisons for five saltbush species grown on a range of saline irrigation regimes, whereas others investigated water use of different tree species under a range of saline conditions (Morris et al. 1998; Slavich et al. 1999; Cramer et al. 1999; Benyon et al. 1999).

For Pakistani conditions, Shah et al. (2000) provided information on crops, salt-tolerant trees, grasses, and saltbush. Khanzada et al. (1998) monitored the water use of *Acacia nilotica*, *A. ampliceps*, and *Prosopis pallida* on 3–5-year-old plantation sites with contrasting soil and groundwater salinity in the Indus Valley in Pakistan. Annual water use by *A. nilotica* was 1248 mm on a severely saline site and 2225 mm on a moderately saline site. This was considerably higher than the annual rainfall, indicating that much of the water was taken up from the saline water tables underlying the sites (20 dS m^{-1} at 1–1.5 m below surface at the saline site and 1.5 dS m^{-1} at 2 m below surface at the moderately saline site). It was concluded that trees can evaporate large volumes of saline groundwater; however, there may be accumulation of salts. Therefore, drainage system might be needed to evacuate the accumulated salts in the root zone to ensure long-term sustainability of this technology.

Fast-growing *Eucalyptus* species known for luxurious water consumption under excess soil moisture conditions are suitable for bioremediation (Shah et al. 2003). *Eucalyptus* species has a higher bioremediation potential as compared to relatively slow bio-drainers like *E. tereticornis* and *Pongamia pinnata* (Zhang et al. 2004). *E. tereticornis* and *Eucalyptus* hybrid are fast bio-drainers primarily due to their ability to display large leaf area (Angrish et al. 2009). Cloned *E. tereticornis* (Mysure gum) is fast growing, goes straight, and thus has low shading effect and has luxurious water consumption where excess soil moisture conditions exist. Because of fast growth rate, good wood properties, and carbon sequestration, world's *Eucalyptus* plantation area has increased to 19 million ha (Mha) (Iglesias and Wilstermann 2009).

In Pakistan, the use of *Eucalyptus* has significantly increased over the last three decades. A rough estimate that *Eucalyptus* plantings cover is about 10,000 ha. Of all *Eucalyptus* introduced in Pakistan to date, *E. camaldulensis* has proven most adaptable under all agroecological zones. Consequently, it has been planted more than other species; it is the prescribed species of all afforestation programs (Bilal et al. 2014). It is especially favored in arid and semiarid plains as single trees, in block and linear plantations, and is raised with or without artificial irrigation. The existing growing stock of *Eucalyptus* is mostly of *E. camaldulensis* and its numerous hybrid forms, occurring naturally. *Eucalyptus* is being grown in irrigated plantations of the government land and on farmlands of Punjab and Sindh. A recent survey shows that about 200 million trees have been planted in the Punjab on farmlands, mostly irrigated, of which *Eucalyptus* is 2.5 % (Shah et al. 2011).

During 1996–1998, the *E. camaldulensis* tree and saltbushes (*Atriplex amnicola* and *Atriplex lentiformis*) species have been successfully used to restore the productivity of about 400 hectare of salt-affected lands in the Faisalabad District of the Punjab province (Shah et al. 2011). This was followed by a UNDP- and AusAID-funded project, which was executed in three districts of the Punjab province. The unique feature of this project was the establishment of Salt Land User Groups (SLUGs) and Women's Interest Groups (WIGs) to encourage involvement of communities in the process of rehabilitation of marginal lands (Shah et al. 2011). During 2006–2010, bioremediation technology was further extended to 80,000 ha with a US$ 13 million financial assistance of the Asian Development Bank. Under this project, land rehabilitation of 45,000 ha was done by adopting an integrated approach, i.e., through gypsum application and tree plantation (Shah et al. 2011). In many other areas, farmers have adopted these species for reclaiming their lands.

Comparison of Bioremediation and Other Reclamation Strategies

Despite above advantages, bioremediation systems have certain shortcomings which can make their adoption restricted under certain circumstances. The efficiency of different plant species for the reclamation of saline-(sodic) soils is highly variable (Qadir and Oster 2004). In general, species with higher biomass production and tolerance against ambient salinity are more efficient in soil reclamation (Ghaly 2002; Kaur et al. 2002). The reduction in sodicity levels through bioremediation treatment was found to be 52 % compared to 62 % through chemical treatments (gypsum). Furthermore, bioremediation works well on coarse- to medium-textured, moderately saline, and saline-sodic soils.

Production systems based on salt-tolerant grasses and forage crops using saline irrigation water are considered more sustainable for bioremediation. If these systems are linked with a livestock production system, economic benefits can increase manifolds, and environmental problems of disposal of saline effluent can be minimized. Therefore, for the success of bioremediation systems, selection of plants capable of producing adequate biomass is very vital. The bioremediation technology is economically beneficially when there is a market for the bioremediation crops for forage and/or for firewood (Barrett-Lennard 2002). However, for the economic analysis of these technologies, one must also consider the value of rehabilitated lands in the long run.

In view of above, bioremediation can effectively contribute to lowering the groundwater table to reduce waterlogging and consequent salinity problems in irrigated and nonirrigated areas. Pakistani experience with bioremediation shows that involvement of communities in rehabilitation process through bioremediation can significantly contribute to rural development and the well-being of rural communities. The bioremediation systems are beneficial in producing timber, fruits, oils, and fuel wood, contributing to carbon sequestration, diminishing the effects of wind erosion, providing shade and shelter, functioning as windbreaks, yielding organic matter for fertilizer, enhancing biodiversity as flora and fauna flourishes, and diminishing air pollution (Heuperman et al. 2002).

One of the disadvantages of bioremediation is its requirement for large proportion of land area for its effectiveness, which might not be possible for an individual farmer to afford. Therefore, in countries like Pakistan where individual land holdings are very small, farmers have to make joint efforts for bioremediation. Bioremediation is also considered less effective in removing salts from the root zone and do not allow controlled drainage. Therefore, these systems need to be complimented with conventional drainage system for removing salts from the root zone. However, combining bioremediation with conventional drainage systems could yield potential results in reclaiming salt-affected and waterlogged soils.

Conclusions

The irrigated agriculture sector of Pakistan is encountering a great challenge in terms of waterlogging and salinity, which deprive farmers of productive resources and threaten their livelihoods. Pakistan has understood the significance of these issues, and efforts were started to overcome these problems through the implementation of waterlogging and salinity control projects. These projects mainly consisted of engineering work to install tube wells and surface and subsurface drains. Although engineering solutions help increase cropping intensities and yields, they fail to stop emergence of similar environmental problems in adjacent areas. For the reclamation of saline-sodic and sodic soils, the government promoted the use of gypsum and other physical methods such as acids and addition of organic matters. For this purpose, gypsum was subsidized by the government to facilitate farmers in the reclamation of troubled soils.

Considering the high installation and operational costs of engineering solutions, thinking of scientists diverted to more inexpensive and sustainable solutions such as biosaline agriculture. In the last 20 years, a wide range of research has been done on saline agriculture through the profitable and integrated use of genetic resources of plants, animals, fish, and insects and improved agricultural practices. For the success of saline agriculture in Pakistan, a number of salt-tolerant crop varieties and the use of improved planting techniques and fertilizers have been introduced. These technologies have shown very encouraging results in reclaiming saline and waterlogged soils. The growth of perennial forage grasses has been quite successful in Pakistan. Rhodes grass (*Chloris gayana*), Tall wheatgrass (*Elytrigia elongata*), and Puccinellia (*Puccinellia ciliata*) are the most popular examples. The incorporation of salt-tolerant trees and saltbushes into agricultural systems of salt-affected and waterlogged lands has the potential to increase crop and animal production and decrease land degradation. Such land improvements combined with improved agricultural practices will ensure that the current unsustainable trends in agriculture are reversed.

The success of saline agriculture necessitates more concerted efforts both at the public and private levels. The major areas of focus should be the enhancement of research capacity on different aspects of engineering, reclamation, and biological approaches. There is also a need to focus on action programs for the most seriously affected areas, capacity building of farmers, introduction of groundwater extraction regulations, and promotion of saline agriculture. Making these priorities in Pakistan is a basis for achieving sustainable irrigated agriculture and improved livelihoods.

The salt management issues in Pakistan are complex; therefore, an integrated approach dealing with irrigation and drainage issues is of paramount importance to sustain and improve the productivity of irrigated agriculture. For the future sustainability of irrigated agriculture, provision of drainage should be considered as a complimentary activity to irrigation because these two are interlinked through (1) over or inefficient irrigation as a cause of waterlogging and (2) the relationship between irrigation management and effluent disposal. Improved irrigation efficiencies and sustainable reuse of drainage water at the farm level can help in minimizing drainage effluent. Where relevant, farmers need to have access to the knowledge and inputs for reclaiming salt-affected lands through physical, chemical, and biological approaches. Timely availability of farm inputs such as good quality water, salt-tolerant germplasm, and promotion of saline agriculture through crop diversification options such as salt-tolerant medicinal and aromatic plant species can improve the capacity of individual farmers to be productive.

References

Angrish RKS, Datta C, Rani VS, Chawla SM (2009) Comparative bioremediation potential of some tree species. In: Asian regional conference, 6–11 Dec, New Delhi, India

Barrett-Lennard EG (2002) Restoration of saline land through revegetation. Agric Water Manage 53:213–226

Benyon RG, Marcar NE, Crawford DF, Nicholson AT (1999) Growth and water use of *Eucalyptus camaldulensis* and *E. occidentalis* on a saline discharge site near Wellington, NSW, Australia. Agric Water Manage 39:229–244

Bhutta MN, Smedema LK (2007) One hundred years of waterlogging and salinity control in the Indus Valley, Pakistan: a historical review. Irrig Drain 56:581–590

Bilal H, Ali SS, Kim MK (2014) Potential of *Eucalyptus* in the remediation of environmental problems: a review. Intern J Innov Sci Res 4(2):136–144

Choudhry MR, Bhutta MN (2000) Problems impeding the sustainability of drainage systems in Pakistan. In: Proceedings and recommendations of the national seminar on drainage in Pakistan, Organized by the Institute of Irrigation and Drainage Engineering and Mehran University of Engineering and Technology. Jamshoro, Pakistan

Cramer VA, Thorburn PJ, Fraser GW (1999) Transpiration and groundwater uptake from farm forest plots of *Casuarina glauca* and *Eucalyptus camaldulensis* in saline areas of southeast Queensland, Australia. Agric Water Manage 39:187–204

FAO (2000) Crops and drops: making the best use of water for agriculture. Food and Agriculture Organization of the United Nations, Rome. Retrieved 2 Nov 2010, from ftp://ftp.fao.org/docrep/fao/005/y3918e/y3918e00.pdf

Ghafoor A, Qadir M, Murtaza G (2004) Salt-affected soils: principles of management. Allied Book Center, Urdu Bazar, Lahore, p 304

Ghaly FM (2002) Role of natural vegetation in improving salt affected soil in northern Egypt. Soil Till Res 64:173–178

Ghassemi F, Jakeman AJ, Nix HA (1995) Salinisation of land and water resources: human causes, extent, management and case studies. CABI Publishing, Wallingford, p 517

Heuperman AF (2000) Biodrainage: an Australian overview and two Victorian case studies. In: Proceedings of the Eighth ICID international drainage workshop, New Delhi, India

Heuperman AF, Kapoor AS, Denecke HW (2002) Bio drainage - principles, experiences and applications. International Programme for Technology and Research in Irrigation and Drainage, Food and Agriculture Organization of the United Nations, Rome, p 89

Hossain MA (2010) Global warming induced sea level rise on soil, land and crop production loss in Bangladesh. In: 19th world congress of soil science, soil solutions for a changing world, Brisbane

Iglesias TG, Wilstermann D (2009) Global cultivated eucalypt forests map. In: GIT forestry consulting's eucalyptologics: information resources on *Eucalyptus* cultivation worldwide. Retrieved from http://www.git-forestry.com

Jeet-Ram, Dagar JC, Lal K, Singh G, Toky OP, Tanwar VS, Dar SR, Chauhan MK (2011) Bioremediation to combat waterlogging, increase farm productivity and sequester carbon in canal command areas of northwest India. Curr Sci 100:1673–1680

Kaur B, Gupta SR, Singh G (2002) Bioamelioration of a sodic soil by silvopastoral systems in northwestern India. Agrofor Syst 54:13–20

Khanzada AN, Morris JD, Ansari R, Slavich PG, Collopy JJ (1998) Groundwater uptake and sustainability of *Acacia* and *Prosopis* plantations in southern Pakistan. Agric Water Manage 36:121–139

Maas EV, Hoffman GJ (1977) Crop salt tolerance - current assessment. Proc Am Soc Civil Eng, Irrig Drain Div, 103(IRZ), pp 115–134

Marcar N, Crawford D, Leppert P, Jovanovic T, Floyd R, Farrow R (1995) Trees for saltland: a guide to selecting native species for Australia. CSIRO, Division of Forestry Canberra, Canberra

Masters DG, Benes SE, Norman HC (2007) Biosaline agriculture for forage and livestock production. Agr Ecosyst Environ 119(3 &4):234–248

Morris J, Mann L, Collopy J (1998) Transpiration and canopy conductance in a eucalypt plantation using shallow saline groundwater. Tree Physiol 18:547–555

Oster JD (2000) Irrigation sustainability. In: Proceedings of the sixth international micro-irrigation congress, 22–27 October 2000. Cape Town. ICID-CIID. Plenary Session, P6. CD Rom. Document Transformation Technologies, POB 560, Lirene, 0062, South Africa

Qadir M, Oster JDV (2002) Vegetative bioremediation of calcareous sodic soils: history, mechanisms, and evaluation. Irrig Sci 21:91–101

Qadir M, Oster JDV (2004) Crop and irrigation management strategies for saline sodic soils and waters aimed at environmentally sustainable agriculture. Sci Total Environ 323(1–3):1–19

Quirk JP (2001) The significance of the threshold and turbidity concentrations in relation to sodicity and microstructure. Aust J Soil Res 39:1185–1217

Qureshi AS (2011) Water management in the Indus Basin in Pakistan: challenges and opportunities. Mountain Res Develop 31(3):252–260

Qureshi AS, Akhtar M (2002) Drought and saline agriculture. In: Proceedings of the SAARC workshop on drought and water management strategies, 16–18 Sept, Pakistan Council of Research and Water Resources, Islamabad, Pakistan

Qureshi RH, Barrett-Lennard ED (1998) Saline agriculture for irrigated land in Pakistan. Monograph no. 50. Handbook published by Australian Centre for International Agricultural Research (ACIAR), Canberra, Australia

Qureshi AS, McCornick PG, Qadir M, Aslam Z (2008) Managing salinity and waterlogging in the Indus Basin of Pakistan. Agric Water Manage 95:1–10. doi:10.1016/j.agwat.2007.09.014

Rhoades JD (1999) Use of saline drainage water for irrigation. In: Skaggs RW, van Schilfgaarde J (eds) Agricultural drainage. ASA-CSSA-SSSA, Madison, pp 615–657

Schulz M (1994) Development of sustainable systems using irrigated salt tolerant plants for productive revegetation of saline land. Final Report NRMS Project V2112. Institute of Sustainable Irrigated Agriculture, Department of Agriculture, Tatura, Victoria, Australia

Shah AH, Sufi AB, Bhutta MN, Patto PM (2000) Prospects for saline agriculture in Pakistan. In: Proceedings of the eighth ICID international drainage workshop, vol IV, Group V Biodrainage. New Delhi, India, pp 61–78

Shah AH, Anwar Ul Haq, Bhutta MN (2003) Success of bio-saline approach for land rehabilitation: Pakistan community development project for rehabilitation of saline and waterlogged land. IWASRI Report No. 2003/22, p 22

Shah AH, Gill AH, Syed NI (2011) Sustainable salinity management for combating desertification in Pakistan. Intern J Water Resources Arid Environ 1(5):312–317

Slavich PG, Smith KS, Tyerman SD, Walker GR (1999) Water use of grazed saltbush plantations with saline water table. Agric Water Manage 39:169–186

Smedema LK (1997) Biological drainage: myth or opportunity? GRID, Magazine of the IPTRID Network, Issue 11, p 3

Smedema LK (2000) Irrigation-induced river salinization. five major irrigated basins in the arid zone. IWMI, Colombo

Sumner ME (1993) Sodic soils: new perspectives. Aust J Soil Res 31:683–750

Tilman D, Cassman KG, Matson PA, Naylor R, Polasky S (2002) Agricultural sustainability and intensive production practices. Nature 418:671–677

WAPDA (2007) Waterlogging, salinity and drainage situation. SCARP Monitoring Organization, Water and Power Development Authority, Lahore

Wichelns D (2009) An economic perspective on the potential gains from improvements in irrigation water management. Agric Water Manage 52:233–248

Wicke B, Smeets E, Dornburg V, Vashev B, Gaiser T, Turkenburg W, Faaij A (2011) The global technical and economic potential of bioenergy from salt-affected soils. Energy Environ Sci 4:2669–2681

Wicke B, Smeets E, Dornburg V, Akanda R, Stille L, Singh RK, Awan AR, Mahmood K, Faaij APC (2013) Biomass production in Agroforestry and forestry systems on salt-affected soils in South Asia: exploration of the GHG balance and economic performance of three case studies. J Environ Manage 127:324–334

Zhang J, Xing S, Li J, Makeschin F, Song Y (2004) Agroforestry and its application in amelioration of saline soils in eastern China coastal region. For Stud China 6(2):27–33

Combating Waterlogging in IGNP Areas in Thar Desert (India): Case Studies on Biodrainage

M.M. Roy, N.D. Yadava, M.L. Soni, and J.C. Tewari

Abstract

Indira Gandhi Nahar Pariyojana/Project (IGNP) in the *Thar Desert* is one of the largest irrigation projects in the world. As a result of this initiative, 1.86 million hectare area has come under cultivation. Although irrigation has greatly increased the agricultural production potential, it has also rendered vast tracts of land under waterlogged category (leading to salinization as well) on account of seepage and deep percolation losses from the irrigation network and lack of a proper drainage system. The concept of biodrainage, removal of groundwater through evapotranspiration, has emerged as an effective way to tackle this problem. Such an approach is economical and eco-friendly. However, more research is required to fully understand the implications, especially at farm level. The policy to control the problems associated with waterlogging and salinization should include biodrainage, reduced water allowance, efficient irrigation, and conjunctive uses.

Introduction

Irrigation of agricultural land has a long and well-documented history. It has a remarkable impact on agricultural development all over the world with increased production and better supply of food, fiber, bioenergy, and bio-industrial feedstock needs. Irrigation with poor drainage creates waterlogging and increases groundwater salinity. The waterlogged areas existing now are the results of negligence and improper drainage management. The history of the Assyrian civilization in Mesopotamia is an example where whole population was forced to abandon a region because of rising groundwater tables and salinity (Jacobsen and Adams 1958).

In India, net irrigated area increased from 22.5 Mha in 1951 to 93.95 Mha in 2002 (Dash et al. 2005). Similarly, the area of degraded land has also changed with time which has been estimated differently by different agencies. In this context, the efforts have been made for integration and streamlining of land based and remotely sensed

M.M. Roy (✉) • N.D. Yadava • M.L. Soni
J.C. Tewari
Central Arid Zone Research Institute,
Jodhpur 342003, Rajasthan, India
e-mail: mmroyster@gmail.com;
narendra_yadava@yahoo.co.in;
soni_ml2002@yahoo.co.in; drjctewari@gmail.com

J.C. Dagar, P.S. Minhas (eds.), *Agroforestry for the Management of Waterlogged Saline Soils and Poor-quality Waters*, Advances in Agroforestry 13, DOI 10.1007/978-81-322-2659-8_7

revised databases in the geographical information system environment so that a harmonized database is made available for use by the planning departments. According to NAAS (2010), the area under salt-affected soils is 6.64 Mha, and waterlogged soil due to surface ponding is 0.88 Mha. The area under subsurface waterlogging (4.75 Mha) has been excluded. Rajasthan is the largest state in India, with 27.24 Mha of cultivable area, of which nearly 60 % is classified as desert (GoR 2012). The ambitious project of the government to bring water in this state through interlinking of rivers Ravi (6.9×109 m^3 of water) and Beas (2.3×109 m^3 of water) and developing storage reservoirs and extensive canal systems has led to intensification of agriculture but with the associated problems of waterlogging and salinity (CAD 2007, 2009; IWMI 2007). In this chapter, an analysis is presented in context of combating these twin problems through biodrainage approach.

Indira Gandhi Nahar Project (IGNP)

IGNP occupies the northwestern and far western parts of the *Thar* Desert in Rajasthan (India). It is one of the biggest projects of its kind in the world aiming at transforming desert wasteland into agriculturally productive area with the objectives of drought proofing, providing drinking water, improvement of environment, development and protection of animal wealth, and increasing agricultural production. The project was conceived as a canal project for irrigating desolate land in the desert area. As per Indus Water Treaty, India got the exclusive right to use waters of the three rivers, namely, Ravi, Beas, and Sutlej. Rajasthan got a share of 10.6 billion cubic meters (BCM) from the surplus water to develop the desert area. Out of 10.6 BCM of water, 9.36 BCM was allocated for meeting the irrigation, drinking purpose, and industrial requirements of the area. The balance 1.24 BCM has been allocated for *Gang*, *Bhakra*, and *Sidhmukh* canal systems. The canal system in IGNP is enormous. The main canal, Indira Gandhi Mukhya Nahar (IGMN), has a total length of 445 km with 9060 km of distribution canals. The project work has been taken up in two stages – Stage I and Stage II (Fig. 1).

IGNP Stage I: It consists of head reach of the project, comprising 204 km long feeder canal originating from the Harike Barrage in Punjab, 189 km long main canal (from Masitawali head to Pugal head), and 3400 km long distribution system.

IGNP Stage II: It consists of lower reaches of the project comprising 256 km long main canal starting from Pugal head of the main canal (Bikaner) up to its tail and near Mohangarh (Jaisalmer). It has a long distribution system of 5780 km with a culturable command area (CCA) of 1.41 Mha at 80 % intensity of irrigation in flow and 60 % intensity under lift systems (CCA of 0.54 Mha).

The area is unique in many aspects like climate, soil, and landforms. Shifting sand dunes, scantly and uneven distribution of rainfall, extreme temperature (<0 °C in winter to>50 °C in summer), and frequent occurrence of drought make the Indian desert one of the most inhospitable for agriculture. Annual mean rainfall varies in between 200 and 300 mm which is highly variable (>45 %) in nature. Daily potential evaporation rate rises to 12 mm day^{-1} in summer, when high wind speed, high temperature, and low humidity produce a desiccating environment. The underground water present in most of the places is brackish, deep (150 m), and unfit for human consumption.

Landforms and Physiography

The area falls in the hot western region of Rajasthan in which about 60 % of the area is covered by sand dunes which range from small, bare, and fairly mobile drifts to a few meters tall to large, semi stable longitudinal, parabolic, and transverse dunes over 50 m tall (Rahmani 1989, Cloudsley-Thompson 1977). Within the IGNP area, five main landform types: (1) younger alluvial plan, (2) flat older alluvial plain, (3) sandy undulating older alluvial plain, (4) sand dunes and inter-dunal plain, and (5) saline depressions were identified (Singh and Kar 1991). Soils of IGNP area fall into two broad geographic zones:

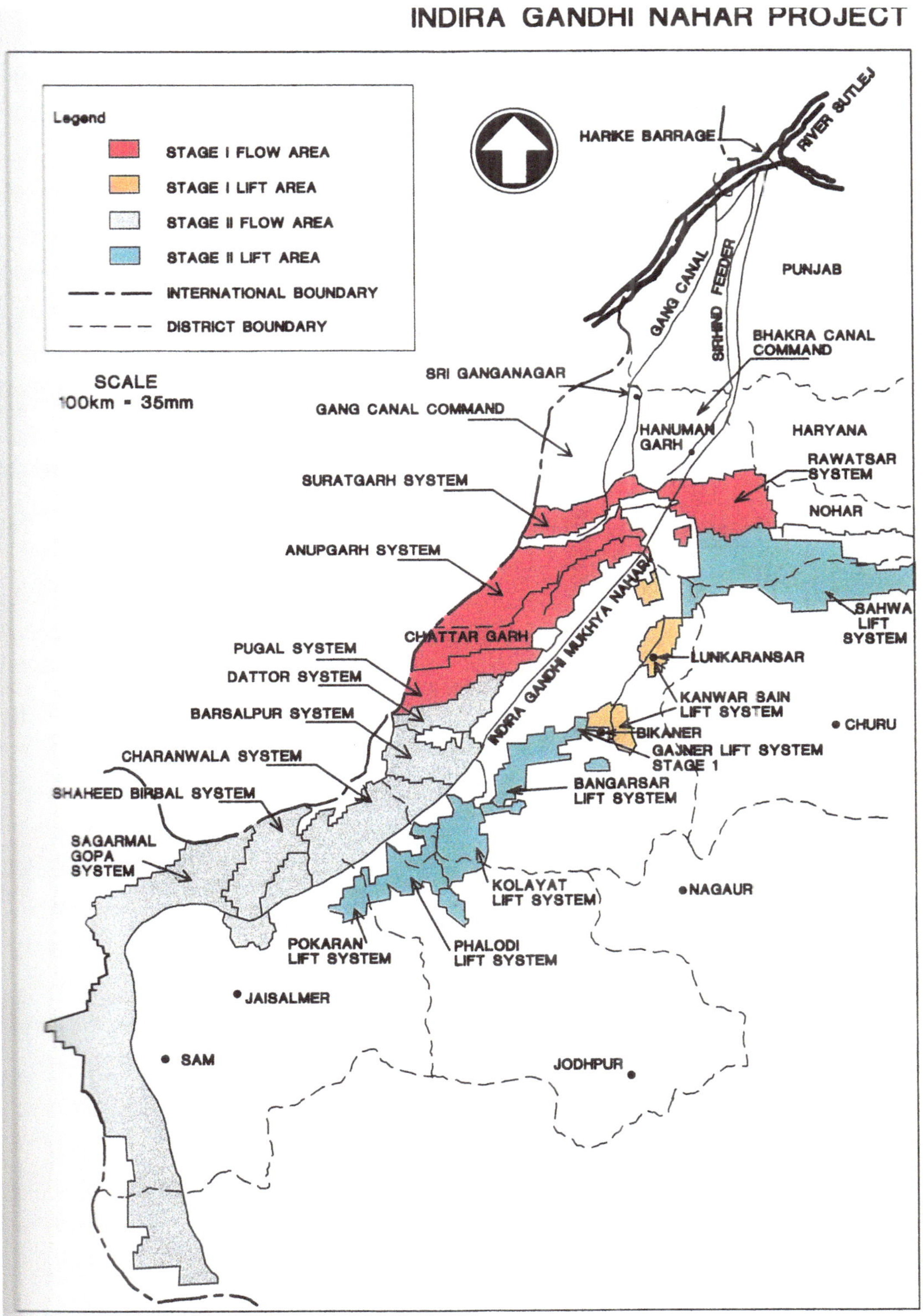

Fig. 1 Layout of IGNP areas in Rajasthan (India)

1. The area north of Pugal branch command, which consists mainly of soils that have developed on late quaternary to recent marine flood plains, with substantial additions of aeolian sand
2. The area south of Pugal branch including Stage II area, the soils of which comprise mainly late quaternary drift sand

The area largely consists of desert plains, sandy plains, and dunes of various types with an underlying impervious layer of calcareous accumulations and rocks. The depth of this layer is up to 15 m below ground level in Stage I and rises toward southwest. In Stage II, it occurs almost on the surface. The most worrying feature is the presence of shale/clay, hard compact friable carbonate nodules, and lime-coated gravel with clay at varying depths with poor infiltration rate and behaving as an impervious barrier. In about 30–35 % of the area under Stage II, the depth down to these hydrological barriers is less than 10 m, being shallower in lift areas and becoming deeper toward the international boundary (CAD 1997, 1999). The hard pan acts as a hydrological barrier to vertical groundwater flow and is not continuous on a regional basis. Perched aquifers are likely to exist where the barrier layer is of particularly low vertical permeability and where it occurs within a shallow depth of ground surface. About 33.4 % flow in command and 76.4 % in lift command (excluding the Sahwa lift area) are prone to waterlogging due to the presence of the hardpan layer.

Causes of Waterlogging in IGNP

IGNP is a boon to the northwestern arid part of India. However, improper and overuse of water coupled with seepage from the canal has created the problem of waterlogging and salinity in the area and has become a major problem to augment agricultural productivity. The conditions those have led to waterlogging in command areas include seepage losses from the conveyance and distribution system, excess water accumulation in lowland, recharge of groundwater table, inadequate drainage, etc. Seepage from the canal, in combination with relatively impervious layers at shallow depth in the profile, resulted in the formation of a perched water body (Kapoor and Denecke 2001). The major causative factors responsible for the development of waterlogging and secondary soil salinization in command area of IGNP have been described by Hooja et al. (1994). These include (1) more incidence of flooding in Ghaggar plains in Rajasthan; (2) inundation of Ghaggar depressions through the use of Ghaggar diversion channel; (3) high water allowances; (4) inefficient lining material; (5) uncontrolled supply of direct outlets from the main and branched canals; (6) over irrigation, surface wild flooding, and growing high water requiring crops by the farmers in the head reaches of the canal systems; and (7) hydrological barrier (hard pan) in subsurface layer. Above changes have induced the recharge component to the groundwater system by various degrees, and in the absence of compensating natural subsurface outflow from the system, the excess water became stagnant on impervious layer resulting in rising water tables. As a result of rising water table, heavy load of salt is left in the profile, rendering the soil saline. Initially the crop yields were marginally reduced, but in the subsequent years, land became completely unproductive.

Water Level Changes Due to Irrigation

Prior to the introduction of canal irrigation, the groundwater levels in the area were very deep varying from 40 and 50 m below the surface. With the introduction of irrigation, the groundwater table started rising. During 1981–1992, the average water table rise in Stage I was 0.92 m year^{-1}. In Stage II of the project, the groundwater table before the advent of irrigation generally ranged between 20 and 100 m below surface. With irrigation it has been rising, though not at the same rate as Stage I. Shrivastava et al. (2013) analyzed the groundwater scenario of IGNP and observed that the earliest groundwater levels during 1952–1972 showed a general rise of 0.42 m year^{-1} which was mainly due to advent of irrigation and groundwater flow from upper reaches, where irrigation was in full swing. During the period 1972–1982, there was a substantial rise in water table up to 1.17 m year^{-1} which was attributed mainly to the return flow of irrigation and filling of depressions through Ghaggar diversion channel in Stage I of IGNP. The net rise in water table reduced to 0.76 m year^{-1} during 1982–1992 in some of the areas, but a large area came under potentially sensitive and critical categories (water level between 1.50 m and 6.00 m below ground level). During the period 1992–1994, the water level depleted drastically due to scanty rainfall and less availability of water in canal system which resulted in the reduction of waterlogged condition significantly. The analysis of groundwater data for the year 1995–1997 revealed rising trend ranging from 0.30 to 0.70 m year^{-1} due to continuous recharge owing to seepage from canal system, return flow of irrigation, etc. The analysis of water level data during the year 1999–2000 revealed that the shallow water table had declined due to scanty rainfall (drought year) and nonavailability of water in the canal.

An analysis of development of waterlogging since 2000–2001 in the IGNP command areas is shown in Table 1. During the year 2002–2005, the average fluctuation of water level showed a declining trend in Stage I and rising trend in Stage II of IGNP command, respectively. However, the water levels in the vicinity of canal had shown declining trend. From the year 2005–2007, the average fluctuation of water level in Stage I as well as Stage II showed an overall rising trend in IGNP command. The water levels in the vicinity of canals had also shown marginally rising trend. During the period 2007–2008, the average fluctuation of water level was −0.09 m in Stage I and −0.02 m in Stage II. This shows an overall marginally depleting trend in IGNP command. However, the fluctuation of water levels in the vicinity of canal of Stage II was −0.12 m year^{-1} showing declining trend (Table 1). The overall increase or reduction in the waterlogged area, critical area, and potentially sensitive area is attributed to the variation in rainfall, availability of water in canal system, and return flow of irrigation.

Impact of Waterlogging

Rise of the water table closer to the surface and inundation of the low-lying areas have caused submergence of agricultural lands, village common lands, villages/habitations besides causing damages to road communication, and public utilities and constraints in the choice of crops and loss of production (Sharma et al. 2009). By the end of year 1997–1998, a total CCA of 514,000 ha (56 % of the total area) in Stage I and about 23,000 ha (about 13 %) in Stage II had become potentially sensitive to waterlogging (CAD 2007). As per estimations of NAAS (2010) in districts of Ganganagar, Hanumangarh, and Jaisalmer, exclusively saline and sodic areas are 36,000 ha and 61,000 ha, respectively.

The analysis of water samples collected from various sources (accumulated seepage lakes (*Tals*), auger bores, and profile pits) of waterlogged and salt-affected soils of IGNP area from different villages near Hanumangarh and Bikaner (29°00′N and 29°35′N latitude and 74°00′E and 74°40′ E longitude) showed a wide variation in electrical conductivity (ECiw); and high to very high ECiw (93.4–102.6 dS m^{-1}) was found in severely affected waterlogged areas of Masitanwali and Bhairusari villages apparently due to prolonged salt accumulation (Mondal and Sharma 2011). The ECiw of water samples collected from profile pits and auger bores (1.7–22.0 dS m^{-1}) and drainage water (3.2–30.6 dS m^{-1}) suggesting its safe reuse after mixing with good quality irrigation water. The transportation of salts from higher to lower elevation through the coarse sandy mass facilitates in development of poor water quality in *Tals* and inter-dunal seepage lakes. The higher EC of water sample (22.2 dS m^{-1}) from profile pit in *Dabli Kalan* village was due to the presence of highly saline *Tal* close to profile pit. The ionic composition of water samples showed significant contribution of sodium (8.2–1935 me L^{-1}), calcium plus magnesium (2.0–370 me L^{-1}) cations for development of high to very high soil salinity. Among anions, chloride and sulfate contributed significantly for the development of salinity. The sandy soil texture in most places enhanced upward movement

Table 1 Status of waterlogging in IGNP command (year wise)

	Total area (000 ha)							
	00–01[a]	01–02	02–03	03–04	04–05	05–06	06–07	07–08
	Potentially sensitive area (water table within 2–6 m)							
Stage I	225.1	179.1	164.3	195.0	168.7	196.8	202.1	181.2
Stage II	12.18	18.3	24.5	13.5	16.0	18.8	15.9	11.8
Total	237.3	197.4	188.9	208.5	184.7	215.4	218.0	193.1
	Critical area (water table within 1–2 m)							
Stage I	13.4	11.3	8.7	9.2	10.6	11.2	16.9	12.5
Stage II	2.2	1.2	0.4	0.3	0.5	1.1	2.6	1.1
Total	15.6	12.6	9.2	9.5	11.1	12.3	19.4	13.6
	Waterlogged area (water table within 0–1 m)							
Stage I	12.7	10.1	5.7	2.5	2.9	3.1	6.8	1.8
Stage II	0.3	0.1	0.0	0.0	0.0	0.5	0.8	0.3
Total	13.0	10.2	5.7	2.5	2.9	3.6	7.6	2.1

Source: CAD (2009)
[a]Denotes years 2000–2001 onwards

of water table and capillary rise of salts to develop secondary soil salinity within the root zone. Such process occurring continuously for longer period produces severe salinization in form of salt efflorescence/salt crust at surface with no agriculture operation. The analysis of soil samples from different pedons showed moderate (ECe 9.3 dS m^{-1}) to very high salinity (ECe 40.3 dS m^{-1}) in IGNP area.

The waterlogging and soil salinity had greatly affected the yield of different crops in the region. According to a World Bank study, India loses 1.2–2.0 million Mg of food grain production every year due to waterlogging (Prasad and Biswas 1999). The average cotton crop yields are low at 1300 kg ha^{-1} as compared to about 1500 kg ha^{-1} in normal soils under Stage I. The same trend was with cluster bean, wheat, mustard, and chickpea. The chickpea yields due to waterlogging and soil salinity reduced up to 50 %. The net returns of cotton and wheat were lowered by about 25 % and 46 %, respectively. In a survey conducted by IWMI (2007), it was emphasized that waterlogging (28 % of respondents) and soil salinity (26 % of respondents) are major problems in IGNP with a lot of area submerged under pools of water. On an average, the additional expenditure due to waterlogging and soil salinity on practices like field preparation enhanced seed rate, and fertilizer applications increased to the tune of INR 1095 ha^{-1}.

Biodrainage in Solving Waterlogging Problems

Many measures for controlling the waterlogging in IGNP area have been planned by the department of IGNP. These includes (a) reduction in excess supplies of water allowance in Stage I till the irrigation in Stage II develops considerably, (b) increase in irrigation area, (c) reduction in number of days of running canal, (d) removal of excess of outlets and correction of over-sized outlets and direct outlets, (e) remedial measures in canal to control seepage through damaged lining, (f) installing shallow skimming wells along with infiltration gallery, and (g) installation of subsurface pipe drainage system.

The usefulness of various techniques in mitigating the problems of waterlogging and salinization or combating the adverse effects of waterlogging has been discussed by many workers (Mann and Chatterji 1978, Chatterji 1985, Shankarnarayan and Sarkar 1985, Chatterji and Saxena 1988). These include construction of open drainage, horizontal subsurface drainage, vertical drainage, soil management, artificial recharge, growing aquatic cash crops, etc. (Sharma 1999). Physical drainage works require expensive capital investment, operation, and maintenance and also generate drainage effluent. Disposal of these effluents is considered unacceptable because downstream users in the catchment rely on these river systems for their water supplies. Greenhouse gas emissions caused during pumping may also be disapproved in a changing climate scenario.

To overcome these issues, biodrainage, i.e., the use of vegetation to manage water fluxes in the landscape has recently attracted interest in drainage and environmental management circles. Following requirements are desired to be fulfilled for effective biodrainage systems: (1) the annual quantity of water removed from the groundwater should equal the quantity of recharge; (2) the quantity of minerals removed annually should be nearly equal to the quantity of mineral import; (3) afforestation or agroforestry should be economically comparable with that from other alternative uses of land. If it is not so, afforestation may still be justified, on considerations of the environmental and drainage benefits; (4) under ideal situation, trees in afforestation area on full development should be able to draw most of their requirement of water from the groundwater table, so that surface irrigation water can be put to other productive uses; (5) when the water table approaches the root zone of trees, the quality of groundwater should be tolerable by the plant species, otherwise the trees would need to be supplied irrigation water; (6) for effective biodrainage system, the groundwater table must be lowered in the irrigated area to a minimum critical depth (say 2 m bg^{-1}) at the farthest point from the edge of the plantation area.

The researchers in India (Hooja et al. 1995, Kapoor 2001, Kapoor and Denecke 2001, Tewari et al. 1997) and abroad (Bhutta and Choudhry 2000, Zhang et al. 1999, Stirzaker et al. 1999) reported that trees could be used to manage rise in water table. When the water table surface comes up sufficiently high and is within the reach of roots of trees in plantations, the trees start drawing water from the groundwater reservoir through the process of transpiration. The root systems of trees could intercept saturated zone or unsaturated capillary fringe above water table to control shallow water table. These plants are known as *phreatophytes*. The unsaturated top soil zones are intercepted by root systems of plants mainly following rainfall or irrigation to remove water from soil profile by controlling recharge. This process of withdrawal of groundwater by plantations is termed biodrainage.

The term biodrainage is relatively new, although the use of vegetation to dry out soil profiles has been known for a long time. One of the earliest documented observations of water table lowering beneath a tree plantation was recorded by Heuperman et al. (1984). The first documented use of the term biodrainage can be attributed to Gafni (1994). Prior to that date, Heuperman (1992) used the term *bio-pumping* to describe the use of trees for water table control. Another term relating to the *bio* aspect of soil water removal is *bio-disposal*, which refers to the use of plants for final disposal of excess drainage water (Denecke 2000). The term biodrainage has attracted interest in drainage and environmental management circles about 10–15 years back due to its eco-friendly nature. The driving force behind the biodrainage concept is the consumptive water use of plants. The rates of transpiration and groundwater uptake by trees underlain by relatively shallow (5–8 m below surface) water tables were very high. It exceeded by a factor 3–6 (1200–2300 mm $year^{-1}$) as compared to pasture (~400 mm $year^{-1}$) (Greenwood et al. 1985).

Lowering of the Raised Groundwater Table

Biodrainage helps in lowering down the groundwater table underneath the plantation area. Large variation in capacity of trees to grow and transpire water has been reported by different workers because of the many factors, influencing the rate of transpiration. Tewari et al. (1997) suggested the suitability of tree species for waterlogged soils on the basis of salinity hazard in IGNP areas. On waterlogged area with lesser salinity, plantations of the recommended tree species, viz., *Eucalyptus camaldulensis*, *Populus casale*, *Phoenix* spp., and *Acacia nilotica*, may be done by making mounds at a spacing of 10 m × 5 m. The width and height of these mounds are recommended as 1.5 m and 1.2 m, respectively. No irrigation is recommended for the plants in such areas. However, in waterlogged area with considerable salinity, narrow plantations with

4 m×4 m are recommended. The suitable tree species which may tolerate high salinity and waterlogging are *Tamarix articulata*, *Prosopis juliflora*, *Acacia nilotica* and *Salvadora oleoides*. Apart from tree species, the shrubs like Saji (*Haloxylon recurvum*) and Lana (*Haloxylon salicornicum*) may also be dribbled for their rehabilitation.

Eucalyptus species are generally considered to be effective for biodrainage purposes. *Eucalyptus camaldulensis* is a hardy tree that grows under a wide range of climatic conditions and soil types in this area. *Eucalyptus* grows fast and has good water consumption capacity when the water is available in sufficient quantity (Alvares 1982). Some provenances of the species tolerate saline conditions quite well. The study conducted in 25 ha waterlogged area in IGNP revealed that in an unplanted area (14.7 ha in 1991), surface water was removed from pools mostly by surface evaporation. Over the period 1991–1997, the water table fell about 10 m. During the same period, the water transpired by the trees was estimated to the tune of 20.6 m (Heuperman et al. 2002).

The impact of block plantations of *E. tereticornis* on reclamation of waterlogged areas was found effective at the Indira Gandhi Nahar Project (IGNP) site in Rajasthan, India (Kapoor 2014). The groundwater table underneath the strip plantations remained lower than that in the adjacent fields. The drawdown of water in IGNP area by the tree plantations established along the canal was reported to the tune of 14 m in 6 years (Kapoor 2001). However, Jeet Ram et al. (2011) working at Puthi research plot in Hisar (Haryana), northwest India, reported that the drawdown in groundwater table was 0.85 m in 3 years. The main reasons for the difference in drawdown of groundwater table at the two sites (IGNP and Puthi research plot) were the design and density of plantations and the sources of recharge of groundwater. IGNP had block plantation of 160 m width with the density of 1000 plants ha^{-1}, and the sources of recharge of groundwater were only rainwater and seepage from canal. Whereas Puthi had strip plantations of 1 m×1 m width on acre line with a density of only 300 plants ha^{-1} and the sources of recharge of groundwater were rainwater, seepage from the canals and irrigation applied to agricultural crops between the strip plantations. Therefore, at the Puthi research plot, the relatively higher recharge of groundwater from many sources and the relatively lower discharge of groundwater due to less number of trees ha^{-1} resulted in less drawdown of the groundwater table.

Besides lowering the water table, *Eucalyptus* fetches high price and provides fuel wood. Strip plantations of clonal *E. tereticornis* also worked as bio-pumps and helped in reducing water table. The drawdown curve of groundwater table due to *E. tereticornis* was similar to the cone of depression of a pumping well (Heuperman 1992, Jeet Ram et al. 2007). Heuperman (1999) estimated annual water use in desert area of Rajasthan by *Eucalyptus* species to the tune of 3.44 m (density of 1900 trees ha^{-1}). Cramer et al. (1999), however, reported that *Casuarina glauca* could extract groundwater more than *Eucalyptus camaldulensis* planted at similar densities. Heuperman and Kapoor (2003) observed the average annual rate of transpiration as high as 3446 mm from a 25 ha mixed plantation (*Eucalyptus camaldulensis, Acacia nilotica, Prosopis cineraria* and *Ziziphus* spp.) in the IGNP area of Rajasthan during 1991–1997. The water removal rate was estimated as equivalent to a vertical drainage network with 500 m well spacing with a 33 m^3 h^{-1} pumping rate. They estimated that a forest cover of 1, 77,000 ha (10 % of the total irrigable area in IGNP) would be able to transpire the estimated annual groundwater recharge of 2.6 BCM. Jeet Ram et al. (2011) reported that transpiration value depends on variation in radiation, temperature, and vapor pressure gradient prevailing during the period. The average rate of transpiration in *E. tereticornis* trees ranged from 44.5 to 56.3 l d^{-1} tree^{-1} in May, 30.5 to 34.0 in July, 24.1 to 28.3 in October, and 14.8 to 16.2 l d^{-1} tree^{-1} in January. The overall average rate of transpiration in the 5-year-old *E. tereticornis* was 30.9 l day^{-1} tree^{-1}, which was 268 mm annum^{-1} by 240 trees ha^{-1} against the mean annual rainfall of 212 mm. The discharge of groundwater by the strip plantations of clonal

E. tereticornis was 1.3 times more than the recharge by rainfall resulting in reclamation of waterlogged areas.

Lowering of water table and associated soil improvement by *Eucalyptus* plantations increased the wheat grain yield by 3.4 times as compared to non-planted agricultural fields and resulted in reclamation of waterlogged areas (Jeet Ram et al. 2011). A close relationship between growth and transpiration rates has been observed by Calder (1992) and Delzon and Loustau (2005). *Eucalyptus* plantation improved overall surface soil properties. Lowering of water table by *Eucalyptus* induced higher root activities and secretion of organic acids in the surface layers. The average steady-state infiltration rate in fields with plantation was 8.3 mm h^{-1}, whereas it was only 5.1 mm h^{-1} in fields without plantation.

Bala et al. (2014) made an attempt to study the removal of excess water from the land through biodrainage and to increase vegetation cover and productivity of a waterlogged area of IGNP. They tried four species on raised bunds (60 cm high, 60 cm wide, and 2 m apart) prepared in waterlogged (inundated water of 15–25 cm) area to provide comfortable root zone for young seedlings. Among the four species tried, the performance of *E. rudis* was best with respect to growth, biomass, transpiration rate, and overall biodrainage potential. *E. rudis* maintained uniform transpiration and photosynthesis rate throughout the year. Groundwater table has receded from 25 cm to 145 cm in *E. rudis* plot compared to 90 cm, 70 cm, and 60 cm in *Corymbia tessellaris, E. camaldulensis* and *E. fastigata*, respectively, within a period of four and half year. Soil working at the site resulted in heavy regeneration of *Eucalyptus camaldulensis*. The regenerated plants were mostly concentrated between 6 and 10 m from the tree trunk of the mother trees situated at the edge of the experimental site. Apart from the planted ones, some species like *Prosopis juliflora*, *Tamarix dioica*, *Saccharum munja* (tall grass) and *Arundo donax* (grass) also have come up in the area. The number of *A. donax* has reduced gradually with recession of groundwater table in the experimental plot. With the lowering of groundwater level, other species started growing in the area as natural succession. Population of *S. munja* was highest followed by *P. juliflora* and *T. dioica*. The total biomass per tree in *P. juliflora* was recorded as 110 kg. The contribution of the roots to the total biomass was 25 %, and *S. munja* and *T. dioica* accumulated a total biomass of 76.5 kg and 73.2 kg per bush, respectively.

Biomass Production and Carbon Sequestration

In a study in IGNP, 5-year-old subirrigated plantations produced dry biomass of 185 Mg ha^{-1}. The utilizable biomass production was 29 Mg ha^{-1} $year^{-1}$. *Acacia nilotica*, *Dalbergia sissoo*, *Tecomella undulata* and *Ziziphus mauritiana* are other species that have performed well in plantations along leaking canals in arid conditions. Under waterlogged saline soils in IGNP area in Lakhuwali, Hanumangarh, Rajasthan, Soni et al. (2012) reported that among the three dominant species in waterlogged salinity, the above ground biomass was highest in *Acacia nilotica* (132.1 Mg ha^{-1}) followed by *Eucalyptus* spp. (77.6 Mg ha^{-1}) and *Acacia tortilis* (40.6 Mg ha^{-1}). Maximum carbon storage was observed by *A. nilotica* (66.5 Mg ha^{-1}) followed by *Eucalyptus* spp. (38.8 Mg ha^{-1}).

Several techniques, viz., double ridge mound (DRM) and circular dished mound (CDM), were developed for the establishment of trees and shrubs in waterlogged saline-alkali soils of IGNP area in Jodhpur district of Rajasthan (Arya et al. 2006). Both mound practices, viz., CDM and DRM, recorded higher survival and growth as compared to control during 36 months of establishment. There was no difference in height of all the three shrub species between CDM and DRM structures. However, for crown diameter, CDM structure exhibited higher growth as compared to DRM structure. Among shrubs, *Suaeda nudiflora* was the best species on the basis of survival and growth followed by *Atriplex lentiformis* and *Atriplex stocksii*. Among trees, *Acacia colei* also showed potential by maintaining 89 % survival on DRM in a drought year.

Salt Balance

Salt balance is one of the most important issues to be addressed before biodrainage can be promoted as an appropriate drainage management technology. The water quality of Indus river system is basically good having low salinity (TSS ~125 ppm). However, due to the large volume of water introduced in the landscape of IGNP system, it increases salt import significantly. When the irrigation water is removed by means of evaporation or by growing plants, the salts remain within the area unless it is removed either by natural or through artificial means. To maintain the salt balance in soil, two mechanisms can be considered in soil-plant systems: (a) salt balance through removal of salts from the vegetation root zone by leaching and (b) removal of salts used by the vegetation through grazing or harvesting of plant matter for its further use. The former is achieved through the concept of leaching fraction on the basis of amount of salt to be leached through root zone. The latter mechanism however needs some elaboration as it is often mentioned in biodrainage related information.

The important aspects to be considered in the salt balance analysis are (1) mineral content in supply (irrigation or ground) water and (2) mineral content in plant biomass. In high-salinity environments, salt uptake by the plants might be negligible in relation to the salts present in the soil. However, under low-salinity scenarios, the salt balance might be achieved between the applied and removal of salt by the plants. This option needs to be critically reviewed. The studies conducted by various researchers showed that the salt uptake by plants is negligible compared to the total salt applied through irrigation water (Hoffman 1990, Chhabra and Thakur 1998, Dash et al. 2007, Heuperman 1999, NIAB 1997, Pessarakli and Szabolcs 1999) and thus leave behind the salts in soil. Schulz (1994) measured average dry matter yields of *Atriplex nummularia* of 0.6 kg plant^{-1} per year^{-1} across a range of applied irrigation salinities (100–10,000 mg L^{-1}, NaCl-dominant water) over a 3-year period. At planting densities of 10,000 bushes ha^{-1}, the salt export in the leaves (considering the major ions) would be between 350 and 433 kg ha^{-1} year^{-1} against the salt application of 1 Mg ha^{-1}year^{-1} and 100 Mg ha^{-1} year^{-1}, respectively, for low- and high-salinity treatment. This suggests that with the low-salinity irrigation water, the plants made a significant contribution to salt removal, but with the higher-salinity values, salt balance control by vegetation was not possible. Lambert and Turner (2000) present data on sodium and chloride accumulation in different components of 22-year-old plantations. The salt balance study on 22 years old plantations of *Eucalyptus grandis* showed an uptake of 14 kg ha^{-1} year^{-1} salt (predominantly Na and Cl). These quantities are small in relation to the salt inputs of 65 kg ha^{-1} year^{-1} in the plantations. As a whole, the potential for export of salt through plant harvesting does not look promising. Salt balance through the removal of vegetation has only been reported for situations with very low salt input/freshwater supplies such as channel seepage.

Conclusions

There are conclusive evidences to suggest that appropriate tree stands have the ability to lower water tables in areas having the problem of rise in water table. This is either on account of reduction in recharge of water or direct extraction of water from the saturated zone or a combination of both. There are also some indications of harvesting of salts from soil by vegetation and plantations. In IGNP areas in Thar Desert, growing of trees in afforestation and other projects have shown its positive impact and supported the concept of biodrainage in combating the problems of waterlogging and associated salinity. Although several studies from IGNP areas in Rajasthan have led to identification of tree species – their plantation density, salt-loving vegetation, etc. – more research efforts on the biodrainage aspects are required for a better understanding and wider adoption. Suggested aspects are:

- Undertaking field measurements of trees in the area and their effect on water table depth and soil and groundwater salinity from farmers' fields

- Pre- and post-monsoon measurement of salt content of plant samples at periodical intervals
- Studies on transpiration rates of planted tree species and its relationship with pan evaporation
- Screening of salt tolerant tree species
- Investigations on the salt sensitiveness of different tree species and salt balance between the tree and the growing medium
- Economic assessments on biodrainage and other approaches to combat waterlogging in specific situations

References

Alvares C (1982) Social forestry: the likely distortions. Bioenergy Res News 1:1–6

Arya R, Choudhary KR, Lohara RR (2006) Studies on mound practices for establishment and growth of various plant species on saline and waterlogged soil in hot arid zone of India. Indian For 132(5):556–564

Bala N, Singh G, Bohra NK, Limba NK, Baloch SR (2014) Biodrainage for restoration of canal command waterlogged area in Indian desert. Indian For 140(5):(in press)

Bhutta MN, Choudhry MR (2000) Biological control of waterlogging. In: Proceedings of the 8th ICID international drainage workshop, vol IV, Group V. New Delhi, India, pp 33–45

CAD (1997) Monitoring of water table in IGNP command. Ground Water Wing, CAD Department, Indira Gandhi Nahar Pariyojana, Bikaner

CAD (1999) Status report: command area development. Ground Water Wing, CAD Department, Indira Gandhi Nahar Pariyojana, Bikaner, Bikaner

CAD (2007) Note on monitoring of water table and status of water logging. CAD Department, Indira Gandhi Nahar Pariyojana, Bikaner

CAD (2009) Report of command area development. CAD Department, Indira Gandhi Nahar Pariyojana, Bikaner

Calder IR (1992) A model of transpiration and growth of *Eucalyptus* plantation in water-limited conditions. J Hydrol 13:1–15

Chatterji PC (1985) Impact of human activities on water resources of arid zone. Proc Indian Nat Sci Acad 51:135–141

Chatterji PC, Saxena SK (1988) Canal irrigation in arid zone of Rajasthan. In: Prakash I (ed) Desert ecology. Scientific Publisher, Jodhpur

Chhabra R, Thakur NP (1998) Lysimeter study on the use of biodrainage to control waterlogging and secondary salinization in (canal) irrigated arid/semi-arid environment. Irrig Drain Syst 12:265–288

Cloudsley-Thompson JL (1977) Man and biology of arid zone. Edward Arnold, London

Cramer VA, Thorburn PJ, Fraser GW (1999) Transpiration and groundwater uptake from farm forest plots of *Casuarina glauca* and *Eucalyptus camaldulensis* in saline areas of southeast Queensland, Australia. Agric Water Manage 39:187–204

Dash CJ, Sarangi A, Singh AK, Dahiya S (2005) Bio drainage: an alternate drainage technique to control water logging and salinity. J Soil Water Conserv (India) 4(3&4):149–155

Dash CJ, Bhattacharya AK, Sarangi A, Singh AK (2007) Salt uptake by *Acacia biflora* saplings from saline root zone. J Agric Eng 44(2):93–98

Delzon S, Loustau D (2005) Age-related decline in stand water use: sap flow and transpiration in a pine forest chronosequence. Agric For Meteorol 129:105–119

Denecke HW (2000) Concept note on biodrainage and bio-disposal systems. http://www.fao.org/iptrid/drainage/bio_drainage/concept_note_bio_drainage.htm

Gafni A (1994) Biological drainage-rehabilitation option for saline-damaged lands. Water Irri 337:33–36

GoR (2012) Basic statistics 2012. Directorate of economics and statistics. Government of Rajasthan, Jaipur, p 194

Greenwood EAN, Klein L, Beresford JD, Watson GD (1985) Differences in annual evaporation between grazed pasture and Eucalyptus species in plantations on a saline farm catchment. J Hydrol 78:261–278

Heuperman AF (1992) Trees in irrigation areas; the bio-pumping concept. Tree Nat Resour 34:20–25

Heuperman AF (1999) Hydraulic gradient reversal by trees in shallow water table areas and repercussions for the sustainability of tree-growing systems. J Agric Water Manage 39:153–167

Heuperman AF, Kapoor AS (2003) Biodrainage status in India and other countries. Indian National Committee on Irrigation and Drainage, New Delhi

Heuperman AF, Kapoor AS, Denecke HW (2002) Bio drainage- principles, experiences and applications. International Programme for Technology and Research in Irrigation and Drainage. Food and Agriculture Organization of the United Nations, Rome, p 89

Heuperman AF, Stewart HTL, Wildes RA (1984) The effect of *Eucalyptus* on water tables in an irrigation area in northern Victoria. In *Water Talk*, Rural Water Commission of Victoria, Vol 52

Hoffman GJ (1990) Leaching fraction and root zone salinity control. ASCE Manual & Report on Engineering Practice, No. 71

Hooja R, Srinivas V, Sharma G (1994). Waterlogging and salinity problems in IGNP, Rajasthan. National seminar on reclamation and management of waterlogging saline soils. CCS, Haryana Agricultural University, Hisar, Haryana, India

Hooja R, Srinivas V, Sharma G (1995). Water and salinity problems in IGNP, Rajasthan. In: Rao KVGK, Agarwal MC, Singh OP, Oosterbaan RJ (eds) Proceedings on national seminar on reclamation and management of waterlogging saline soils. India, CSSRI, Karnal, Haryana, India, pp 141–159

IWMI (2007) Study on groundwater externalities of surface irrigation transfers under National River Linking Project – Indira Gandhi Nahar Pariyojana (IGNP). Final Report. IWMI, New Delhi, p 97

Jacobsen T, Adams RM (1958) Salt and silt in ancient Mesopotamian agriculture. Science 128(3334):1251–1258

Jeet Ram, Dagar JC, Lal K, Singh G, Toky OP, Tanwar VS, Dar ISR, Chauhan MK (2011) Biodrainage to combat waterlogging, increase farm productivity and sequester carbon in canal command areas of northwest India. Curr Sci 100(11):1673–1680

Kapoor AS (2001) Biodrainage - a biological option for controlling waterlogging and salinity. Tata McGraw Hill Publishing Co. Ltd, New Delhi, 315 p

Kapoor AS (2014) Managing groundwater for irrigated agriculture- the relevance of biodrainage model. Rawat Publications, Jaipur, p 341

Kapoor AS, Denecke HW (2001) Biodrainage and biodisposal: the Rajasthan experience. Grid (IPTRID Network Magazine) 17:3–4

Lambert MJ, Turner J (2000) Commercial forest plantations on saline lands. CSIRO Publishing, Collingwood

Mann HS, Chatterji PC (1978) Effect of land use change on the groundwater conditions in canal irrigated area of western Rajasthan and their management. In: Proceedings of land and water management in the Indus Basin, pp 252–268

Mondal AK, Sharma RC (2011) Delineation and characterisation of waterlogged salt affected soils in IGNP using remote sensing and GIS. J Indian Soc Remote Sens 39(1):39–50

NAAS (2010) Degraded and wastelands of India – status and spatial distribution. National Academy of Agricultural Sciences, New Delhi

NIAB (1997) The Australian Centre for International Agricultural Research (ACIAR). For Newsletter, pp 1–4

Pessarakli M, Szabolcs I (1999) Soil salinity and sodicity as particular plant/crop stress factors. In: Pessarakli M (ed) Handbook of plant and crop stress, 2nd edn. Marcel Dekker, New York, pp 1–16

Prasad RN, Biswas PP (1999) Soil resources of India. In: Singh GB, Sharma BR (eds) 50 years of natural resource management. Indian Council of Agricultural Research, New Delhi

Rahmani AR (1989) The uncertain future of Desert National Park in Rajasthan. Environ Conserv 16(3):237–244

Jeet Ram, Garg VK, Toky OP, Minhas PS, Tomar OS, Dagar JC, Kamra SK (2007) Biodrainage potential of *Eucalyptus tereticornis* for reclamation of shallow water table areas in north-west India. Agroforest Syst 69:147–165

Schulz M (1994) Development of sustainable systems using irrigated salt tolerant plants for productive revegetation of saline land. Final report NRMS project V2112. Institute of Sustainable Irrigated Agriculture, Department of Agriculture, Tatura, Victoria, Australia

Shankarnarayan HS, Sarkar TK (1985) Integration of arid land irrigation with other resources development. In: Irrigated agriculture in arid areas. WAPCOS, New Delhi, pp 142–178

Sharma KD (1999) Rajasthan canal—a boon or environmental disaster? Survey Environ 99:101–105

Sharma BR, Rao KVGK, Sharma G (2009) Groundwater externalities of large surface irrigation transfers: lessons from Indira Gandhi Nahar Pariyojana, Rajasthan, India. In: Strategic Analyses of the National River Linking Project (NRLP) of India Series 5. Proceedings of the Second National Workshop on Strategic Issues in Indian Irrigation, New Delhi, India, 8–9 Apr 2009, International Water Management Institute (IWMI), Colombo, Sri Lanka, pp 107–121. doi:10.3910/2010.202

Shrivastava M, Sharma IK, Sharma DD (2013) Ground Water Scenario in Indira Gandhi Nahar Pariyojna (IGNP) in parts of Sri Ganganagar, Hanumangarh, Churu, Bikaner, Jaisalmer Jodhpur and Barmer Districts, Rajasthan. Memoir Geological Soc India 82:16–35

Singh S, Kar A (1991) Morphological characteristics of land forms and associated problems. In: Abrol IP, Venkateswarlu J (eds) Prospects of Indira Gandhi Canal Project. Publication and Information Division, ICAR, New Delhi

Soni ML, Beniwal RK, Garg BK, Tanwar SPS, Burman U, Yadav P, Yadava ND (2012) Above ground biomass production and carbon storage during restoration of water logged saline soils of IGNP area. Crop Improv (Special Issue): 555–556

Stirzaker RJ, Cook FJ, Knight J (1999) Where to plant trees on cropland for control of dryland salinity: some approximate solution. Agric Water Manag 39:115–133

Tewari VP, Arrawatia ML, Kishan Kumar VS (1997) Problems of soil salinity and waterlogging in Indira Gandhi Canal Area of Rajasthan State. Ann Biol 13:7–13

Zhang L, Dawes WR, Slavich PG, Meyer WS, Thorburn PJ, Smith DJ, Walker GR (1999) Growth and ground water uptake responses of Lucerne to changes in ground water levels and salinity: lysimeter, isotope and modelling studies. Agric Water Manage 39:265–282

Agroforestry to Rehabilitate the Indian Coastal Saline Areas

J.C. Dagar and P.S. Minhas

Abstract

About 35 % of Indians live within 100 km of the Indian coastline measuring 7517 km consisting of parts of the mainland India, Andaman-Nicobar Islands and Lakshadweep Islands. The coastal and island ecosystems have a vide variability in climate and topographical and edaphic conditions. These support diverse cultivated crops as well as natural vegetation ranging from tropical rainforests to coastal mangroves. The area is environmentally disadvantaged both on anthropogenic activities and weather adversities. Paddy is the predominant crop except the plantation trees mainly in homesteads. The soil salinity and waterlogging problems arise with intrusion of seawater, and these are expected to become severe with rise in sea level due to global warming. However, these ecosystems offer immense scopes and opportunities of increasing productivity through integration of agroforestry with livestock and aquaculture particularly in mangrove areas. Some of the possible strategies for their reclamation and management through sustainable agroforestry systems are discussed. These systems should further improve the livelihood security of the coastal population.

Introduction

Agricultural prosperity in India is evidenced by almost fivefold increase in food production from 50.8 in 1950–1951 to 247.6 million tons in 2012–2013 (ICAR 2014). This needs to be enhanced further to cope up with the increasing demand of population and that too in the wake of challenges like climate change, degradation of land and water resources, shrinking of land and scarcity of water available for agriculture. Already about 121 Mha land in India is suffering with various

J.C. Dagar (✉)
Central Soil Salinity Research Institute,
Karnal 132 001, Haryana, India
e-mail: dagarjc@gmail.com

P.S. Minhas
National Institute of Abiotic Stress Management,
Baramati, Pune 413115, Maharashtra, India
e-mail: minhas_54@yahoo.co.in

J.C. Dagar, P.S. Minhas (eds.), *Agroforestry for the Management of Waterlogged Saline Soils and Poor-quality Waters*, Advances in Agroforestry 13, DOI 10.1007/978-81-322-2659-8_8

kinds of degradation (NAAS 2010). Agroforestry is probably the only means for getting the desired tree cover and is not only a technique for rehabilitation of degraded lands but is also a source of renewable energy resource and a means of reducing the risk of environmental degradation and climate change. The livelihood security through agroforestry and its potential in meeting basic needs (food, fuel, fodder, timber and employment generation) are well known. The importance of agroforestry is more relevant in coastal areas which are ecologically more disadvantaged and at a greater risk due to the anthropogenic activities and weather adversities. Trees have to play crucial role in conserving and protecting vast coast line particularly by mangroves and associate halophytes. About 35 % of Indians live within 100 km of the Indian coastline measuring 7517 km consisting of parts of nine states on mainland, Andaman-Nicobar Islands and Lakshadweep Islands (India 2013). The coastal and island ecosystems have a wide variability in climate, topographical and edaphic conditions, and they support diverse cultivated crops as well as natural vegetation ranging from tropical rainforests to coastal mangroves. The coastal ecology offers immense opportunities of commercial use not only for wide varieties of fish and fruit and vegetable crops but also plantation crops, spices and medicinal and aromatic plants, which may be blended in multistorey plantation-based farming systems and home gardens. The extent of land degradation due to waterlogging and salinity and rehabilitation of these areas through farming systems and agroforestry approach are discussed in this chapter.

Extent of Waterlogging and Salinity

Broadly the entire agroclimatic zones of the coastal region of India come under four agro-ecological regions and are spread across nine states and three union territories. The west coast covers parts of five states, namely, Gujarat (13 districts), Maharashtra (5 districts), Goa (2 districts), Karnataka (3 districts) and Kerala (14 districts), and union territories (Pondicherry, Daman and Diu and Dadar and Nagar Haveli); and east coast covers parts of four states, namely, Tamil Nadu (13 districts), Andhra Pradesh (9 districts), Orissa (11 districts) and West Bengal (3 districts). Two groups of islands—Andaman-Nicobar and Lakshadweep islands—are situated in the Bay of Bengal and the Arabian Sea, respectively. Entire coastal region of India except north Gujarat receives a normal annual rainfall of more than 1000 mm; in the west coast it is as high as 2500 mm. However, more than 80 % of the annual rainfall is received during June–September except coastal Tamil Nadu, which receives maximum rainfall during October and November. The major source of water in coastal regions is rainfall and groundwater. Inefficient rainwater management and over-exploitation of groundwater for irrigation in the west and south coastline have caused acute shortage of water. On the other hand, the major part of the east coast suffers from severe waterlogging problems due to flat topography and low hydraulic conductivity. All along the coastline, groundwater quality is being threatened due to the surface or subsurface seawater intrusion. In coastal regions, the areas affected by salinity and waterlogging are about 3 and 7 Mha, respectively (Table 1). The total area of groundwater within 3 m has been reported to be 7016 thousand ha, out of which 1.4 % is most critical (<1 m), 24.4 % critical (1–2 m) and 74.2 % less critical (2–3 m) for post-monsoon session during the year 2003–2005. However, during post-monsoon, this area increased to 2.25 times (15.76 Mha), out of which 8.5 % to be most critical, 39.9 % critical and 51.7 % less critical (RRSSC & CWC 2009).

Under the threat of climate change, increased levels of salinity arising from sea level rise and coastal flooding will pose a serious problem to the rural inhabitants of the coastal area. The predicted negative impacts of climate change are likely to bring new challenges in addition to magnifying existing problems, particularly in the areas like Sundarbans, where rural community already has limited capacity to adapt to these changes (Sujana-Dhar 2011). In coastal areas, the excess of water is stagnated during rainy season, and draining is very slow leading to waterlogging.

Table 1 Extent of coastline, mangrove areas, saline and waterlogged soils of maritime states and union territories of India

State/union territory	Coastal length (km)[a]	Mangrove area (000 ha)[b]	Saline area (000 ha)[c]	Waterlogged area (000 ha)[d]
West coast region				
Gujarat	1214.7	105.8	714	2602.3
Maharashtra	652.6	18.6	64	931.2
Karnataka	280.0	0.3	86	369.9
Kerala	569.7	0.6	26	116.4
Goa, Daman and Diu	160.5	2.4	18	5.3
Pondicherry	30.6	0.1	1	
East coast regions				
West Bengal	157.5	215.5	820	151.7
Orissa	476.4	22.2	400	754.7
Andhra Pradesh	973.7	35.2	276	1379.6
Tamil Nadu	906.9	3.9	100	704.7
Island regions				
Andaman-Nicobar	1962.0	61.7	15	NA
Lakshadweep	132.0	NA	Na	NA
Total	7516.6 (say 7517)	466.3	2520+466=2986	7015.8

Source:
[a]India (2013)
[b]FSI (2011)
[c]Yadav et al. (1983)
[d]RRSSC & CWC (2009)

Further, during summer season, the low rainfall leads to exploitation of groundwater for agriculture. Over a period of time, the groundwater table has decreased resulting in intrusion of seawater into the groundwater. Subsequently, the salinized groundwater used for irrigation has led to the accumulation of salt on the surface of the soil. The intrusion of seawater towards mainland/inland may vary from depth of water table, amount and distribution of rainfall. The distance may not be uniform throughout the coast line and vary due to climatic factors, topography, amount and distribution of rainfall, depth of water table, cropping intensity with use of groundwater etc. In Kerala, the lands are mostly waterlogged due to the lagoons, estuaries and wetlands. The intrusion of seawater is more frequent along east coast. However, certain areas where the estuaries and other forms of land with negative mean sea level (MSL) are the possible reasons for seawater intrusion in western coast.

Natural Vegetation and Cropping Systems

The coastal regions are very rich in biodiversity both of flora and fauna. The natural vegetation is composed of evergreen forests, semi-evergreen forests, deciduous elements and littoral species including mangroves. Tree species which are commonly seen include *Acacia auriculiformis*, *Acacia planiformis*, *Achras zapota*, *Ailanthus triphysa*, *Albizia amara*, *Anacardium occidentale*, *Annona ramosa*, *A. squamosa*, *Artocarpus chaplasha*, *Bombax ceiba*, *Borassus flabellifer*, *Clerodendron inerme*, *Cocos nucifera*, *Dalbergia latifolia*, *Elaeis guineensis*, *Emblica officinalis*, *Erythrina indica*, *Eucalyptus tereticornis*, *Gliricidia sepium*, *Hernandia peltata*, *Hibiscus tiliaceus*, *Jatropha curcas*, *Leucaena leucocephala*, *Mangifera indica*, *Morinda citrifolia*, *Moringa oleifera*, *Musa paradisiaca*, *Pandanus leram/fascicularis*, *Pongamia pinnata*, *Psidium guajava*, *Punica granatum*, *Ricinus communis*, *Salvadora persica*, *Samanea saman*, *Sesbania grandiflora*, *Tamarindus indica*, *Thespesia populnea*, *Vitex negundo*, *V. trifoliata*, etc. Quite good numbers of trees are fruit bearing.

Among mangroves *Acanthus ilicifolius*, *A. volubilis*, *A. ebracteatus*, *Aegialitis rotundifolia*, *Aegiceras corniculatum*, *Avicennia marina*, *A. officinalis*, *Bruguiera gymnorrhiza*, *B. parviflora*, *B. cylindrica*, *B. sexangula*, *Ceriops tagal*, *C. decandra*, *Cynometra ramiflora*, *C. iripa*, *Excoecaria agallocha*, *Heritiera fomes*, *H. littoralis*, *Kandelia candel*, *Lumnitzera racemosa* (*L. littorea* in Andamans), *Nypa fruticans*, *Phoenix paludosa*, *Rhizophora apiculata*, *R. mucronata*, *R. stylosa*, *R. lamarckii* (only in Andamans), *Scyphiphora hydrophyllacea*, *Sonneratia alba*, *S. apetala*, *S. caseolaris*, *S. ovata*, *Xylocarpus gangeticus*, *X. granatum* and *X. moluccensis* and associated species such as *Acrostichum aureum*, *Barringtonia asiatica*, *B. racemosa*, *Caesalpinia bonduc*, *C. crista*, *Calophyllum inophyllum*, *Casuarina equisetifolia*, *Cerbera floribunda*, *Cordia subcordata*, *Cycas rumphii* (only in Andaman-Nicobar Islands), *Dalbergia monosperma*, *Derris scandens*, *Dolichandrone spathacea*, *Erythrina indica*, *E. variegata*, *Guettarda speciosa*, *Hernandia peltata*, *Hibiscus tiliaceus*, *Intsia bijuga*, *Licuala spinosa*, *Manilkara littoralis*, *Mimusops elengi*, *Morinda citrifolia*, *Ochrosia oppositifolia*, *Pandanus* spp., *Pongamia pinnata*, *Scaevola taccada*, *Premna corymbosa*, *Syzygium samarangense*, *Tabernaemontana crispa*, *Terminalia catappa*, *Thespesia populnea*, *Tournefortia ovata* and *Vitex negundo* are common. Many climbers and epiphytes such as *Asplenium nidus*, *Canavalia maritima*, *Clitorea ternatea*, *Dischidia bengalensis*, *D. nummularia*, *Finlaysonia maritima*, *Hoya parasitica*, *Hydnophytum formicarum* and many orchids are conspicuous throughout mangrove stands. *Ipomoea pes-caprae*, *Launaea sarmentosa*, *Wedelia biflora*, *Mapania cuspidata*, *Scirpus littoralis*, *Triumfetta repens* (only seen in Nicobars), *Canavalia maritima*, *Clitorea ternatea* and *Vigna marina* (wild pulse) are quite frequent on sandy beaches. *I. pes-caprae* shows its allegiance due to its evergreen shining leaves and violet flowers and found trailing over sandy beaches throughout coastal areas.

Mangrove forests bear net of aerial prop roots and thick stand (if undisturbed) along coast (Fig. 1) protecting the coastal areas from cyclonic tidal waves. These also provide an important habitat for young stages of commercially important fish and shrimps, breeding grounds for fish, shellfish, turtles and home for variety of wild life (Dagar et al. 1991; Dagar 1995b, 1996, 2003; Dam Roy 2003, 2015; Goutham-Bharathi et al. 2014).

The major cropping/agroforestry systems in these areas are coconut (*Cocos nucifera*) and *Casuarina*-based and homestead gardens. In the Bay Islands areca nut (*Areca catechu*, *A. triandra*), banana (*Musa* spp.) and *Artocarpus* spp. are also common. Screw pine (*Pandanus* spp.) and cashew nut (*Anacardium occidentale*) are quite frequent in Andamans and along Orissa coast. The rice during rainy season and vegetables and pulses during post-monsoon season are major crops. Fish cultivation along rice is common practice. Farming systems involving livestock, aquaculture, poultry, plantations, vegetables and rice cultivation are now frequently adopted. Despite of sufficient rainfall, the post-monsoon season faces scarcity of water due to non-harvesting of water during rainy season. Modern technologies (discussed later) of *in situ* and *ex situ* water harvesting are being developed.

Afforestation/Agroforestry Systems

Agroforestry systems provide alternatives for restoring soil health and amelioration of salt-affected coastal soils for their productive use. Location-specific multipurpose tree species (MPTs) can be grown mixed or separately planted for various purposes such as fruits, vegetables, wood, fodder, soil protection and soil amelioration. Amelioration of waterlogged salt-affected coastal soils with the domesticated agroforestry trees will reduce the pressure on the productive lands to fulfil the food needs of the growing population and environmental concerns. Agroforestry is the only alternative to meet the country's target

Fig. 1 Mangroves protecting shores from natural disasters like cyclones

of increasing forest cover besides improving the livelihood of rural people. A detailed account of many tree and shrub species suitable for planting in soils of different degrees of salinity is given in chapter "Global Perspectives on Agroforestry for Management of Salt-affected Soils" (Dagar and Minhas) of this publication. Mounding is essential on sites prone to waterlogging for establishing seedlings of trees and shrubs, with mounds positioned over the rip lines. In Western Australia, the seedlings planted in a 0.5–0.9 m wide trough between mound ridges and with a mound height of between 0.5 and 1 m had the highest survival rates. This mound design is preferable to a single ridged or flat-top mound. A single-ridge mound tends to shed rainfall and wet up from the water table or the furrows beside the mound, causing salt to accumulate in the mound. In contrast, the trough of a double-ridge mound tends to collect rainfall, favouring percolation and leaching of salts from the seedling root zone (Ritson and Pettit 1992). In areas underlain by palaeochannel, irrigated tree plantations are unlikely to have a significant effect on the local water table but may help to reduce subsurface drainage from the palaeochannel system if the leaching fraction beneath the trees is less than for other crop types. An additional advantage may be gained if the trees can derive a portion of their water requirement directly from the water table (Ayars et al. 2005). Groundwater pumping tests by Smith et al. (2005) further confirm the water table drawdowns beneath trees, planted over the palaeochannel system.

The drainage problems of coastal areas are related essentially to monsoon, caused by high rainfall and inadequate field tertiary and main drainage. Discharge from inland catchment also adds to the problem. Waterlogging and flooding conditions in most of the coastal lowlands are such that essentially only rice is grown during the wet (*kharif*) season. Trees can also influence groundwater tables by consuming groundwater 'directly'. This can occur in two ways: (a) extraction of water from the capillary fringe and (b) extraction of water from within the saturated zone. The capillary fringe is the area immediately above the water table in which groundwater is drawn up by capillary forces. The capillary fringe may be saturated close to the water table, but its water content decreases with increasing distance from the water table and is consequently well aerated. As such, trees with roots penetrating to the capillary fringe can readily utilize this water. Transpiration of the capillary water leads to a continuous movement of groundwater into the capillary fringe. Groundwater extraction can take place by trees specifically adapted to transpiring water when their roots are in saturated soil. Greenwood (1986) cited experimental evidence of Van Hylckama (1974) which showed decreasing water uptake of the phreatophyte *Tamarix pentandra* with increasing depth. Reasons for groundwater uptake decreasing with depth include decreasing root density, increasing soil bulk density (affecting root penetration), decreased oxygen level and greater gravitational potential difference (i.e. more effort required to lift the water against gravity).

The presence of 'ultrafilter' in roots of mangrove species of family Rhizophoraceae enables only selective absorption of ions. They may retain a low internal salinity by means of salt-excluding mechanisms in the roots. In this type, sodium and chloride concentrations are higher in xylem sap and do not reach the metabolic cellular environment. Another mechanism of salt regulation in mangroves is salt excretion. In species of *Avicennia*, *Aegialitis* and *Acanthus*, NaCl concentration in the excreted solution exceeds that of seawater. The same holds true for the halophytic grass *Aeluropus lagopoides* and tree *Tamarix articulata*. The ionic excretion of salt glands is constituted mainly of sodium and chloride. Under high-temperature conditions (25–30 °C), excretion rate is usually accelerated and the excretion process is dependent on light. Tree plantations in Sundarbans of West Bengal can dramatically lower the water table through decreasing local groundwater replenishment and increasing transpiration. In areas where the aquifer has small transmissivity, tree plantations may be effective at lowering the local water table, but this advantage may not be sustainable in the long term due to salt accumulation in the root zone (Sujana-Dhar 2011). Depending upon the need and situation, the following methods of planting have been adopted:

Row planting: Usually shelterbelts having 50–100 rows of trees are planted to minimise the wind speed.

Line planting: Line planting in coastal islands helps in reducing the wind force and stabilising the sand dunes.

Cluster planting: The latest method which is becoming popular with the farmers is cluster planting. In this method 2–3 seedlings are planted at each spot at a wider spacing of 2 m or 3 m.

On the coastlines wherever sulphur-containing sediments accumulate in tidal marshes or swamps or highly degraded mangrove ecosystems, acid sulphate soil formation takes place. These formations are quite frequent in Peninsular India and Andaman-Nicobar Islands, and rice is a major crop in these areas with very low productivity. However, under such conditions, it also suffers from an excess of water-soluble iron and particularly of aluminium, and toxic effect of these is a consequence of the strongly acidic pH. The application of lime mitigates the adverse and harmful effects on plants, but it is a costly preposition as the requirement of lime is very high. Therefore, alternative land use systems (agroforestry systems) need to be evolved.

Plantations in the Areas Closer to the Sea

The areas lying closer to the sea are flooded regularly with seawater and, therefore, have high salinity. The tidal areas protected against high wind velocity and waves of high intensity (as in case of many creeks, lagoons and estuaries) form a suitable situation for cultivation of mangroves. For cultivation of mangroves, we need seedlings of appropriate size. Many mangrove genera such as *Rhizophora*, *Ceriops*, *Aegialitis*, *Aegiceras*, *Bruguiera*, *Kandelia* and *Camptostemon* are viviparous in nature, and seed germinates when the fruit is intact to branch. The radical falls when mature and develops roots when it touches the muddy substratum. These mature radicals can be collected from the mangrove stands and planted directly in mangrove habitats or along protected shores as nursery in polybags. When seedlings are of proper size, these may be planted directly in tidal zone. Earlier attempts at restoration, regeneration and afforestation have been undertaken with direct planting of mangrove propagules or seeds (Karim et al. 1984; Hamilton and Snedaker 1984; Kogo 1985; Kogo et al. 1986). Untawale (1993) attempted rehabilitation of denuded mangrove areas via raising mangrove nursery in tidal zone on slightly raised platforms and supporting these with split bamboos to prevent possible drifting of polybags. Better results were obtained in areas with freshwater influence. After observing critically the natural zonation pattern (Dagar 1982, 1996, 2003; Dagar et al. 1991, 1993; Dagar and Singh 1999), it was recommended that *Rhizophora* may be planted facing the sea followed by belts of *Bruguiera*, *Kandelia*, *Ceriops*, *Avicennia*, *Sonneratia* and *Excoecaria* in middle zone towards land and species of *Ceriops*, *Aegiceras*, *Aegialitis* and *Camptostemon*, grown towards border with associate mangroves such as *Thespesia populnea*, *Pongamia pinnata*, *Terminalia catappa* and *Calophyllum inophyllum*. Species such as *Rhizophora mucronata, R. apiculata*, *Avicennia marina*, *A. officinalis*, *Bruguiera gymnorrhiza* and *B. parviflora* can be grown in highly saline sandy-clay substratum; while species such as *R. stylosa*, *Ceriops tagal*, *Aegiceras corniculatum* and *Sonneratia alba* are found more predominantly on silty clay substratum. *Sonneratia caseolaris*, *Xylocarpus granatum* and *Excoecaria agallocha* prefer low salinity and silty substratum and found distributed towards land. *Avicennia marina* is most tolerant to biotic stress and may be planted widely in all kinds of mangrove habitats. *Nypa fruticans*, a mangrove palm, is more predominant in muddy substratum along creeks. This can be propagated from suckers. *Terminalia catappa* (coastal almond), *Pandanus* spp., *Calophyllum inophyllum*, *Salvadora persica* and *Pongamia pinnata* are useful oil-yielding trees and can be commercially explored in areas bordering mangroves, but their nursery cannot be raised in tidal zone. The natural zonation pattern of mangrove stands of the area gives perfect understanding where to plant a particular species. The natural zonation pattern of Andaman-Nicobar Islands studied by Dagar (1982), Dagar et al. (1991) and Dagar and Singh (1999) is given in Fig. 2.

The alcohol yielding *Nypa fruticans* is more frequent in muddy creeks of Andamans and Sunderbans and may be cultivated as commercial crop. Several projects have later been implemented for rehabilitation of mangroves in Indian subcontinent, and quite sizeable area has been planted with mangroves particularly in Korangi-Phitti creek and Indus Delta in Pakistan and Goa and Pichavaram in Tamil Nadu. Some denuded areas have been rehabilitated with suitable mangrove species along Goa and Tamil Nadu coasts of India. The sandy beaches along Orissa coast have been planted successfully with *Casuarina glauca*, *C. equisetifolia*, *Pandanus*, cashew nut (*Anacardium occidentale*), coconut and *Acacia auriculiformis* (Dagar 2014). Due to *tsunami* land got elevated in North Andaman and in many uplifted mangrove areas due to nonavailability of tidal water mangroves have died and soils have turned acidic. Dam Roy and Krishnan (2005) reported critical analysis on effects of *tsunami* at different locations of Andamans. *Avicennia marina* and *Sonneratia alba* were least affected showing their adaptability to cyclones. To rehabilitate the uplifted areas for the choice of species, we need careful analysis of soil and natural succession of vegetation. Species bordering mangroves such as *Clerodendron inerme*, *Thespesia populnea*, *Terminalia catappa*, *Salvadora persica*, *Casuarina glauca*, *Pandanus* spp. and *Pongamia pinnata* may find a place.

Brackish Water Aquaculture in Mangrove Areas or behind Mangroves

The seawater near mangroves is resource for a wide variety of species of pelagic, demersal and oceanic fish. The pelagic resources include anchovies, sardines, mackerels, carangids, ribbon, seer, neritic tunas and barracuda; the demersal resources constitute perches, silver bellies,

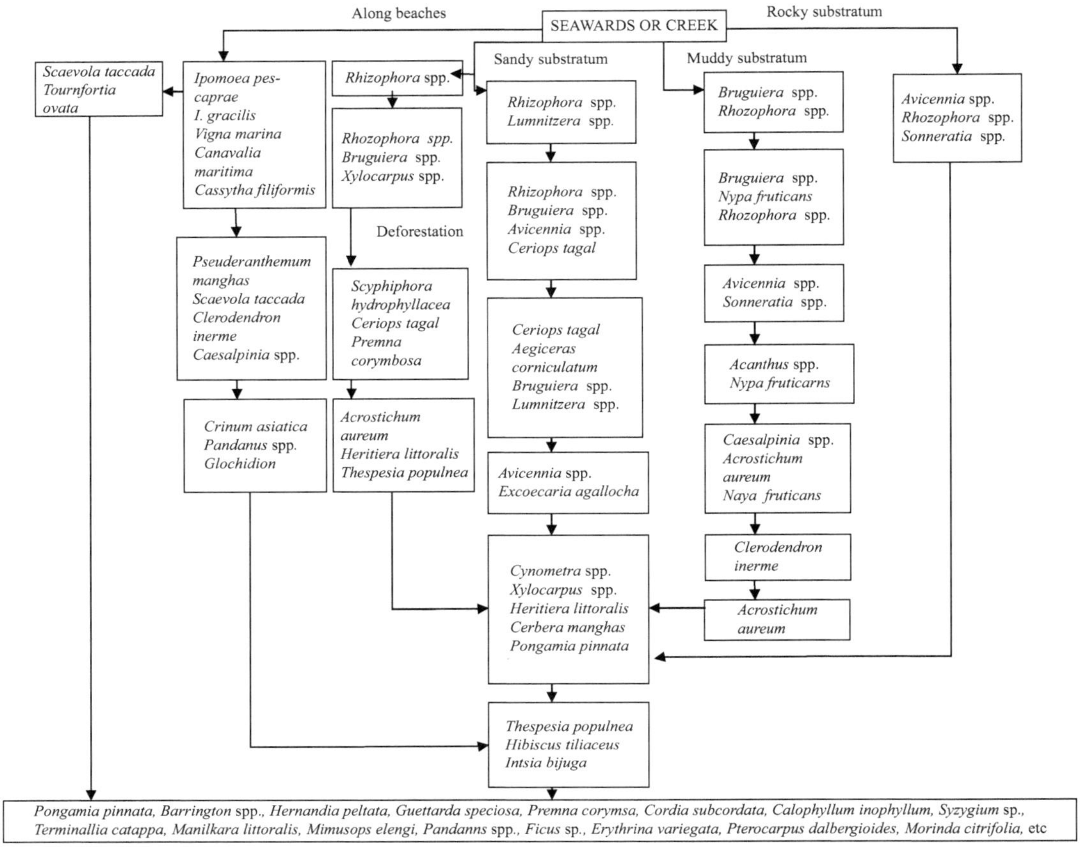

Fig. 2 Zonation pattern of mangrove forests in Andaman-Nicobar Islands

pomfrets, scads, sciaenids, nemipterids, shrimps and lobsters; and the oceanic resources are skipjack, tuna, sail, marlin and sword fish, and pelagic shark (Dam Roy et al. 2015). High-value crabs, prawns and lobsters are in plenty. These resources are dwindling as mangrove stands are facing tremendous anthropogenic pressure. Reclamation of backwater areas for aquaculture acid sulphate soils is constrained due to acidification of exposed mud with drying process. Any dyke construction from this mud thus results in acid leaching with rainwater to reduce pH resulting in stress or even death to the cultured stock. The iron carried into pond water is oxidized and later precipitated out as insoluble ferric hydroxide that clogs the gill filaments of aquatic organisms (fish) leading to asphyxiation (Dam Roy et al. 2005). The iron in pond interferes with smooth osmotic exchange of gases between organisms and water resulting in poor excretion. The carbonates and bicarbonates in pond water deplete as they react with acid resulting in soft shelling and poor moulting response of shell fish. These chemical reactions also negatively affect composition of bacteria-plankton–meiobenthos-microphytes affecting the pond food chains. The reclamation of saline acid sulphate soils for brackish water aquaculture can be achieved only through a model, which nullifies or reduces the abovementioned problems. Shrimp or fish farming is one of the viable commercial alternatives to agriculture in these areas. The development of brackish water aquaculture especially shrimp farming has been one, and it has been found to have substantial economic gains.

Aquaculture keeping mangroves intact is most feasible and sustainable option for promotion of aquaculture in inundated areas. In a preliminary study, mullet, prawn tilapia and fish culture could be made feasible and by connecting these ponds with brackish water behind mangroves particularly in association with *Avicennia* communities. The shrimp fry (*Penacus* and *Metapenaeus*) could be produced with rice. *Nypa fruticans*, a mangrove palm (frequent in Sunderbans and Andaman-Nicobar Islands), is cultivated in the Philippines and Bangladesh as a commercial crop and can have potential for Indian conditions also. Vannucci (1989) reported 15,000 l ha^{-1} $year^{-1}$ alcohol production by this palm. As it is a predominant palm in Sunderbans, the feasibility of its commercial cultivation must be explored which will generate employment in poor coastal population. Thus, fish/prawn culture, keeping mangroves intact, is very viable and useful system particularly in areas where freshwater streams merge with seawater. Bee humps are natural in Sunderbans; hence, beekeeping and duck-based poultry can be blended with fish culture associated with mangroves particularly along creeks. In the post-*tsunami* scenario, in South Andaman alone, due to the subduction of the land by about 1.25 m, the level of submergence due to tidal influence has also increased. Approximately 4000 ha areas of agricultural farmlands was submerged, out of which 630 ha suited for coastal aquaculture. On dykes of ponds, coconut and areca nut can successfully be cultivated for extra income.

Central Island Agricultural Research Institute (CIARI) took an initiative in collaboration with the Forest Department to look into the mangrove ecosystem as a livelihood source for the coastal fisherman, and a mangrove-based agro-aqua farming system was developed and demonstration adjacent to the creek within the brackish water farm complex of the institute in Sippighat, Port Blair (Dam Roy 2015). It demonstrates sustainable aquaculture where stocking and harvesting will be done round the year without supplementary feed, and the same is advocated for the areas adjacent to the mangroves in Andaman. Mangroves contribute about 3.7 Mg of litter ha^{-1} $year^{-1}$ (Dam Roy 2015), which decompose into detritus and made available as nutrients by microorganisms to marine animals. In protected stands, the litter quantity may exceed this amount as Dagar and Sharma (1991, 1993) reported 8.4 to 11.0 $Mgha^{-1}$ $year^{-1}$ litter fall in some protected mangrove stands of Andamans. Further, Dam Roy (2015) reported that mangroves contribute 18–42 Mg C $ha^{-1}year^{-1}$, approximating the distribution of the tropical rainforests and 10 times higher than primary production in the open ocean. Dagar and Singh (1999) gave details of structure, standing biomass, productivity of mangrove water and nutrient component of leaves of different mangrove species of Andamans. Ghoshal et al. (2006) reported 426 and 397 μg g^{-1} microbial biomass carbon in surface and subsurface soil at relatively undisturbed mangrove site in Andamans showing richness of microbial activities in the substratum. Besides other animals, mangrove waters are very rich in fish fauna. Rajan and Dam Roy (2003) reported 239 species of fish from mangrove ecosystem of Bay Islands, 68 % being of commercial importance while the rest of other types. This shows the potential of fish culture keeping mangroves intact.

Agroforestry in Inundated Areas Due to Seawater Flooding

The 2004 Indian Ocean *tsunami* severely damaged the coastal ecosystems of India. Velmurugan et al. (2015) conducted experiments to assess the impact of bunding and broad bed and furrow (BBF) systems in restoring the productivity of inundated areas of South Andaman (Fig. 3). They found that bunding of agricultural land leached out the salts by impounding of rainwater with significant reduction in electrical conductivity (ECe), sodium absorption ration (SAR) and exchangeable ions (Na^{+}, Ca^{2+}, Mg^{2+}, Cl^{-} and SO_4^{2-}). The BBF system installed in low-lying waterlogged areas improved the drainage of the beds, harvested the rainwater (4476 m^3ha^{-1}), prevented entry of tidal and runoff water into the furrow and reduced the overall salinity. In addition microbial biomass C was significantly improved (193–210 mg kg^{-1} soil) and the soil under BBF systems adequately drained. The

Fig. 3 Broad bed furrow system in Andamans (Source: CIARI, Port Blair)

depth of submergence (R^2=0.798) and soil salinity (R^2=-0.787) were found correlated with rainfall amount. Consequently BBF systems enabled a higher cropping intensity (218 %), increased fish productivity (INR 47.36 m^{-3}) and enhanced employment generation (213 man-days). These results can easily be upscaled elsewhere at least in Southeastern Asia.

By adopting this technology, farmers of Andamans could earn INR 1,17,532 per ha by growing vegetables [okra (*Abelmoschus esculentus*)+*Amaranthus* sp. - okra)] in beds; brinjal (*Solanum melongena*)+*Moringa oleifera*+banana (*Musa paradisiaca*) in border areas; and rice+*Azolla*+fish in furrows (Dam Roy et al. 2015). Trees such as *Casuarina*, noni (*Morinda citrifolia*), *Moringa oleifera*, coconut (*Cocos nucifera*), banana (*Musa* spp.) and areca nut (*Areca catechu*, *A. triandra*) can successfully be grown on bunds, vegetables and pulses on broad beds and fish and rice in furrows. Farmers in Andamans have four types of land holdings—hilly, slopping hilly uplands, medium upland valley and low-lying valleys. Low-lying valleys may be cultivated following BBF system; medium upland valleys may be planted with banana, coconut, areca nut and spices like clove (*Syzygium aromaticum*) or cinnamon (*Cinnamomum zeylanicum*) and black pepper (*Piper nigrum*) as climber; slopping hilly uplands may be utilized growing fodder grasses and legumes, while hilly lands may be retained with forest trees and planted with multipurpose trees (Dagar 1995a).

Need-based integrated land improvement approach comprising of six different methods, viz. broad bed and furrow, rice-fish cultivation, three-tier farming, farm pond, paired bed and furrow and pond-nursery systems, was developed and implemented in degraded coastal areas, and more than 500 farmers were benefited, which led to the crop diversification and livelihood security in islands.

Different land-shaping techniques for improving drainage facility, rainwater harvesting, salinity reduction and cultivation of plantation and crops and fish (freshwater and brackish water fish) for livelihood and environmental security were tested on about 400 ha degraded and low-productive land in disadvantaged areas in Sundarbans region of Ganges delta (West Bengal) and *tsunami*-affected areas in Andaman-Nicobar Islands covering 32 villages in 4 districts (South 24 Parganas and North 24 Parganas districts in West Bengal and South Andaman and North and

Fig. 4 Land shaping for deep furrow and high-ridge cultivation in coastal saline areas of West Bengal (Source: CSSRI, Karnal)

Middle Andaman districts in Andaman-Nicobar Islands) during 2010–2014. The soil in the study area was affected by high level of soil salinity (ECe up to 18 dS m^{-1}) and water salinity (EC up to 22 dS m^{-1}) that limits the choice and options of growing crops in the area. The following land-shaping technologies were tried on farmers' fields in coastal and islands areas (Burman et al. 2013):

Land shaping for deep furrow and high-ridge cultivation: The 50 % of farm land was sloped into alternate furrows (3 m top width × 1.5 m bottom width × 1.0 m depth) and ridges (1.5 m top width × 3 m bottom width × 1 m height). The ridges remained relatively free from drainage congestion and low in soil salinity build-up. These could be successfully used for raising plantations (fruits) or vegetables during both *rabi* and *kharif* and furrows for rainwater harvesting (to be used as life-saving irrigation in *rabi* season) and cultivation of rice and fish along with remaining original field. During dry season, the remaining field was also used for cultivation of low-water-requiring crops.

Land shaping for shallow furrow and medium ridge cultivation: About 75 % of the farm land was shaped into furrows (2 m top width × 1 m bottom width × 0.75 m depth) and medium ridges (1 m top width × 2 m bottom width × 0.75 m height) with a gap of 3.5 m between two consecutive ridges and furrows (Fig. 4). In wet season, the furrows could be used for rice and fish culture with rest of the field and rainwater harvesting. In dry season these could be used for rice cultivation. The ridges are planted with fruit trees or cultivated with vegetables or pulses throughout the year. The remaining original land could be used for low-water-requiring crops.

Land shaping for farm ponds: The 20 % of farm land was converted into on-farm reservoir (OFR) for in situ conservation of excess rainwater used during dry season, supplemental irrigation in *kharif* and freshwater aquaculture. The dugout soil was used to raise land to be used for crop cultivation. The dykes of the pond may also be planted with fruit trees.

Land shaping for paddy-cum-fish culture: Trenches of about 3 m width × 1.5 m depth

Table 2 Average depth of standing water (ADSW) and soil properties under different land situations created through land-shaping techniques in Sunderbans and Andaman group of islands

Land situation	ADSW (cm) in *kharif*	ECe (dS m^{-1})	pH	Org-C (%)	MBC (μg g^{-1} dry soil)	Avail N (kg ha^{-1})	Avail P (kg ha^{-1})	Avail K (kg ha^{-1})
Low land (without LS)	40–50	15.5	7.2	0.61	187.7	195.8	15.4	673.8
Low land (LS)	30–40	13.2	7.5	0.80	244.0	234.0	17.1	628.4
Medium land (LS)	15–20	7.3	7.4	1.10	279.0	238.1	18.9	480.3
High land (LS)	0	6.6	7.3	1.20	280.5	251.7	22.4	430.5

Source: Burman et al. (2015)
LS land-shaping technique

were dug around the field with a ditch of 6 m × 6 m × 3 m (depth) at one corner. The excavated soil was used for making dykes of about 3 m width and 1.5 m height to protect the fish cultivated with paddy. During wet season paddy and fish were grown on original land and vegetables/fruits on dykes. During summer low-water-requiring crops and vegetables were grown on dykes (in case fruits are not grown), and low-water-requiring crops on original land and lifesaving irrigation were given from water harvested in furrows. The original land in some cases was used for brackish fish culture. In that case at the end of summer season, brackish water was drained out along with monsoon rains, and the land was again used for paddy-*cum*-fish culture. Due to the creation of different land situations and following cultivation of crops round the year organic C, available N, P and K and biological activities (microbial biomass C) in surface soil were increased under land-shaping techniques compared to land without land shaping (Table 2).

About 1950 water storage structures were created under different land-shaping techniques, and 13,05,000 m^3 rainwater was harvested annually in these structures in the study area, and with this harvested rainwater, about 260 ha areas which were earlier under monocropping with rice due to shortage of irrigation water were brought under irrigation for growing multiple crops round the year. The cropping intensity increased up to 240 % from a base level value of 100 % due to implementing the land-shaping techniques (Table 3). These land-shaping techniques are very popular among the farmers of both Sundarbans and Andaman-Nicobar Islands as these have increased the employment and income of the farm family by manifolds compared to baseline value. Average net income per ha of farm land has been increased from INR 22,000 to 1,23,000 in Sundarbans and INR 22,400 to Rs 1,90,000 in Andaman-Nicobar Islands.

Homestead-based Integrated Farming Systems

In Bay Islands before penal settlement, Nicobari tribes indulged in community farming under *tuhet* system, wherein they practised a sustainable type of coconut based farming in a traditional way, which is prevalent even these days. Coconut trees are allowed to grow and propagate as wild forming a coconut forest. No spacing is followed; fertilizers are never applied. The interspaces (if any) are utilized for growing tuber crops. *Amorphophallus paeoniifolius*, *Coleus* spp., *Colocasia esculenta*, *Dioscorea alata*, *D. glabra*, *D. pentaphylla*, *D. bulbifera*, *Ipomoea batatas*, *Manihot esculenta*, *Maranta arundinacea*, *Tacca leontopetaloides* and *Xanthosoma sagittifolium* are common tribal tuber food in islands, which have several wild varieties/cultivars. Bhargava (1981, 1983), Dagar (1989), Dagar and Dagar (1989, 1991, 1999) and Dagar and Singh (1999)

Table 3 Enhancement in cropping intensity, employment generation and net income under different land-shaping techniques in Sundarbans and Andaman and Nicobar Islands

Land-shaping technologies	Cropping intensity (%)		Employment generation (man-days hh^{-1*} $year^{-1}$)		Net return (000 INR ha^{-1} $year^{-1}$)	
	Before	After	Before	After	Before	After
Farm pond	114[a], 100[b]	193[a], 200[b]	87[a], 8[b]	227[a], 22[b]	22[a], 10[b]	140[a], 148[b]
Deep furrow and high ridge	114[a]	186	87	218	22[a]	102[a]
Paddy-*cum*-fish	114[a], 100[b]	166[a], 200[b]	87[a], 8[b]	223[a], 35[b]	22[a], 24[b]	127[a], 148[b]
Broad bed and furrow	100[b]	240[b]	9[b]	48[b]	24[b]	212[b]
Three tier	100[b]	220[b]	10[b]	42[b]	30[b]	221[b]
Paired bed	100[b]	240[b]	9[b]	54[b]	24[b]	216[b]
Brackish water aquaculture	0/100	–	25[a]	100[a]	–	146[a]

Source: Burman et al. (2015)
Note: Costs and returns at current price of 2012–2013
*hh^{-1}: per household
[a]av. holding was 0.35 ha in Sundarbans
[b]av. holding of implementation was 0.20 ha in Andaman and Nicobar Islands

have reported the uses of several plants as food, shelter, canoe-making and drugs in the life of aborigines of Andaman-Nicobar Islands.

The community with their wise wisdom go for a shifting cultivation, but they do not burn the vegetation. The Nicobarese are of the habit of rearing pig (Nicobari pig is endemic to Nicobar group of islands), poultry (the bird is also endemic to Nicobars) and of recent cattle and goat. Keeping in view the dietary habits, limitation of land and markets for perishable produce, CIARI has developed homestead-based farming system for them comprising home garden (400 m^2), backyard poultry (20 nos.) and goat farming (2 nos.), and composting was developed for improving productivity. Consumption of fruits, vegetables, eggs and meat increased significantly, and a total of 95 man-days were generated from the system (Swarnam et al. 2014).

In other parts of islands, coconut and areca nut plantations are raised in coastal areas, while spices like black pepper, cinnamon, clove, Bay leaf, nutmeg and turmeric are cultivated in interspaces. In some areas forage grasses are cultivated. In low-lying areas, salt-tolerant rice varieties such as CARI Dhan-4, CARI Dhan-5 and CSR-36 are cultivated. *Gliricidia sepium* and *Sesbania* are cultivated for green manuring. *G. sepium* is also used as support for black pepper in many localities. Now farmers have also gained knowledge of mushroom cultivation on agricultural wastes and by-products available. On saline lands fruit species such as *Khaariphal* (*Ardisia solanacea* and *A. andamanica*), *Phoenix paludosa*, pond apple (*Annona glabra*) and noni (*Morinda citrifolia*) are grown successfully (Dam Roy et al. 2015). Experiments on grafting cultivated nutmeg (*Myristica fragrans*) on related *Knema andamanica* of the same family have given positive results (Rema et al. 2006). Other related species include *Endocomia macrocoma* syn. *M. prainii*, *Horsfieldia glabra* syn. *M. glabra*, *H. irya* syn. *M. irya*, *H. macrocarpa* var. *canarioides* syn. *M. canarioides*, *Knema conferta* syn. *M. conferta*, *K. glaucescens* syn. *M. glaucescens* and *M. glauca*. Similarly there are numerous wild cultivars of banana under *Musa acuminata* and *M. textilis* and of mango under *Mangifera andamanica*, *M. indandamenasis*, *M. camptosperma*, *M. indica* and *M. sylvatica* species, which can be explored for developing disease-resistance cultivars. Different integrated farming system (IFS) models for different micro-farming situations in hilly upland (plantation + dairy + backyard poultry), medium upland (crop + cattle + fish + poultry + goat) and valley areas (rice + vegetables + fish) at farmers' fields could increase farm income up to INR 3, 90, 000 ha^{-1} $year^{-1}$ besides additional employment generation of 163 man-days ha^{-1} y^{-1} (CIARI 2014).

Afforestation of Land Impregnated with High Salinity

Most of the area in Rann of Kutchh along Gujarat and Pakistan coast comes under this category. Because of low rainfall and high evapotranspiration, and saline shallow underground water, the problem becomes more severe. In many areas, natural salt is prepared in evaporation saltpans. MPTs such as *Prosopis juliflora*, *Salvadora persica*, *Tamarix articulata*, *T. troupii*, *Arthrocnemum indicum* and many halophytes are found growing naturally in these areas with stunted growth. Gururaja-Rao et al. (2004, 2013) advocated that *Salvadora persica* can be cultivated on highly saline soils (ECe>55 dS m^{-1}). Even nursery of this plant could be raised using saline water of ECiw 15 dS m^{-1}. Plant started bearing seeds during second year producing about 725 kg seeds per ha. During the fourth and fifth years, it could produce 1580 and 1838 kg seeds per ha, respectively. Fruit trees such as ber (*Ziziphus mauritiana*), pomegranate (*Punica granatum*), sapota (*Achras zapota*) and banana (*Musa paradisiaca*) can successfully be cultivated in saline black soil as well as coastal sandy saline soils of Gujarat. Many seed spices such as cumin, fennel, coriander, dill and fenugreek in isolation or with forest and fruit trees are suitable for saline black Vertic Haplustepts soils (Dagar and Tomar 1998; Gururaja-Rao et al. 2000, 2004, 2013; Gururaja-Rao 2004). These may be irrigated with saline water.

A viable and productive silvopastoral system can be developed incorporating forage trees with suitable salt-tolerant forages such as species of *Atriplex*, *Kochia indica*, *Aeluropus lagopoides*, *Chloris gayana*, *C. barbata*, *Dichanthium annulatum*, *Leptochloa fusca*, *Echinochloa colonum* and *Sporobolus helvolus*. Oil-yielding species such as *Salvadora persica*, *Salicornia bigelovii*, *Pongamia pinnata* and *Terminalia catappa* and firewood trees like *P. juliflora*, *Acacia nilotica* and *Casuarina glauca* can be raised in furrows and abovementioned grasses in interspaces. In coastal sandy areas particularly along beaches of Orissa, *Casuarina equisetifolia* is successfully grown. At many places, plantations of *Eucalyptus*, cashew nut (*Anacardium occidentalis*), soapnut (*Sapindus trifoliatus*), *Acacia leucophloea*, *A. auriculiformis* and *Tamarindus indica* are found raised successfully. More species to be grown with saline water have been discussed in detail in chapter "Saline Irrigation for Productive Agroforestry Systems".

The silvopastoral system usually refers to land use system in which pasture (grazing land) and livestock production are integrated with woody perennials on the same land management unit and grazing is major component but a more productive and efficient concept covers broadly 'cut-and-carry' fodder production practices also. These land use systems are generally characterized by higher productivity on account of the vertical stratification of the shoot and root systems of different components. The trees in managed species have a great potential for efficient cycling of plant nutrients. Growing of nitrogen fixing trees has additional advantage as these help in fixing the atmospheric nitrogen into the soil which in turn is utilized by the associated field crops.

Mathew et al. (1992) in Kerala revealed that growth and yield of fodder species are significantly influenced by tree components only after tree canopy formation. The fodder species such as *Pennisetum purpureum*, *Panicum maximum*, *Brachiaria ruziziensis* and *Euchlaena mexicana* grown in association with *Casuarina equisetifolia* and *Ailanthus malabarica* recorded comparatively higher forage yield even after canopy formation. However, forage yield in association with *Acacia auriculiformis* and *Leucaena leucocephala* was relatively lower. The forage grasses are cultivated in the order of *P. purpureum*>*P. maximum*>*B. ruziziensis*>*E. mexicana* producing mean biomass of 74.5, 59.0, 42.5 and 23.9 Mg ha^{-1}, respectively. Among other shade trees *Acacia auriculiformis*, *Ailanthus triphysa*, *Casuarina equisetifolia* and *Leucaena leucocephala* can successfully be utilized along with grasses. Overall *Casuarina* among abovementioned trees and hybrid napier (*Pennisetum purpureum*) and guinea (*Panicum maximum*) among forage crops (other forage crops were congo-signal (*B. ruziziensis*) and teosinte (*Zea mexicana*) performed better than others. Opportunities exist for growing salt-tolerant fodder trees along banks of wetlands in coastal areas of West Bengal. Forage grasses such as *Coix lacryma-jobi*, *Brachiaria mutica* and *Echinochloa* spp. could be successfully cultivated and produced 41.3, 31.1 and 24.4 Mg ha^{-1}, respectively, forage biomass from five cuts during *kharif* season (Biswas 1994). Further 100 kg N ha^{-1} was sufficient for optimal yields of these forages.

In canal command areas of Karnataka, *Acacia nilotica* and *Casuarina equisetifolia* are found effective in controlling seepage along canals. The grasses in between complimented the effects. The water table receded significantly underneath the plantation while increased significantly outside plantation area. Two to four rows of *A. nilotica* with a spacing of 4 m × 2 m parallel to canal (5 m away from canal) helped in controlling canal seepage in Tug Bhadra irrigation command area (Vishwanath et al. 2013). Among other trees *Acacia auriculiformis*, *A. ferruginea*, *Albizia lebbeck*, *Gliricidia sepium* and *C. equisetifolia* performed well under saline (ECe 10–12 dS m^{-1}) and high water table conditions. All these species also helped in ameliorating the soil in terms of organic matter and nutrient pool. Among fruit trees pomegranate (*Punica granatum*) and *ber* (*Ziziphus mauritiana*) maintained steady growth in low salinity and shallow water table, while wood apple (*Feronia limonia*) was promising under relatively high salinity but deep water table conditions. Jamun (*Syzygium cuminii*) and sapota (*Achras zapota*) were better performers under shallow water table conditions.

Along the 880 km long coastal area of Pakistan, the soils are light-textured grazing lands and inhabited by *Prosopis juliflora* vegetation. There are a few well-known halophytes that are grazed, the most outstanding being *Arthrocnemum indicum*, *Salsola drummondii*, *Bienertia cycloptera* and *Zygophyllum simplex* (Qureshi et al. 1993). These are highly palatable to camels. Taxa such as *Tamarix articulata*, *T. indica*, *T. stricta*, *Salvadora persica* and *S. oleoides* are other valuable fodder sources for desert livestock and may be cultivated with saline water. Other potential species identified for this region include *Leucaena leucocephala*, *Prosopis cineraria*, *Acacia nilotica* and *Sesbania sesban* among woody species; species of *Atriplex* and *Maireana* among shrubs; and *Leptochloa fusca*, *Echinochloa crus-galli*, *Cenchrus ciliaris* and *Dactyloctenium sindicum* among grasses.

Subsurface Freshwater Skimming System (Improved Doruvu Technology) in Coastal Areas with Sandy Soil along Sea Coast

Out of the total 82 thousand ha of sandy soils along the sea coast of Andhra Pradesh, 18 thousand ha have shallow water table (0.5–3.0 m). These soils occur in 10 km wide and 972 km long strip along the coast extending from Ichapuram in Srikakulam district in the north to Tada in Nellore district in the south. Although the area receives 855 mm annual rainfall, irrigation water at critical growth stages of crops is the major impediment for obtaining optimal crop yield. With very high permeability of coastal sands, almost all the rainwater percolates in to the soil, and this water having lesser density floats over the subsurface saline water, which itself is underlain by impervious soil layer. Traditionally farmers dug a conical pit known as *doruvu* in local language and draw good-quality water manually and irrigate limited crops using pitchers. To overcome the drawbacks of traditional method, a subsurface freshwater skimming system (SSFWSS) was evolved by Raghu Babu et al. (1999) known as 'Improved Doruvu technology', which works on the principal of rapid phreatic flow in sandy soils under the influence of vertical recharge. The lateral flow is collected in a sump constructed on an impermeable layer. One or two parallel collector lines are imbedded for the collection of lateral flow by digging a horizontal trench and connecting these to the sump. Experiences show that each SSFWSS can supply sufficient good-quality water to raise plantations in 4–5 ha area using drip irrigation. Plantations such as red oil palm (*Elaeis guineensis*), coconut, banana, guava and pomegranate can successfully be grown in wider spaces. In the interspaces groundnut, pulses and vegetables can be grown successfully.

Agroforestry on Waterlogged Soils Intruded with Seawater

Cramer et al. (1999) showed that *Eucalyptus camaldulensis* intercepted deep groundwater, while *Casuarina glauca* relied on shallower unsaturated zone. Keeping this fact into consideration, Roy Chowdhury et al. (2011) conducted experiments in coastal deltaic Orissa where problem of waterlogging was both due to seawater intrusion and due to topographical depression. *Casuarina* was more efficient in discharging saline groundwater than *Eucalyptus*. The deltaic Orissa on an average experiences annual rainfall more than 1400 mm, during monsoon. As a result, for about 10–12-week period, the plantations experience above-ground waterlogged condition and remain so till the end of monsoon. Therefore, in present scenario, the scope for assessment of efficacy of biodrainage plantation has been limited to only in post-monsoon season. The effect of planted tree species on underlain water table was monitored by them through observation wells monitoring systems. The lowest water level at Patna (Orissa) was 1.02 m below ground which declined to 1.18 m in 3 years. Similarly decline from 1.27 to 1.52 m and from 1.69 to 1.85 m at other sites was observed. Thus, it is evident that at phreatic surface, there has been a clear drawdown in level of water table underneath tree plantations.

This accelerated drainage has helped the farmer to advance *rabi* cultivation by a period of 15–20 days. There was an advantage of growing watermelon (*Citrullus vulgaris*) with *Casuarina*, as the latter caused 21–34 % reduction in incident of photosynthetically active radiations over watermelon crop in comparison to control, and there was 34 and 39 % increase in yield in cultivars Mokassa and Arka Manik, respectively, in association with *Casuarina* as compared to control (Roy Chowdhury et al. 2012). Through this process, the cultivation of watermelon as intercrop inside *Casuarina* vegetation could get additional benefit of about INR 15,000 per ha for the farmer due to better market price of the crop as well as avoiding the market glut. In *kharif* season, rice was taken as intercrop inside *Casuarina* vegetation at one site. The final average yield of the paddy obtained during 4 years was 1.8 Mg ha^{-1}. The yield under *Eucalyptus* ranged from 2.3 to 3.5 Mg ha^{-1} (average 2.6 Mg ha^{-1}) during the same period. At another site, the net return of watermelon under *Casuarina* plantation in *rabi* season was INR 30,000 with B/C ratio of 2.14. Similarly under *Eucalyptus* from groundnut, net return was INR 21,000 with a B/C ratio 2.10, and from watermelon, net return was INR 62,500 and with B/C ratio of 3.67. Many other crops were also tried and showed promise as intercrops with these plantations. Aquaculture intervention in the biodrainage field was also initiated during the first week of June 2007 using a dugout pond of 400 m^2 of water surface area with *Casuarina* plantation. After carrying out standard pond preparation protocol, air-breathing fish like magur (*Clarias batrachus*) and koi (*Anabas testudineus*) were cultivated. A yield of 1.25 Mg ha^{-1} composite yield of fish was obtained within 10 months with a B/C ratio of 2.5. *Eucalyptus* clones 'JK Super' developed by JK Paper Ltd., Rayagada, Orissa, were planted and found effective. A stretch of a canal of 0.63 km length and 3.0 m width was under observation with growing of *Eucalyptus tereticornis* and banana varieties, viz. Poovan, Karpuravalli, Rastali and Monthan. Half of the canal length was planted with *Eucalyptus tereticornis* and the remaining length was planted with banana varieties. The survival of *Eucalyptus tereticornis* was much higher (Masilamani et al. 2003). The average transpiration rate (measured by sap flow) of groundwater by *Eucalyptus* plantations ranged from (L d^{-1} tree^{-1}) 44.5–56.3 in May to 14.8–16.2 in January. The annual transpiration rate was 268 mm (Roy Chowdhury et al. 2011). Optimizing micro-water resource design and integrated farming system approach for enhancing productivity of waterlogged area, Jena et al. (2006, 2011) modified the land by excavating ponds for storing excess water and created soil platforms for raising high transpiration trees such as *Acacia mangium* and *Casuarina equisetifolia*. This land was highly acidic, low in organic carbon and available nutrients and high in iron contents. The growth of trees was far superior and remunerative in modified land configuration and helped in lowering down the water table for growing intercrops. Similarly Mohanty et al. (2006) found feasibility of growing cowpea (*Vigna unguiculata*) and turmeric (*Curcuma domestica*) intercrops with drip irrigated banana plantation in waterlogged situation where there were additional benefits from the turmeric irrigated by microtubes, and extension tubes were Rs 24,700 and INR 24,200 ha^{-1} per season, respectively. This benefit with cowpea irrigating with microtubes and in-line drippers was lesser INR 11,000 and 7200 ha^{-1} $year^{-1}$, respectively.

Storage of rainwater in ponds for developing aquaculture-based integrated farming systems involving fish, poultry and halophytic crops gave net return of INR 69 thousand ha^{-1} $year^{-1}$ on 15 year basis in Orissa (Mohanty et al. 2004). The horticultural plants included banana, papaya, pineapple, mango and areca nut. From another enterprise involving poultry, fish and plantation on dykes, they could get net returns of more than INR 200 thousand per ha. The water productivity also of integrated farming system was far superior to paddy-based or sugarcane-based and vegetable-based systems. Therefore, in coastal waterlogged areas integrated farming, system approach is most profitable and feasible approach. A multi-enterprise model based on an integrated farming system and multiple water-use approach involving components of crops, fisheries, dairying, horticulture, vegetables, beekeeping, poultry, duck-based poultry, cow-dung-based plant, solar heating system etc. was developed on 2 ha reclaimed sodic land, to provide regular income, employment and livelihood to small farmers. The preliminary results indicated that the field crops (rice and wheat) gave a net income of about INR 52 thousands, berseem INR about 46 thousands and bottle gourd (*Lagenaria siceraria*) INR about 62 thousands ha^{-1}. Fish worth about INR 13 thousands was sold during the year from 0.2 ha fish pond (DARE/ICAR 2008–2009). Behera and Mahapatra (1999) reported that with the following integrated farming system approach in coastal areas of Orissa, a farmer with holding of 1.25 ha could get net return of INR 58,360 and generate 573 man-days in a year. Recently, Gangwar and Ravisankar (2014) reported that with the following farming system approach, the per-day profit of marginal and small households can be increased by 69 % through low-cost interventions such as using improved varieties, balanced recommended nutrient application, integrated pest management, good-quality fodder supply to cattle and the following farming system approach.

In Tamil Nadu, the coastal wetlands are highly saline. Since agriculture is not sustainable, silviculture is practised in river beds, on river and canal banks and in farm holdings and homesteads. Trees such as teak (*Tectona grandis*), *Prosopis juliflora*, *Eucalyptus* and bamboos (*Bambusa* spp.) are grown in small woodlots on patches of land in the wetlands for domestic use, but comparatively dry lands are preferred for raising commercial plantations of *Casuarina*. *C. equisetifolia*, *Thespesia populnea*, *Calophyllum inophyllum*, *Madhuca indica*, *Azadirachta indica*, *Pongamia pinnata*, *Borassus flabellifer*, *Lannea coromandelica* and *Bambusa bambos* are commonly grown on field boundaries and also on small landscapes. There is quite change in attitudes of tree growers from past to present seeing their sustainability and commercial use particularly in industry (Harikrishnan 1993). *Casuarina junghuhniana*, *C. equisetifolia*, *C. glauca*, *Eucalyptus*, *Melia dubia*, *Leucaena leucocephala*, *Prosopis juliflora* and bamboos are viable source of feedstock for combustible-type biomass power plants. Some major companies established their units in Ramanathapuram, Pudukottai, Sivaganga, Virudhunagar, Tirunelveli and Tuticorin districts of Tamil Nadu with the intention of using *Prosopis juliflora* which is seen almost everywhere in Tamil Nadu, as their main biomass material is feeling the pinch now with the high cost and non-availability of feedstock. However, the rural people utilize this wood for the production of charcoal by traditional methods to sustain their livelihood.

Shelterbelts and Shore Protection

In coastal areas high winds also carry salt with them and damage crops. Besides mangroves and other littoral species mentioned earlier, many trees and shrubs such as *Casuarina equisetifolia*, *Acacia auriculiformis* and *Gliricidia sepium* may play very important role in reducing the speed of the winds and protect the crops from injury. These also help in soil amelioration. Most of the coastal areas are prone to damage caused by cyclones and even *tsunamis*. Hence attempts should be made to conserve and restore the mangrove-degraded areas by planting suitable species. Besides mangroves, littoral species such as *Pandanus* spp., *Thespesia populnea*, *Scaevola taccada*, *Tournefortia ovata*, *Hibiscus tiliaceus* and *Salvadora persica* may also play important role in protecting the shores and beaches. MPTs such as *Calophyllum inophyllum*, *Pongamia pinnata*, *Heritiera littoralis*, *Terminalia catappa* and *Manilkara littoralis*, which are found growing luxuriously along beaches of Andamans (Fig. 5), can also be raised on degraded low-lying areas. These belts protect the shores/beaches, provide valuable forest products and also give shelter to wild life.

Domestication of Halophytes

Halophytes are naturally evolved salt-tolerant plants that represent at most 2 % of terrestrial plant species (Flowers and Colmer 2008). They have the ability to complete their life cycle in salt-rich environment where almost 99 % of salt-sensitive species die because of NaCl toxicity and thus may be regarded as a source of potential new crops (Glenn et al. 1991; Jaradat 2003; Dagar 2003) particularly for coastal areas where if necessary these may be irrigated with seawater. While halophytes since long have been in the diet of the people and are utilized in variety of ways in routine life, their scientific exploration as crops developed only in the latter half of the twentieth century (reviewed by Rozema et al. 2003 and Panta et al. 2014). Many halophytes have been evaluated for their potential use as crop plants (Miyamoto et al. 1996; Barrett-Lennard 2003; Reddy et al. 2008; Ruan et al. 2008; Qadir et al. 2008; Flowers et al. 2010; Tomar et al. 2010; Rozema et al. 2003; Dagar et al. 2013, 2014; Zhang 2014). Species such as *Distichlis palmeri*, *Chenopodium quinoa*, *C. album*, *Pennisetum typhoides*, *Salicornia bigelovii*, *Diplotaxis tenuifolia* and many others

Fig. 5 Littoral forest protecting the natural beach of Havelock, South Andaman

have been established as food crops, are being explored commercially and can be cultivated using seawater for irrigation. Similarly species of *Atriplex* and *Maireana*; grasses *Leptochloa fusca*, *Chloris gayana*, *Aeluropus lagopoides*, *Brachiaria mutica* and *Paspalum conjugatum*; and many other are constituents of silvopastoral systems developed on waterlogged salt lands in different agroclimatic regions of the world. At least 50 species of seed-bearing halophytes are potential sources of edible oil and proteins. *Salicornia bigelovii*, *Terminalia catappa*, *Suaeda moquinii*, *Kosteletzkya virginica*, *Batis maritima*, *Chenopodium glaucum*, *Crithmum maritimum* and *Zygophyllum album* are a few examples. A number of species including the halophytes *Tamarix chinensis*, *Phragmites australis*, *Spartina alterniflora* and species of *Miscanthus* have been evaluated as biofuel crops for ethanol production in the coastal zone of China (Liu et al. 2012), while many species such as *Halopyrum mucronatum*, *Desmostachya bipinnata*, *Phragmites karka*, *Typha domingensis* and *Panicum turgidum* are grown in coastal regions of Pakistan as source of bioethanol (Abideen et al. 2011). In addition, sugar beet (*Beta vulgaris*), mangrove palm (*Nypa fruticans*) and *kallar* grass (*Leptochloa fusca*) are identified as a source of liquid and gaseous fuel (Jaradat 2003). Screw pine (*Pandanus fascicularis*), quite predominant along Indian coast, is rich in methyl ether of beta-phenylethyl alcohol and is used as a perfume and flavouring ingredient (Dutta et al. 1987). Many woody and succulent halophytes are used for turf production for golf and landscape development, paper industry, medicinal use and other commercial purposes. As stated earlier mangroves are unique resources for tidal zone, which must be protected and multiplied in degraded areas. Therefore, more efforts are needed to domesticate these useful resources, particularly in coastal areas in agroforestry mode. For more details also see chapter "Saline Irrigation for Productive Agroforestry Systems".

Conclusions

The coastal areas are usually struck by disasters like cyclone and other climate vagaries incurring heavy productivity losses and associated salinity and waterlogging problems. These fragile and resource poor areas also face several other socio-economic constraints and thus demand for a holistic approach to raise their agricultural productivity. Opportunities exist because of the rich biodiversity and high availability of rainwater (>1000 mm) and thus paving ways for agroforestry-based strategies. The site-specific farming systems can combine forest and fruit trees, plantation crops, spices, forages, vegetables and halophytic plants. Integrated farming systems involving fish, shrimps, all kinds of aquaculture, multistoreyed plantation-based cropping systems, duck- and chick-based poultry, high-value medicinal and aromatic plants and spices can be highly remunerative. In low-lying areas, land shaping has helped in utilizing salt-affected waterlogged areas for increasing farm productivity; hence, such programmes must be undertaken at larger scale with support of different agencies involved in agriculture.

Mangroves are unique ecosystems which are nursery ground of several aquatic species, which stand a scope of commercial exploration through all kinds of aquaculture keeping mangroves intact. These will not only act as life-support system but also will protect the shores from natural disasters and act as carbon sink. Thus, agroforestry land use system should be of great relevance to the coastal and island ecologies particularly in the scenario of climate change. Unfortunately natural mangrove stands are in most dwindled stage and need to have special attention under mission-mode approach. These habitats must be replanted and developed at all possible sites.

Research efforts are needed in developing and domestication of high-value halophytic woody and herbaceous species. In coastal saline areas, leguminous species such as *Vigna marina*, *Clitorea ternatia* and *Canavalea* spp. are found natural giving a room to develop stress-tolerant pulses by inculcating the potential genes. Many other wild relatives of fruit and spice trees and orchids may be utilized for developing disease-resistance high-value crops. Seaweeds and marine wealth can successfully be explored in food, cosmetic, fertilizer and drug industries. Many ornamental fish species can be explored for ecotourism. Thus, though vulnerable to climate change and other natural disasters, the coastal and island regions have tremendous opportunities for increasing agricultural and related productivity through integrated farming system mode involving agroforestry science.

References

Abideen Z, Ansari R, Khan MA (2011) Halophytes: potential source of ligno-cellulosic biomass for ethanol production. Biomass Bioenergy 35:1818–1822

Ayars JE, Christen EW, Soppe RWO, Hornbuckle JW, Schoneman RA (2005) In-situ crop water use from shallow ground water to reduce drainage and save irrigation water. In: Proceedings of the international salinity forum, managing saline soils and water, riverside, 25–27 Apr 2005, pp 25–28

Barrett-Lennard EG (2003) Saltland pastures in Australia-a practical guide (Second Edition). State of Western Australia and CRC for Plant-based management of dryland salinity, South Perth, Western Australia, p 176

Behera UK, Mahapatra IC (1999) Income and employment generation of small and marginal farmers through integrated farming systems. Indian J Agric Sci 44(3):431–439

Bhargava N (1981) Plant in folk life and folk lore in Andaman and Nicobar Islands. In: Jain SK (ed) Glimpses of Indian ethnobotany. Oxford/IBH Publishing Co Ltd, New Delhi, pp 329–344

Bhargava N (1983) Ethnobotanical studies of the tribes of Andaman and Nicobar Islands. I. Onge. Econ Bot 37:110–119

Biswas CR (1994) Studies on feasibility and yield of forage crops in coastal saline areas of Sundarbans (West Bengal). J Indian Soc Coastal Agric Res 12:91–94

Burman D, Bandyopadhay BK, Mandal S (2013) Land shaping-a unique technology for improving productivity of coastal land. Technical Bulletin 02/2013. CSSRI Regional Research Center, Canning Town, West Bengal, India

Burman D, Mandal S, Bandopadhyay BK, Maji B, Sharma DK, Mahanta KK, Sarangi SK, Mandal UK, Patra S, De S, Patra S, Mandal B, Maitra NJ, Ghoshal TK, Velmurugan A (2015) Unlocking production potential of degraded coastal land through innovative land management practices: a synthesis. J Soil Salinity Water Qual 7(1): (in press)

CIARI (2014) CIARI-the Journey and the Road Ahead. ICAR-Central Island agricultural Research Institute, Port Blair, p 24

Cramer VA, Thornburn PJ, Fraser DW (1999) Transpiration and ground water uptake from farm forest plots of *Casuarina glauca* and *Eucalyptus camaldulensis* in saline areas of south east Queensland, Australia. Agric Water Manage 39:187–204

Dagar JC (1982) Some ecological aspects of mangrove vegetation of Andaman and Nicobar Island. Sylvatrop Philippines. For Res J 7:177–216

Dagar HS (1989) Plant folk medicines among Nicobarese tribals of Car Nicobar Island, India. Econ Bot 43:215–224

Dagar JC (1995a) Agroforestry systems for Andaman and Nicobar Islands in India. Int Tree Crop J 8(2&3):107–128

Dagar JC (1995b) Ecology of halophytic vegetation in India: a review. Int J Ecol Env Sci 21:273–296

Dagar JC (1996) Rehabilation of Coastal saline lands by planting with suitable species. J Indian Soc Coastal Agric Res 14:185–191

Dagar JC (2003) Biodiversity of Indian saline habitats and management & utilization of high salinity tolerant plants with industrial application for rehabilitation of saline areas. In: Alsharhan AS, Wood WW, Goudie AS, Fowler A, Abdellatif EM (eds) Desertification in the third millennium. Swets & Zeitlinger Publishers, Lisse, pp 151–172

Dagar JC (2014) Greening salty and waterlogged lands through agroforestry systems for livelihood security and better environment. In: Dagar JC, Singh AK, Arunachalam A (eds) Agroforestry systems in India: livelihood security & environmental services-advances in agroforestry, vol 10. Springer, New Delhi/Dordrecht/New York, pp 273–332 un, India, p 122

Dagar HS, Dagar JC (1989) Ethnobotanical observations among the Onge tribe of Little Andaman. Bull Medico Ethno Bot Res 10(1–2):1–10

Dagar HS, Dagar JC (1991) Plant folk medicines among the Nicobarese of Katchal Island, India. Econ Bot 45(1):114–119

Dagar JC, Dagar HS (1999) Ethnobotany of aborigines of Andaman-Nicobar Islands. Surya International Publications, Dehra Dun

Dagar JC, Sharma AK (1991) Litterfall beneath Rhizophora apiculata in mangrove forests of Andamans, India. Trop Ecol 32(2):231–235

Dagar JC, Sharma AK (1993) Litterfall beneath *Bruguiera gymnorrhiza* in mangrove forests of South Andamans, India. Indian J For 16(2):157–161

Dagar JC, Singh NT (1999) Plant resources of the Andaman & Nicobar Islands. Bishen Singh Mahendra Pal Singh, Dehra Dun, p 987

Dagar JC, Tomar OS (1998) Suitable agroforestry systems for coastal saline regions: case study. J Range Manage Agrofor 19:165–172

Dagar JC, Mongia AD, Bandyopadhyay AK (1991) Mangroves of Andaman and Nicobar Islands. Oxford/ IBH Publishing Co Ltd, New Delhi

Dagar JC, Singh NT, Mongia AD (1993) Characteristics of mangrove soils and vegetation of Bay Islands in India. In: Lieth H, Al Masoom A (eds) Towards the rational use of high salinity tolerant plants, vol 1. Kluwer, Dordrecht, pp 59–80

Dagar JC, Tomar OS, Minhas PS, Mukesh K (2013) Lemongrass (*Cymbopogon flexuosus*) productivity as affected by salinity of irrigation water, planting method and fertilizer doses on degraded calcareous soil in a semi-arid region of northwest India. Indian J Agric Sci 83(7):734–738

Dagar JC, Pandey CB, Chaturvedi CS (2014) Agroforestry: a way forward for sustaining fragile coastal and island agro-ecosystems. In: Dagar JC, Singh AK, Arunachalam A (eds) Agroforestry systems in India: livelihood security & environmental services-advances in agroforestry, vol 10. Springer, New Delhi/ Dordrecht/New York, pp 185–232

Dam Roy S (2003) A compendium on mangrove biodiversity of Andaman and Nicobar Islands. Central Agricultural Research Institute, Port Blair, p 196

Dam Roy S (2015) Mangroves ecosystem of the Islands: significance and management options. In: Dhyani SK, Newaj R, Alam B, Dev I (eds) Agroforestry: present status and way forward. Biotech Books, New Delhi, pp 395–401

Dam Roy S, Krishnan P (2005) Mangrove stands of Andamans vis-à-vis tsunami. Curr Sci 89(1):1800–1804

Dam Roy S, Soundararajan R, Krishnamoorthy V, Krishnan P (2005) Tsunami impact on fisheries sector, rehabilitation required and strategies to be followed in Andaman and Nicobar Islands. In: Paper in national symposium "The Seventh Indian Fisheries, Forum" 8–12 Nov 2005, Bangalore, India

Dam Roy S, Velmurugan A, Zamir-Ahmed SK, Nagesh-Ram K-S, Sankaran RM, Gautam RK, Swarnam TP, Singh S (2015) Island agriculture: perspective and strategies. In: Souvenir-national seminar on harmonizing biodiversity and climate change: challenges and opportunities. Andaman Science Association, Port Blair, pp 1–21

DARE/ICAR (2008–2009) Annual report 2008–2009. Department of Agricultural Research and Education, Indian Council of Agricultural Research, New Delhi, India

Dutta PK, Saxena HO, Brahman M (1987) Kewda perfume industry in India. Econ Bot 41:403–410

Flowers TJ, Colmer TD (2008) Salinity tolerance in halophytes. New Phytol 179:945–963

Flowers TJ, Galal HK, Bromham L (2010) Evaluation of halophytes: multiple origins of salt tolerance in land plants. Funct Plant Biol 37:604–612

FSI (2011) India State of Forest Report 2011. Forest Survey of India (FSI), Ministry of Environment & Forests, Govt. of India, Dehra Dun, p 286

Gangwar B, Ravisankar N (2014) Farming systems approach for managing soil, water and crops in coastal areas. J Indian Soc Coastal Agric res 32(1):1–6

Ghoshal CS, Raja R, Jayakumar V (2006) Biochemical and molecular characterization of the soils of mangrove forests of Andamans. Annual Report. CARI, Port Blair, pp 14–15

Glenn EP, O'Leary JW, Watson MC, Thompsom TL, Kuehl RO (1991) *Salicornia-bigelovii* Trr – an oilseed halophyte for seawater irrigation. Science 251:1065–1067

Goutham-Bharathi MP, Dam Roy S, Krishnan P, Kaliyamoorthy M, Immanuel T (2014) Species diversity and distribution of mangroves in Andaman and Nicobar Islands, India. Bot Mar 57(6):421–432

Greenwood EAN (1986) water use by trees and shrubs lowering saline groundwater. Reclam Reveg Res 5:319–341

Gururaja-Rao G (2004) Diversified cropping systems and socioeconomic prospects in coastal saline soils of Gujarat state. J Indian Soc Coastal Agric Res 22(1&2):178–184

Gururaja-Rao G, Nayak AK, Chinchmalatpure AR (2000) Dill (Anethum graveolens) a potential species for salt affected black soils. Technical Monograph No.1, Regional Research Station Anand of Central Soil Salinity Research Institute, p 10

Gururaja-Rao G, Nayak AK, Chinchmalatpure AR, Nath A, Babu VR (2004) Growth and yield of *Salvadora persica* a facultative halophyte grown on saline black soil (vertic haplustept). Arid Land Res Manage 18:51–61

Gururaja-Rao G, Chinchmalatpure AR, Sanjay-Arora, Khandelwal MK, Sharma DK (2013) Coastal saline soils of Gujarat: problems and their management-technical bulletin 01/2013. CSSRI Regional Station, Bharuch, Gujarat, India, p 60

Hamilton LS, Snedaker SC (1984) Handbook of mangrove area management. UNEP and East-West Centre, Environment and Policy Institute, Hanolulu

Harikrishnan M (1993) Sustainable management of coastal ecosystems. In: Swaminathan MS, Ramesh R (eds) Coastal agroforestry and its sustainable utilisation. Proceedings of the workshop held in Madras in October, 1991, pp 71–83

ICAR (2014) http://www.icar.org.in/en/node/7665

India (2013) India 2013-a reference manual. Publication Division, Ministry of Information & Broadcasting, Government of India, New Delhi, p 1206

Jaradat AA (2003) Halophytes for sustainable farming systems in the Middle East. In: Alsharhan AS, Wood WW, Goudie AS, Fowler A, Abdellatif EM (eds) Desertification in the third millennium. Swets & Zeitlinger Publishers, Lisse, pp 187–204

Jena SK, Sahoo N, Roy Chowdhury S, Mohanty RK, Kundu DK, Mohanty M (2006) Optimizing microwater resource design and integrated farming system approach for enhancing productivity of waterlogged areas. J Indian Coastal Agric Res 24(1):184–187

Jena SK, Sahoo N, Roy Chowdhury S, Mohanty RK, Kundu DK, Behera MS, Kumar A (2011) Reclamation of coastal waterlogged wasteland through biodrainage. J Indian Coastal Agric Res 29:57–62

Karim A, Hossain Z, White KJ (1984) Study of the growth of mangrove plants in relation to edaphic factors in coastal afforestation plantation of Chittagong. In: Proceeding of thesymposium on Mangrove environment research & management, Kuala Lumpur, Malaysia

Kogo M (1985) Mangrove research & recommendations of afforestation in Pakistan. A report for UNDP/UNESCO Mangrove Project by Al Gurm Research Centre, Tokyo

Kogo M, Kaminura D, Miyagi T (1986) Research for rehabilitation of mangroves in Truck Islands. In: Mangroves of Asia & the Pacific-Status and monument. Technical Report. UNDP/4NESCO Regional Mangrove Project, New Delhi

Liu XZ, Wang CZ, Su Q, Lib CK (2012) The potential resource of halophytes for developing bio-energy in China coastal zone. Herald J Agric Food Sci Res 1:44–51

Masilamani P, Santhana Bosu S, Annadurai K (2003) Bio drainage for degraded drainage problem lands. Leisa India 5:19

Mathew T, Kumar BM, Suresh-BabuKV UK (1992) Comparative performance of some multipurpose trees and forage species in silvo-pastoral systems in the humid regions of southern India. Agrofor Syst 17:205–218

Miyamoto S, Glenn EP, Olsen MW (1996) Growth, water use and salt uptake for four halophytes irrigated with highly saline water. J Arid Environ 32:141–159

Mohanty RK, Sahoo N, Jena SK, Roy Chowdhury S, Verma HN (2004) Water productivity and economic evaluation of aquaculture-based integrated farming system approach in waterlogged ecosystem. J Indian Soc Coastal Agric Res 22:298–303

Mohanty S, Srivastava RC, Singandhupe RB (2006) Feasibility of intercrops in drip irrigated banana in coastal Orissa. J Indian Coastal Agric Res 24(1):184–187

NAAS (2010) Degraded and wastelands of India: status and spatial distribution. National Academy of Agricultural Sciences, New Delhi, p 158

Panta S, Flowers T, Lane P, Doyle R, Haros G (2014) Halophytes agriculture: success stories. Environ Exp Bot 107:71–83

Qadir M, Tubeileh A, Akhtar J, Larbi A, Minhas PS, Khan MA (2008) Productivity enhancement of salt-affected environments through crop-diversification. Land Degrad Develop 19:429–453

Qureshi RH, Aslam M and Rafiq M (1993) Expansion in the use of forage halophytes in Pakistan. Davidson N, Galloway R (eds) Productive use of saline land. ACIAR Proceedings No. 42, Australian Centre for International Agricultural Research, Canberra, Australia, pp 12-16

Raghu Babu M, Prasad PRK, Subbaiah GV, Mujeeb-Khan, Minhas PS (1999) Subsurface fresh water skimming system (Improved Doruvu Technique), Technical Bulletin 1/99 Central Soil Salinity Research Institute, Karnal, India, p 19

Rajan PT, Dam Roy S (2003) Mangrove fishes of Andaman and Nicobar Islands. In: Dam Roy S (ed) A compendium on mangrove biodiversity of Andaman and Nicobar Islands. CARI, Port Blair, pp 123–135

Reddy MP, Shah MT, Patolia JS (2008) *Salvadora persica*, a potential species for industrial oil production in semiarid saline and alkali soils. Ind Crop Prod 28:273–278

Rema J, Krishnamoorthy B, Mathew PA, Johnson George K (2006) Conservation of wild nutmegs of Andamans at IISR. Spice India, December, pp 21-22

Ritson P, Pettit N (1992) Research to improve reforestation techniques for saline groundwater discharge control in water resource catchments. Surface Water Branch, Water Authority of Western Australia, Rep. No. WS25, 44 pp

Roy Chowdhury S, Kumar A, Brahmanand PS, Ghosh S, Mohanty RK, Jena SK, Sahoo N, Panda GC (2011) Application of bio-drainage for reclamation of waterlogged situations in deltaic Orissa. Research Bulletin 53. Directorate of Water Management (ICAR), Bhubaneswar, p 32

Roy Chowdhury S, Brahmanand PS, Ghosh S, Jena SK, Mohanty RK, Kumar A (2012) Photosynthesis performance and yield of watermelon grown under biodrainage (*Casuarina equisetifolia*) vegetation. J Indian Coastal Agric Res 30(2):34–40

Rozema J, Muscolo A, Flowers T (2003) Sustainable cultivation and exploitation of halophyte crops in a salinising world. Environ Exp Bot 92:1–3

RRSSC & CWC (2009) Assessment of Waterlogging and Salt and/or Alkaline Affected soils in the commands of all major and medium irrigation projects in the country using satellite remote sensing. Regional Remote Sensing Service Centre, Indian Space Research Organization, Jodhpur; and Central Water Commission, New Delhi, p 165

Ruan CJ, Li H, Guo YQ, Qin P, Gallgher JL, Seliskar DM, Lutts S, Mahy G (2008) Kosteletzkya virginica, an agroecoengineering halophytic species for alternative agricultural production in China's east coast: ecological adaptation and benefits, seed yield, oil content, fatty acid and biodiesel properties. Ecol Eng 32:320–328

Smith AJ, Pollock DW, Salama RB, Palmer D (2005) Ivanhoe plain aquifer pumping trial, July 2003–April 2005: Stage 1 Ord River irrigation area, Kununurra, Western Australia, CSIRO land and water technical report, 24 May 2005

Sujana-Dhar (2011) Impact of climate change on the salinity situation of the Piyali River, Sundarbans, India. J Water Resour Protect 3:495–503

Swarnam TP, Velmurugan A, Zacharia-George RN, Sai TP, Dam Roy S (2014) Integrated farming system for sustainable livelihood in tribal areas of Nicobar Islands. J Andaman Sci Assoc 19(1):19–22

Tomar OS, Dagar JC, Minhas PS (2010) Evaluation of sowing methods, irrigation schedules, chemical fertilizer doses and varieties of *Plantago ovata* Forsk to rehabilitate degraded calcareous lands irrigated with saline water in dry regions of north western India. Arid Land Res Manage 24:133–151

Untawale AG (1993) Development of an intertidal mangrove nursery and afforestation techniques along the Indian Coast. In: Lieth H, Al Masoom A (eds) Towards the rational use of high salinity tolerant plants, vol 1. Kluwer, Dordrecht, pp 371–378

Van Hylckama TEA (1974) Water use by saltcedar as measured by the water budget method. Professional paper 491E. United States Geological Survey, Washington, DC, 30 pp

Vannucci M (1989) The mangroves and us. A synthesis of insight. Indian Association of the Advancement of Science, New Delhi

Velmurugan A, Swarnam TP, Lal R (2015) Effect of land shaping on soil properties and crop yield in tsunami inundated coastal soils of Andaman Islands. Agric Ecosyst Environ 206:1–9

Vishwanath J, Hebbara M, Karegoudar AV, Rajkumar RH, Anand SR, Ambast SK, Gupta SK (2013) Groundwater quality and management of poor quality water for agriculture in Karnataka. Technical Bulletin 2013/1, AICRIP on management of salt-affected soils and use of saline water in agriculture, Gangavati, University of Agricultural Sciences, Raichur, Karnataka, India, p 47

Yadav JSP, Bandyopadhya AK, Bandyopadhya BK (1983) Extent of coastal saline soils in India. Indian Soc Coastal Agric Res 1(1):1–6

Zhang J (2014) Agroforestry systems of combating saline soils. In: Zhang J (ed) Coastal saline soil rehabilitation and utilization based on forestry approach in China. Springer, Berlin, pp 41–54

Saline Irrigation for Productive Agroforestry Systems

J.C. Dagar and P.S. Minhas

Abstract

Land irrigation is playing a major role in enhancing food and livelihood security in the world over. Nevertheless, a typical scenario in the ground water-irrigated regions has emerged; the areas characterized by water scarcity also usually have underlying aquifers of poor quality. Though possibilities have now emerged to safely use waters otherwise designated unfit if the characteristics of water, soil, and intended usages are known. But it is neither feasible nor economical to use highly saline waters for crop production, especially on lands that are already degraded. Best land use under such situations is to retire these areas to alternative uses through agroforestry. With growing scientific and social recognition of many diversified uses of salt-tolerant plants, the research efforts have led to the development of planting technique and other practices for critical management of salts and water in root zone so as to rehabilitate the degraded lands successfully using saline irrigation. Several salt-tolerant plants have been identified with benefits as fuelwood, greening, fruits, forage, and medicinal and aromatic uses in addition to allowing for crop production activity underneath these (agroforestry). Many halophytes also have potential to be used as traditional food, forages and animal feeds, oil seeds, and energy crops.

Introduction

Land irrigation is playing a major role in enhancing food and livelihood security in the world over. About two-fifth of the world's total food and fiber output is now contributed by irrigated agriculture, although its area is only 17 %. In fact, the productivity of irrigated areas in arid and semiarid regions largely depends upon the ability to enlarge this

J.C. Dagar (✉)
Central Soil Salinity Research Institute,
Karnal 132 001, Haryana, India
e-mail: dagarjc@gmail.com

P.S. Minhas
National Institute of Abiotic Stress Management,
Baramati, Pune 413115, Maharashtra, India
e-mail: minhas_54@yahoo.co.in

J.C. Dagar, P.S. Minhas (eds.), *Agroforestry for the Management of Waterlogged Saline Soils and Poor-quality Waters*, Advances in Agroforestry 13, DOI 10.1007/978-81-322-2659-8_9

resource base by better rainwater management and/or development of ground water. Globally, the aquifer withdrawal has increased manifold during the second half of the last century. For example, in the United States, the share of ground water used for irrigation increased from 23 % in 1950 to 42 % in 2000. In the Indian subcontinent, ground water use soared from 10–20 km^3 in 1950 to 240–260 km^3 during 2000. Nevertheless, a typical scenario in the ground water-irrigated regions has emerged; the areas characterized by water scarcity also usually have underlying aquifers of poor quality. These areas often have the greatest need for economic development, public welfare, and more food to supply the growing populations and regional conflicts over water and environmental degradation. But, driven by the pressure to produce more, even the brackish ground water is being increasingly diverted to irrigate agricultural lands. The use of such saline or alkali water to produce many conventional grain, forage, and feed crops as well as salt-tolerant plants and trees is prevalent particularly in Bangladesh, China, Egypt, India, Iran, Pakistan, Syria, and the United States (Tanwar 2003; Minhas 2012). The overexploitation of good-quality water in many developing countries and the alarming rate of decline in ground water levels are also putting aquifers at risk of contamination from adjoining poor-quality aquifers. Moreover, irrigation efficiency in most of the world's irrigated areas is of the order of 50 %, suggesting substantial secondary salinization from seeped water. About 20 % of the globally irrigated area is afflicted with varying levels of secondary salinity and sodicity (Ghassemi et al. 1995). The most technical method to combat irrigation-induced salinity being installation of expensive drainage systems and large amounts of drainage effluents of poor-quality are produced in areas covered with subsurface/surface drainage systems. In addition, recent trends in climate change and salt water intrusion suggest the influence of even greater volumes of the sea waters in agricultural production in coastal areas in the coming years.

Indiscriminate use of saline waters in the absence of proper soil–water–crop management strategies poses grave risks to soil health and environment (Minhas and Gupta 1992; Minhas and Bajwa 2001; Minhas and Samra 2004; Minhas 2012; Minhas et al. 2015). Development of salinity, sodicity, and toxicity problems in soils not only reduces crop productivity and quality but also limits the choice of crops. Its management signifies those methods, systems, and techniques of water conservation, remediation, development, application, use, and removal that provide for a socially and environmentally favorable level of water regime to agricultural production systems at the least economic cost (Hillel 2000). Though possibilities have now emerged to safely use waters otherwise designated unfit if the characteristics of water, soil, and intended usages are known (Minhas and Gupta 1992), and this has led to replacement of too conservative water-quality standards with site-specific guidelines, where factors like soil texture, rainfall, and crop tolerance have been given due consideration. The increased scientific use of these "degraded" waters such as brackish ground water, saline drainage water, and treated wastewaters therefore offers opportunities to address the current and future shortage (O'Connor et al. 2008). However, in some cases, it is neither feasible nor economical to use highly saline waters for crop production, especially on lands that are already degraded. Best land use under such situations is to retire such areas to alternative uses (through agroforestry), where salt-tolerant forest and fruit trees, crops, forage grasses, and medicinal and aromatic and other high value crops can be equally remunerative. Besides providing fuel, fodder, and timber, afforestation will also lead to bio-amelioration of salty lands. Afforestation of these lands should not only help in ecological and environmental considerations but also be useful in relieving pressure on traditionally cultivated lands and forests. These can be specifically beneficial for areas where periodic droughts along with extensive pressures from overgrazing and uncontrolled fuelwood gathering are substantially contributing to desertification and land degradation. Thus, this chapter briefly outlines several remedial management actions at the tree, root zone, and farm and irrigation system level strategies available for alleviating the hazards of highly saline waters and establishing tree-based systems.

Irrigation Management Strategies

Successful establishment of trees with saline irrigation requires appropriate planting techniques and site preparation for planting and post-planting irrigation management. Earlier researchers (Armitage 1984; Tomar and Gupta 1985; Yadav 1991; Gupta et al. 1994; Stevens et al. 1999) supported that the establishment of tree saplings is the most critical stage for raising plantations under the conditions of water and salt stress. However, there is general lack of information on the basic questions as how much and how long to irrigate tree plantations established in arid areas with saline waters. The traditional approach for sustaining the use of saline waters is to irrigate the arable crops more frequently and provide adequate leaching requirements. Nevertheless, such practices demand the application of additional quantities of saline water and thereby also result in enhancement of salt loads of soils (Minhas 1996). In fact, frequent irrigation is usually advocated for shallow-rooted crops in arid environments mainly because the added salts get pushed beyond the rooting zone. But in deep-rooted tree plantations, the additional salts going into the soil through enhanced frequency of irrigations may rather aggravate the problem as these are likely to persist within their expanding rooting zones and subsequently hinder the growth of trees. Though the longer-term data with saline irrigation are not available, but Fig. 1 (from Bandara et al. 2002) illustrates how salt concentrations vary with depth at plantation sites varying in soil conditions, depth to water table, and irrigated water salinities. No salt accumulation was observed at the Cobram site, which was irrigated with good-quality water, and peak concentration was relatively low at Nathalia (irrigated with 2.0–2.5 dS m^{-1} water) and at Mangalore, where the salinity of water applied was 1–1.5 dS m^{-1}. However, at the Tatura and Timmering sites, irrigated with water of 5 and 10 dS m^{-1}, respectively, there is clear evidence of salinity buildup in the upper 3 m of soil. Soils at Tatura and Timmering were also heavier textured than those at Nathalia and Mangalore. Thus, the impacts on plantation growth should be more pronounced where saline water is applied to clay soils rather than to sandy soils.

For proper irrigation scheduling in terms of frequency of irrigation and the amount of water applied each time, aim should be for maximizing productivity while conserving water and ensuring that irrigation systems are environmentally and economically sustainable. Decision on scheduling requires a good knowledge of crop water demand, tolerance to salinity, and soil water characteristics and must account for the type of irrigation method used. Theiveyanathan et al. (2000) predicted annual irrigation demand under flood irrigation to be 14–16 ML ha^{-1}. These volumes far exceed the normal operational rates of irrigation in that region, generally in the range 4–8 ML ha^{-1} $year^{-1}$. Understanding the relationship between the volumes of irrigation applied and growth is fundamental to developing efficient irrigation strategies. In Australia, the growth of trees was linearly related to irrigation rates up to the maximum of 8 ML ha^{-1} $year^{-1}$. The regression equation shows that diameters increased at the rate of 4.3 mm per ML of water applied, whether through irrigation or the sum of irrigation and rainfall.

Tree saplings with their roots located where these can access soil water of low salinity or lateral subsurface flows of salts would achieve better establishment and growth than those with more or less uniform salinity. Therefore, irrigation with saline waters should aim to create favorable niches for the better establishment of saplings and also eliminate the excessive salinity buildup in rooting zone. This could be achieved by using subsurface planting and furrow irrigation technique irrigating only the limited area under furrows planted with tree saplings (Minhas et al. 1996). In this technique, furrows (15–20 cm deep and 50–60 cm wide) are created at 3–5 m intervals and tree saplings are transplanted during rainy season (July-August) at the sill of furrows. The technique is now known as subsurface planting

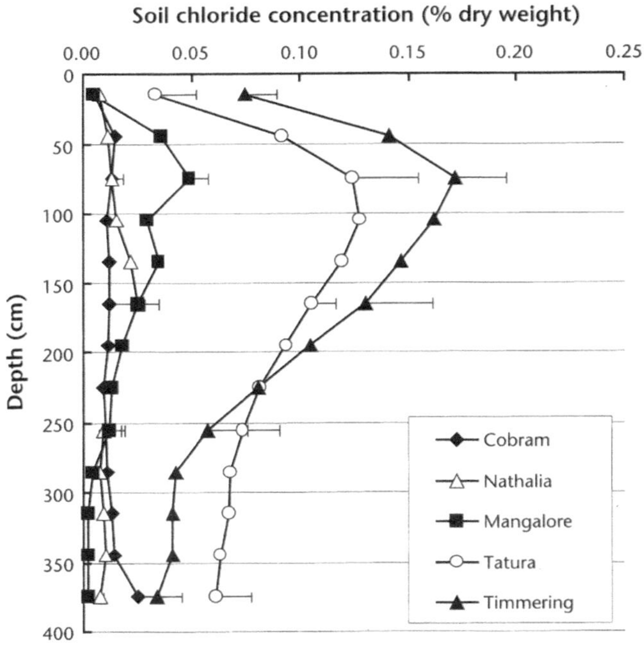

Fig. 1 Soil chloride concentrations at five plantation sites irrigated with water of different salinities, 7–9 years after planting. (Bars represent standard errors) (Source: Bandara et al. 2002)

and furrow irrigation system (SPFIS). Depending upon tree row spacing, furrow vis-a-vis irrigated area occupies only one to fifth to one to tenth of total land area depending upon the space of planting rows. The success of the system was attributed to both the reduced salt load with irrigation only a portion of land, creation of better soil moisture regimes in rooting medium during dry periods (October-June) and significant leaching and lateral distribution of salts by the concentration of rainwater through runoff into these furrows (Fig. 2). It was further reported that irrigation quantities equaling 10 % of open pan evaporation, though saline, sufficed for optimal growth of *Acacia nilotica* and *Dalbergia sissoo* plantations on a highly calcareous soil with little subsoil–water storage.

Though the initial irrigation to plantations may allow for their better establishment and promote growth as compared with un-irrigated, but objective should be to discontinue saline irrigation that may impact the tree growth with accumulation of salts. In a semiarid monsoonal climate, there was two- to four-fold improvement in establishment and growth of *Acacia nilotica* and *Dalbergia sissoo* tree saplings with saline irrigation (Minhas et al. 1997), and irrigation support for at least 2 years seemed necessary. From later experiments (Tomar et al. 2003b), it was concluded that for tolerant trees, saline irrigation may be provided for initial 3 years of transplanting, and thereafter irrigation may be applied once during winter to safeguard against the frost damage. Others (Rogers 1985; Morris et al. 1994) have also recommended a period of at least 5 years for irrigation.

Thorburn et al. (1995) suggested that if salinity in tree root zones increases for extended periods above salinity levels tolerated by particular species, roots are likely to become confined to an increasingly shallow part of the upper profile above the zone of salt accumulation thereby reducing plantation productivity. The long-term salinity dynamics need to be systematically studied but with adequate leaching, on soils with good hydraulic conductivity/lighter textured soils, salts can be moved out of the root zone. However, leaching of salts may be restricted when (Marcar and Morris 2005) (1) the root zone is deep — trees, depending on their growth rates, can exploit much of the soil-available water, and hence large volumes of water may be required to

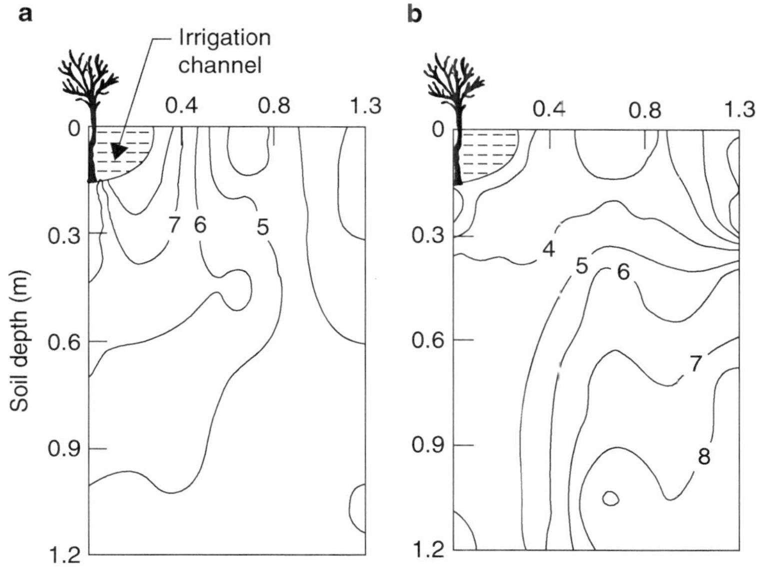

Fig. 2 Contour of soil salinity (**a**) before and (**b**) after monsoon rains in a tree plantation established with sub-surface planting and furrow irrigation (SPFIM) method (Source: Minhas et al. 1996)

leach salts; (2) a low infiltration rate in heavy-textured soils limits the volume of water that can be applied in an irrigation event; and (3) leaching of salt below the root zone may raise the water table and increase the likelihood of upward capillary rise of salt.

Tree Species for Saline Irrigation

Since the saline ground waters mostly exist under water scarcity zones, for sustaining viable wood-producing enterprise, tree species should both be tolerant to salinity and drought as well as well adapted to the local agro-climate. Drought-tolerant species, those develop deep, and extensive root systems are usually successful on long-term basis, although their early growth rates may be slower. Some species may show better growth when under irrigation but may slow down with the cessation of irrigation after establishment. Sufficient information is available about the tolerance of tree species to salinity from pot and short-term field studies (Schofield 1992; Yadav 1991; Ahmad and Ismail 1993a, b; Marcar et al. 1993; Davidson and Galloway 1993; Gupta et al. 1995; Marcar and Khanna 1997; Tomar and Minhas 1998; Barrett-Lennard 2003; Jaradat 2003; Dagar 2003; Dagar and Singh 2007), but knowledge of soil salinity tolerance of specific tree species under field conditions over a longer period of time is limited. With limited information due to lack of longer period field trials over a range of climate, cultural practices, soil types, and soil conditions (e.g., calcareousness goes along with aridity), it is very difficult to draw conclusions about the performance of individual tree species. In addition to salt tolerance as the selection criteria for specific sites, the socioeconomic and ameliorative role of trees has to be accorded due consideration in afforestation programs.

The information on establishment of trees and their subsequent growth under saline-irrigated conditions is mainly available from the Indian subcontinent. Ahmad et al. (1985) observed the faster growth of *Melia azedarach* than *Azadirachta indica* with saline irrigation (ECiw 4.5–14.0 dS m^{-1}), while the performance of *Prosopis juliflora*, *Acacia nilotica*, *Terminalia arjuna*, *Syzygium cuminii*, *Albizia lebbeck*, *Pongamia pinnata*, *Cassia articulata*, and *Cassia siamea* was better with waters ranging from EC 4.0–6.1 dS m^{-1} (Chaturvedi 1984, 1985). Jain et al. (1983, 1985) reported that *P. juliflora* and *Tamarix articulata* tolerated irrigation water salinity of 8 dS m^{-1}, and *Eucalyptus hybrid* and *Leucaena leucocephala* were moderately tolerant to saline irrigation (ECiw ~ 6 dS m^{-1}). Ahmad et al. (1987) and Ahmad and Ismail (1993a, b) observed that certain species of fuelwood and salt bushes show luxuriant growth at sandy strata when irrigated with saline water of oceanic strength. In trees such as *Azadirachta indica*, *Casuarina equisetifolia*, and *Eucalyptus camaldulensis*, three-fourth of the potential biomass could be obtained with waters having ECiw 15 dS m^{-1}; in *Prosopis juliflora*, it was 20 dS m^{-1}, while in *Tamarix articulata*, only beyond 30 dS m^{-1}. It is reported from Pakistan that tree species such as *Prosopis cineraria, P. juliflora, Acacia nilotica, Tamarix articulata, T. indica, T. stricta, Salvadora persica, S. oleoides*, and *Leucaena leucocephala* could be successfully utilized for revegetation of desert land (Thal, Thar, Choliston, and Chaki-Kharan) using waters from saline aquifers ECiw 4 to 18 dS m^{-1} (Qureshi et al. 1993a, b). Hussain and Gul (1993) evaluated 35 tree species on sodic to saline sodic waterlogged sites in Peshawar region and found that local species such as *Tamarix articulata, Acacia modesta*, and *A. nilotica* performed well but the exotics *Acacia stenophylla, A. ampliceps, Casuarina obesa, Eucalyptus camaldulensis,* and *Prosopis chilensis. P. siliquastrum* and *P. alba* exhibited better performance showing their suitability for saline environment.

Systematic efforts were made by Tomar et al. (2003b) to evaluate long-term performance of 31 tree species when irrigated initially with saline water (ECw ~ 8.6–10 dS m^{-1}) on a highly calcareous soil (*Typic Haplustalf*) in a semiarid part of north-west India (rainfall 499 mm $year^{-1}$; PAN-E 1888 mm $year^{-1}$). After 8 years of planting, the best performing tree rated on the basis of survival, vigor, and biomass yields was *Tamarix articulata* (73.5 Mg ha^{-1}) followed by *Acacia nilotica* (22.4 Mg ha^{-1}), *Prosopis juliflora* (20.2 Mg ha^{-1}), and *Eucalyptus tereticornis* (14.8 Mg ha^{-1}). Even after 20 years of growth, the highest biomass-yielding trees continued to be *Tamarix articulata* followed by *Acacia nilotica,A. tortilis, Eucalyptus tereticornis, Prosopis juliflora*, and *Azadirachta indica* (Fig. 3). In addition to fuelwood, the other benefit was improvement in organic matter of soils (>5 g kg^{-1} soil), e.g., *Acacia nilotica, A. tortilis, Azadirachta indica, Eucalyptus tereticornis, Feronia limonia, Tamarix articulata*, and *Guazuma ulmifolia* species from the original (3 g kg^{-1}). Working on the same site, Dagar et al. (2004, 2005, 2006, 2012) reported successful raising of *Salvadora persica, Catharanthus roseus, Cordia rothii*, and *Adhatoda vasica.*

Fruit and Oil-Yielding Tree Species

Most of the fruit trees with the exception of date palm (*Phoenix dactylifera*), pistachio (*Pistachio vera*), pomegranate (*Punica granatum*), olive tree (*Olea europaea*), Ber (*Ziziphus mauritiana*), and a few others are relatively sensitive to salinity stress. The most efficient method of using saline waters in widely spaced fruit trees is the drip irrigation. Enough information is available on cultivation of fruit trees with saline irrigation (Bernstein 1980; Bielorai et al. 1983, 1985; Mass 1986; Hoffman et al. 1988; Aronson et al. 1989; Picchioni and Miyamoto 1990; Banuls and Primo-Millo 1995; Aksoy et al. 1997; Boman 2000; Levy and Syvertsen 2004; Wiesman et al. 2004; Weissbein et al. 2008; Zeinadini et al. 2009; Kamiab et al. 2012) which helps in understanding the mechanism and preliminary concepts of salt tolerance. Mass (1986) rated date palm (*Phoenix dactilifera*) as tolerant; fig (*Ficus carica*), jujube (*Ziziphus mauritiana*), olive (*Olea europaea*), papaya (*Carica papaya*), pineapple (*Ananas comosus*), and pomegranate (*Punica granatum*) as moderately tolerant; grape (*Vitis vinifera*) as moderately sensitive; and almond (*Prunus dulcis*), apricot (*P. armeniaca*), sweet cherry (*P. avium*), sand cherry (*P. bessey*), peach (*P. persica*), plum (*P. domestica*), apple (*Malus sylvestris*), avocado (*Persea americana*), boysenberry (*Rubus ursinus*), cherimoya (*Annona cherimola*), grape fruit (*Citrus paradisiaca*), lemon (*C. limon*), lime (*C. aurantifolia*), orange (*C. sinensis*), pummel (*C. maxima*), tangerine (*C. reticulata*), loquat (*Eriobotrya japonica*), mango (*Mangifera indica*), passion fruit (*Passiflora

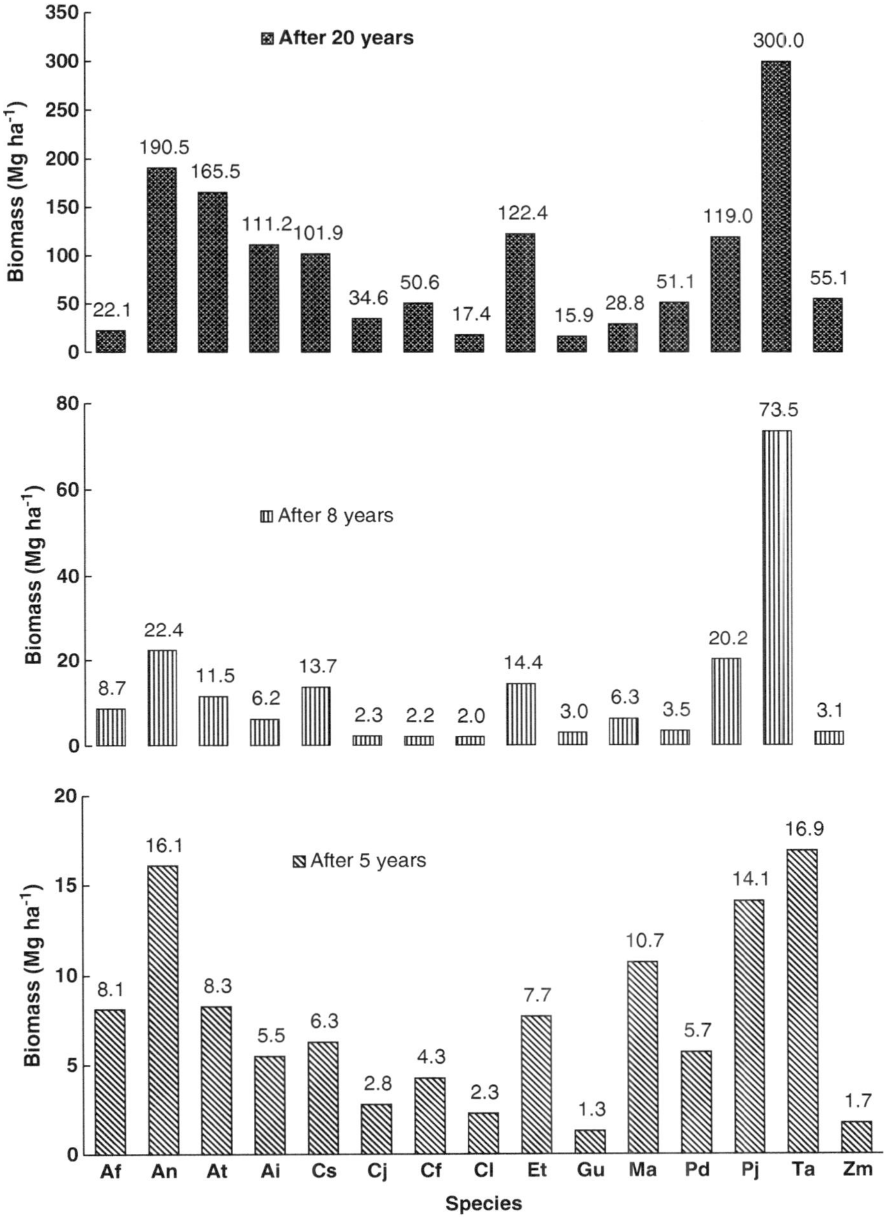

Fig. 3 Biomass of trees harvested after 5, 8, and 20 years of growth. Depictions: Af = *Acacia farnesiana*, An = *Acacia nilotica*, At = *A. tortilis*, Ai = *Azadirachta indica*, Cs = *Cassia siamea*, Cj = *C. javanica*, Cf = *C. fistula*, Cl = *Callistemon lanceolatus*, Et = *Eucalyptus tereticornis*, Gu = *Guazuma ulmifolia*, Ma = *Melia azedarach*, Pd = *Pithecellobium dulce*, Pj = *Prosopis juliflora*, Ta = *Tamarix articulata*, Zm = *Ziziphus mauritiana* (Source: Tomar et al. 2003b and Dagar et al. 2015, in review)

edulis), pear (*Pyrus communis*), persimmon (*Diospyros virginiana*), raspberry (*Rubus idaeus*), sapota white (*Casimiroa edulis*), and rose apple (*Syzygium jambos*) as sensitive to saline irrigation. Mostly olives are grown on well-drained soils and can tolerate moderate salinity of irrigation water. Generally, it is believed that high salinity levels reduce olive yields (Gucci and

Tattini 1997), but Klein et al. (1994) reported higher olive yield in Manzanillo trees irrigated with waters of ECiw 7.5 dS m^{-1}. Wiesman et al. (2004) reported average annual olive yield of Barnea trees to be 15.1, 18.5, and 15.5 Mg ha^{-1} when irrigated with water of ECiw 1.2, 4.2, and 7.5 dS m^{-1}, respectively, indicating yields enhancement under moderate salinity.

Weissbein et al. (2008) observed that in semiarid Israeli Negev area, saline irrigation treatment at 4.2 dS m^{-1} demonstrated only a low rate of retardation effect on growth and yield of olive trees compared with water at 1.2 dS m^{-1}. From the observations of many workers (Munns 1993; Klein et al. 1994; Gucci and Tattini 1997; Wiesman et al. 2004; Aragues et al. 2005; Weissbein et al. 2008), it could be concluded that different cultivars behaved differently, and most of these clearly showed the advantage of using water of ECw 4.2 dS m^{-1} for optimization of the horticultural performance. Further, leaching methodology based on drip irrigation enabled reduction of the salt level in the developing root zone to a level lower than 6 dS m^{-1}, which is essential for normal olive development. Because of salinity tolerance, the cultivation of pistachio is common in California, Arizona, southern New Mexico, and far west Texas. Picchioni and Miyamoto (1990) consider it to be potentially an alternative to salt-sensitive pecan (*Carya illinoensis*) and almond (*Prunus amygdalus*), and *Pistacia atlantica, P. terebinthus*, and *P. integerrima* are the major rootstock of the domestic pistachio industry. Interestingly, most pistachio plantations in all over the world are on saline soils (ECe>6 dS m^{-1}) and irrigated with low quality and saline water (Kamiab et al. 2012). In recent years, however, due to increasing salinity of soil and water, the pistachio yields have reduced. This negative influence was significant at concentrations exceeding 100 Mm salinity. The data reported by Kamiab et al. (2012) also showed that the mechanism of tolerance to salinity in *P. vera* cv. Badami–Zarad rootstock might be attributed to the better K^+-Na^+ discrimination and accumulation of osmolytes (proline and reduced sugars) under salt stress condition. The characteristics of Badami pistachio could be introduced as a mechanism for increasing to salt tolerance. Ahmad et al. (1987) and Ahmad and Ismail (1993a, b) observed that in some fruit trees such as date palm (*Phoenix dactylifera*), *Achras zapota, Grewia asiatica*, and *Ziziphus mauritiana* reduction in growth and fruit yield only with saline irrigation of ECiw 8–10 dS m^{-1}. Further, Dagar et al. (2008) and Dagar (2014) evaluated *Carissa carandas*, *Aegle marmelos*, *Emblica officinalis, Ziziphus mauritiana*, and *Feronia limonia* to be the most suitable fruit trees in a semiarid part of north-west India, and these could perform satisfactorily with saline water of ECiw ~ 10 dS m^{-1}. But there is need to evaluate proper rootstocks and cultivars of most of the salt-tolerant species of fruit trees in India.

At least 50 species of seed-bearing halophytic plants are potential sources of edible or industrial oil and proteins which can successfully be cultivated using sea water or highly saline ground water. Among the most important include *Salicornia bigelovii* (Glenn et al. 1991), *Terminalia catappa* and *Pandanus* spp. (Dagar 2003), *Suaeda moquinii* (Weber et al. 2001), *S. aralocaspica* (Wang et al. 2012), *Kosteletzkya virginica* (Gallagher 1985), *Salvadora persica* (Dagar 2003; Gururaja-Rao et al. 2004; Reddy et al. 2008), *Batis maritima* (Marcone 2003), *Crithmum maritimum* and *Zygophyllum album* (Zarrouk et al. 2003), *Nitraria sibiria, Suaeda salsa, Chenopodium glaucum*, and *Descurainia sophia* (Yajun et al. 2003). Importantly, in all these cases, seeds are relatively salt-free although there may be significantly higher concentrations of salts in other parts of the plant (Jaradat 2003).

Agroforestry Systems

Nonconventional crops (including halophytic crops) are now seen as alternative for farming in saline water-irrigated areas. Many of these crops along with salt-tolerant glycophytes may form constituents of viable and sustainable agroforestrysystems including silvopastoral systems. A perennial salt grass *Distichlis palmeri* (fiber content 8.4 %), used for making biscuits and bread, performs well in flooded and hypersaline conditions producing 1.25 Mg ha^{-1} grains (Pearlsteina et al. 2012). In coastal areas, it can be grown as an understory crop with trees like coconut palm (*Cocos nucifera*), coastal almond (*Terminalia catappa*), pongam (*Pongamia pinnata*), *Casuarina glauca*, and *Pandanus* spp. Quinoa (*Chenopodium quinoa*) is another food crop that can tolerate ECe up to 40 dS m^{-1} (Adolf et al. 2013) and is sold as a nutritious food item at a premium price. Similarly, psyllium (*Plantago ovata*) having medicinal value can be cultivated with saline water (ECw < 10 dS m^{-1}) with trees like *Acacia nilotica, A. tortilis, Tamarix articulata*, and *Feronia limonia* in dry regions of India (Tomar et al. 2010). Many mangrove species such as *Avicennia marina* and A. *germinans* are also used as food (Leith et al. 2000). A typical example of agri-horticultural system has been reported by Dagar et al. (2015b) where some of the crops like pearl-millet (*Pennisetum typhoides*), cluster bean (*Cyamopsis tetragonoloba*), barley (*Hordeum vulgare*), and mustard (*Brassica juncea*) could be taken on calcareous soil in low rainfall (<500 mm) areas in India along with fruit trees like Karonda (*Carissa carandas*), Indian Goose Berry (*Emblica officinalis*), and Bael (*Aegle marmelos*) with irrigation waters of (ECiw 8.5–10.0 dS m^{-1}). After 5 years, all fruit trees (Karonda, Goose berry, and Bael) started bearing fruits and recorded yields of 1.6, 0.5, and 3.5 Mg ha^{-1} fruits, respectively. The crop yields at different stages and growth of fruit trees are shown in Table 1.

Table 1 Yield (Mg ha^{-1}) of intercrops grown alone or under agri-horticultural (AH) systems in north-western India

	Mean for 5 years (2003–2007)		Mean for 4 years (2008–2011)	
Treatment	Barley*	Cluster bean*	Mustard**	Cluster bean**
Control (without trees)				
Low	3.55	1.41	1.58	0.87
Low/high	3.45	1.32	1.68	0.82
High	3.25	1.98	1.58	0.78
Carissa carandas (AH)				
Low	3.43	1.36	1.41	0.77
Low/high	3.32	1.28	1.33	0.71
High	2.99	1.21	1.18	0.69
Emblica officinalis (AH)				
Low	3.56	1.38	1.73	0.83
Low/high	3.29	1.27	1.66	0.78
High	3.04	1.16	1.58	0.73
Aegle marmelos (AH)				
Low	3.27	1.30	1.26	0.78
Low/high	3.08	1.25	1.21	0.72
High	2.78	1.14	1.11	0.66

Source: Dagar et al. (2015)
*Average of first 4 years
**Average of last 3 years. Salinity of irrigation water low (ECiw 5 dS m^{-1}), high (ECiw 8–10 dS m^{-1}), low/high alternate irrigation with low and high salinity water

Silvopastoral Systems/Forage Grasses

Some salt-tolerant plants/halophytes have long been utilized as fodder crops, e.g., Le Houerou (1994) reported that about 0.1 million ha of saline land was planted with *Atriplex* species in the Mediterranean Basin. The other possible candidate is *Leptochloa fusca* (Aslam et al. 1993). The soils with sandy substratum are especially suited to halo-xeric forages with saline irrigation (Ahmad and Ismail 1993a, b). Ahmad et al. (1987) and Ahmad and Ismail (1993a, b) reported that three-fourth of the potential yields could be obtained in *Sporobolus arabicus*, *Panicum turgidum*, and *Thinopyrum ponticum* when irrigated with water of ECiw 10–15 dS m^{-1}, while in *Leptochloa fusca* the ECiw for similar yields was 20 dS m^{-1}.

In Pakistan, about 11 million ha land resources are desert (Thal, Thar, Choliston, and Chaki-Kharan), and 31.7 million ha in India, consisting of great tracts of sand dunes, which in places, are interspersed with sparsely vegetated clay flats and ground water is highly saline ranging from ECiw 4 to 18 dS m^{-1}. These areas can be brought under silvopastoral system utilizing the local vegetation as well as saltbushes consisting of trees (*Prosopis cineraria, P. juliflora, Acacia nilotica, Tamarix articulata, T. indica, T. stricta, Salvadora persica, S. oleoides, Leucaena leucocephala*) and forages and grasses (*Atriplex* spp., *Maireana* spp., *Leptochloa fusca, Echinochloa crusgalli,*

Cenchrus ciliaris, *Arthrocnemum indicum*, *Salsola drummondii*, *Bienertia cycloptera*, *Indigofera oblongifolia*, and *I. cordifolia*) using saline aquifers (Qureshi et al. 1993a, b; Tewari et al. 2014). Further, Abdullah et al. (1993) tested 13 species of *Atriplex* and 8 of *Maireana* for their suitability in Cholistan desert with saline irrigation (~5 dS m^{-1}) and found that species of *Atriplex* (especially *amnicola, bunburyana, halimus,* and *lentiformis*) were most promising as compared to species of *Maireana*. Hanjra and Rasool (1993) reported a dry biomass of 23 Mg ha^{-1} in *L. leucocephala*, 8.0 Mg ha^{-1} in *Atriplex amnicola*, 7.5 Mg ha^{-1} in *Sesbania sesban*, 4.6 Mg ha^{-1} in *Cenchris ciliaris*, 4.3 Mg ha^{-1} in *Leptochloa fusca*, and 3.2 Mg ha^{-1} in *Panicum antidotale* showing their potential with saline irrigation as silvopastoral species in dry regions.

In the USA and Israel, the plantations of *Atriplex* raised with saline water are highly productive ranging from 12 to 20 Mg ha^{-1} dry matter per ha (Watson et al. 1987; Aronson et al. 1988). Malik et al. (1986) reported that *kallar* grass (*Leptochloa fusca*) grown on salty soil and irrigated with brackish water could produce 50 Mg ha^{-1} $year^{-1}$ fresh biomass. Hades and Frankel (1982) showed that the use of brackish irrigation water increases the rate of infiltration into a saline sodic soil. However, Aslam et al. (1993) observed that the application of brackish water did not cause any change in soil properties. In contrast, the roots of *kallar* grass were able to penetrate to depth creating vertical fine channels accelerating the leaching of salts down below 3 m in depth and increasing the hydraulic capacity of the soil. Thus, the cultivation of salt-tolerant plants like *kallar* grass also initiates a soil improvement process by providing soluble Ca^{+2} to the soil through dissolution of native $CaCO_3$ which lowers the pH. Rashid et al. (1993) demonstrated in Peshawar valley that *Atriplex lentiformis* (accession 159) was the most productive of the 20 saltbushes tested irrigating with brackish water. The other promising accessions were *A. amnicola* (971), *A. lentiformis* (178), *A. cineraria* (524), *A. undulata* (471), and *A. amnicola* (573). These saltbushes along with productive salt-tolerant grasses and forage trees may form ideal silvopastoral system on these degraded lands. Qadir et al. (1995) reported the potential of forage biomass production of 32.3 Mg ha^{-1} by *Sesbania aculeata*, 24.6 Mg ha^{-1} by *Leptochloa fusca*, 22.6 Mg ha^{-1} by *Echinochloa colona*, and 5.4 Mg ha^{-1} by *Eleusine coracana* in saline-sodic environment, and these species helped in soil amelioration in terms of reducing soil pH and salinity and increasing nitrogen in the order S. *aculeata*>*L. fusca*>*E. colona*>*E. coracana*. In semiarid part of India, Tomar et al. (2003a) observed forage grasses like *Panicum laevifolium* and *P. maximum*

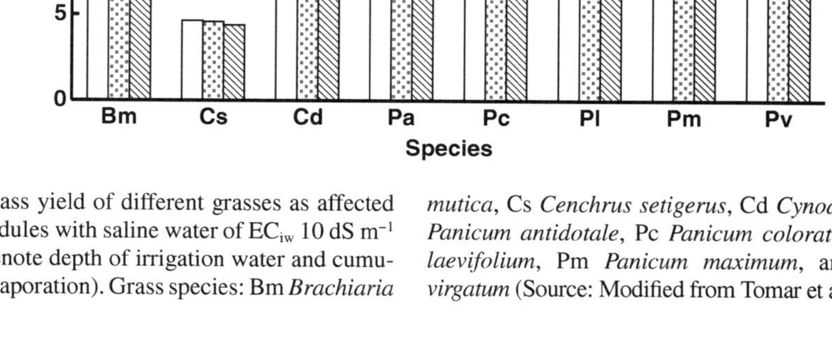

Fig. 4 Dry biomass yield of different grasses as affected by irrigation schedules with saline water of EC_{iw} 10 dS m^{-1} (Diw and CPE denote depth of irrigation water and cumulative open pan evaporation). Grass species: Bm *Brachiaria mutica*, Cs *Cenchrus setigerus*, Cd *Cynodon dactylon*, Pa *Panicum antidotale*, Pc *Panicum coloratum*, Pl *Panicum laevifolium*, Pm *Panicum maximum*, and Pv *Panicum virgatum* (Source: Modified from Tomar et al. 2003a)

were most suitable species producing annually 14–17 Mg ha^{-1} dry forage with saline irrigation (Fig. 4). About 25–30 % of total forage was also available during lean period of summer when most of the nomads move to other areas in search of forage for their cattle. These grasses along with native *Cenchris setigerus* and *C. ciliaris* could also be raised with trees like *Acacia nilotica, A. tortilis, A. ampliceps, A. farnesiana, A. modesta, Azadirachta indica, Feronia limonia, Prosopis juliflora, P. cineraria, Tamarix articulata, Cordia rothii, Salvadora persica*, and *Cassia siamea* (Dagar et al. 2008). Despite regular forage during rainy season, about 3–4 Mg ha^{-1} biomass from protected local grazing stands may be obtained if provided irrigation with saline water (ECiw ~10 dS m^{-1}) during the lean period.

On saline vertisol soils, grasses such as *Aeluropus lagopoides, Leptochloa fusca, Brachiaria mutica, Chloris gayana, Dichanthium annulatum, Bothriochloa pertusa*, and species of *Eragrostis, Sporobolus*, and *Panicum* are most suited to saline irrigation. Oil-yielding bush *Salvadora persica* was grown in combination with forage grasses such as *Leptochloa fusca, Eragrostis* sp., and *Dichanthium annulatum* on heavy textured saline vertisol (pH 7.2–8.9; ECe 25–70 dS m^{-1}). The underground water was 0.5–2 m from surface with ECiw 55–60 dS m^{-1}. These grasses with saline irrigation could produce on an average 3.72, 1.0, and 1.8 Mg ha^{-1} of forage, respectively. During fourth year, the seed yield of *Salvadora persica* ranged from 1.84 to 2.65 Mg ha^{-1} with oil contents ranging from 576 to 868 kg ha^{-1} (Gururaja-Rao et al. 2003).

Other Nonconventional Crops

A large number of salt-tolerant species, besides those mentioned above, can be used as agroforestry or sole crops with saline irrigation; however, these exhibit large differences in salt tolerance based on number of factors, including life cycle, frost tolerance, soil type, and climatic factors. Many important crops which have not mentioned earlier include tubers and foliage crops such as *Eleocharis dulcis, Sesuvium portulacastrum, Beta vulgaris*, and *B. maritima*; fruit yielding trees like *Achras zapota, Manilkara hexandra, Morinda citrifolia*, and *Borassus flabellifer*; pickles (*Capparis decidua* and *Cordia rothii*); liquid fuels *Beta vulgaris* and *Nypa fruticans*; gums, oils, and resins yielding many species of *Acacia, Sesbania*, and *Grindelia*; oil-like sperm whale (*Simmondsia chinensis*); source of natural rubber (*Parthenium argentatum*); and bioactive derivative-yielding plants (*Calophyllum inophyllum, Balanites roxburghii, Azadirachta indica, Catharanthus roseus*) are some interesting crops already adopted in many parts of the world and grown with saline irrigation (NAS 1990; Jaradat 2003; Dagar 2003, 2014; Dagar et al. 2009). Kefu et al. (1995) reported utilization of halophytes in China as source of starch and protein (species of *Zostera, Chenopodium, Atriplex*), oil (*Salicornia, Suaeda*), food and therapeutic value (*Limonium bicolor*), fiber (*Apocynum venetum*), medicine (*Ephedra sinica, Lycium barbarum, Kochia scoparia, Xanthium sibiricum, Glycyrrhiza uralensis, Artemisia stelleriana*), essential oil (*Aster, Artemisia*), and valuable fodder for domestic animals (*Agropyron sibiricum, A. mongolicum, Pennisetum alopecuroides, Spartina anglica, Nitraria sihirwa, Elaeagnus angustifolia, E. umbellata*), which are cultivated in agroforestry systems in saline environments. A number of species including *Tamarix chinensis, Phragmites australis, Miscanthus* spp. and *Spartina alterniflora* were evaluated as biofuel crops for ethanol production in the coastal zone of China (Liu et al. 2012). Halophytic perennial grasses along coastal area of Indian subcontinent such as *Halopyrum mucronatum, Desmostachya bipinnata, Phragmites karka, P. australis, Typha domingensis*, and *Panicum turgidum* are also known to be highly suitable for bioethanol production (Abideen et al. 2011). *Panicum virgatum* like corn (*Zea mays*) grown extensively as a conventional food crop as well as ethanol production (Hendricks and Bushnell 2008). In addition, sugar beet (*Beta vulgaris*), nipa palm (*Nypa fruticans*), and *kallar* grass (*Leptochloa fusca*) are identified sources of liquid and gaseous fuel (Jaradat 2003). Species of *Tamarix* which are highly biomass-producing

Table 2 Impact of different irrigation schedule on fresh yield (Mg ha^{-1}) of lemon grass when irrigated with water of different salinity

	Irrigation schedule (Diw/CPE ratio)				
ECiw (dS m^{-1})	0.2	0.4	0.6	0.8	Mean
Low	10.85	11.93	13.03	14.28	12.52
Low/high	7.65	8.63	9.97	11.53	9.45
High	4.31	7.12	8.51	10.03	7.49
LSD (p=0.05) ECiw 2.37 Irrig. Sch. 0.98 Interaction: NS					

Source: Dagar et al. (2013)

halophytic species are also source of bioethanol production (Panta et al. 2014).

Lemongrass (*Cymbopogon flexuosus*) has also been found promising crop with saline irrigation (Dagar et al. 2013). The average fresh foliage yield was found to be 12.5, 7.5, and 9.5 Mg ha^{-1}, respectively, when irrigated with water of low salinity (ECiw 4.0 dS m^{-1}), high salinity (ECiw 8.6 dS m^{-1}), and alternately with two waters. There was increase in yield with increase in irrigation water (Diw/CPE ratio) average yield ranging from 7.60 to 11.95 Mg ha^{-1} (Table 2). Furrow planting was superior to other methods. The yield with furrow, flat, and top of bund planting was 9.1, 5.7, and 3.1 Mg ha^{-1}, respectively (Dagar et al. 2013).

There lie varietal differences in salt tolerance of many of these species. For example, in lemongrass varieties, OD-58 and RRL-16 could produce 10–20 times biomass as compared to other tested varieties (Dagar et al. 2013), and among psyllium (*Plantago ovata*) varieties, JI-4 was found far superior followed by Sel-10 and Niharika when cultivated with saline water of ECiw 10 dS m^{-1} (Tomar et al. 2010). In another experiment, the aromatic grasses such as vetiver (*Vetiveria zizanoides*), lemon grass, and palmarosa (*Cymbopogon martini*), when irrigated with saline water (EC 8.5 dS m^{-1}) could produce on an average of 90.9, 10.4, and 24.3 Mg ha^{-1} dry biomass, respectively (Tomar and Minhas 2004a). Medicinal *Aloe barbadensis* was also equally tolerant and could produce 18 Mg ha^{-1} fresh leaves under partial shade. *Ocimum sanctum* yielded 910 kg ha^{-1} dry shoot biomass, while dill (*Anethum graveolens*), taramira (*Eruca sativa*), and castor (*Ricinus communis*) yields equaled 931, 965, and 3535 kg seeds per ha, respectively (Dagar et al. 2008). *Cassia senna* and *Lepidium sativum* could also be raised successfully. Among seed spices, fennel (*Foeniculum vulgare*) could be cultivated successfully producing 1.56 Mg ha^{-1} seed without showing any yield penalty when irrigated with water of ECiw 8.6 dS m^{-1} (Meena et al. 2014).

Floriculture

Skimina (1992) conducted a trial on about 100 ornamental species grown in 2.8 L containers and showed the feasibility of growing plants with recycled water without any yield reduction. Experiments conducted in Israel (Shillo et al. 2002), in California (Carter et al. 2002; Carter and Grieve 2008; Grieve et al. 2008a, b), and elsewhere (Niu and Cabrera 2010) proved the feasibility of growing ornamental plants and cut flowers even when the EC of irrigation water was not optimal for growth of other crops. These workers identified species of *Limonium, Dianthus, Celosia, Chrysanthemum, Antirrhinum, Gypsophila*, and *Matthiola* suitable for reuse system. Tomar and Minhas (2002, 2004a, b) and Dagar et al. (2008) reported that most of these species could be raised with water of ECiw 8.6 dS m^{-1}, and yield was higher when established with good-quality canal water. *Chrysanthemum indicum* and medicinal and aromatic oil yielding *Matricaria chamomilla* were the most tolerant. Recently, Cassaniti et al. (2013) have analyzed the potential use of brackish water for growing ornamentals. They have highlighted the effects of salt stress on floricultural crops, halophytic floriculture, importance of genetic traits in developing new crops, and management practices. Many halophytes have significant economic potential in terms of ornamental purpose. For example, *Maireana sedifolia* is used for its cut branches and exported to Europe from Israel. Ornamental species such as *Clianthus formosus, C. puniceus*, and many cultivars of *Portulaca oleracea* can successfully be grown using water of EC 5–15 dS m^{-1}. Interestingly, Cassaniti et al. (2013) identified 42 suitable halophytic species under 24 botanical genera found naturally in the Mediterranean environments and may be grown successfully as ornamental plants. Species such as *Limonium sinuatum, L. perezii, Trachelium caeruleum, Eustoma grandiflorum, Hippeastrum hybridum, Ornithogalum arabicum*, and several others have been identified suitable for saline irrigation having ECiw ranging from 2.5 to 11.5 dS m^{-1} (Grieve et al. 2008a, b).

Conclusions

In most of the arid and semiarid regions, good-quality water is scarce, and utilization of poor-quality water for agricultural purposes is inevitable. For that, we need to use adaptable, sustainable, viable, and affordable techniques and suitable salt-tolerant plants. Alternate land uses/agroforestry is ideal preposition for such areas involving hardy salt-tolerant forest and fruit trees, forage grasses, and low water requiring conventional and nonconventional crops including ornamental plants. There are opportunities to increase the salt tolerance of existing crops using conventional plant breeding and molecular approaches. There is need to produce low water (stress)-requiring and more salt-tolerant plant types. In the scenario of climate change, we also must conserve all the salt-tolerant, stress-tolerant, and also submergence-tolerant land races. Human resource development at all levels and strengthening extension network is as important as technology development. Recently, the attention is being paid toward commercial forestry, raising block plantation of commercial trees and also of biodiesel-yielding plantations like *Pongamia pinnata, Jatropha curcas* and *Euphorbia antisyphilitica*. Moreover, biomass from fast-growing trees like *Prosopis* and agricultural wastes may be used to generate electrical energy. Species such as *Tamarix articulata* and *Prosopis juliflora* producing huge biomass under saline conditions prove them ideal candidates for energy plantations. This approach will change the economic scenario by reducing the import of fossil fuels. Adopting biosaline agroforestry and silvopastoral systems, the nomadic behavior of large population will be checked in dry regions. This will have a tremendous social impact. Institutes already working in salinity-related problems may act as nodal institutes and may play more important role in promoting biosaline agriculture.

References

Abdullah M, Akram M, Khan WA, Davidson NJ (1993) Selecting halophytic shrubs for the Cholistan desert. In: Davidson N, Galloway R (eds) Productive use of saline land. ACIAR proceedings No. 42, Australian Centre for International Agricultural Research, Canberra, Australia, pp 45–48

Abideen Z, Ansari R, Khan MA (2011) Halophytes: potential source of ligno-cellulosic biomass for ethanol production. Biomass Bioenergy 35:1818–1822

Adolf VI, Jacobsen SE, Shabala S (2013) Salt tolerance mechanisms in quinoa (*Chenopodium quinoa* Willd.). Env Exp Bot 92:43–54

Ahmad R, Ismail S (1993a) Provenance trials in Pakistan: a synthesis. In: Davidson N, Galloway R (eds) Productive use of saline land. ACIAR proceedings No. 42, Australian Centre for International Agricultural Research, Canberra, Australia, pp 62–65

Ahmad R, Ismail S (1993b) Studies on selection of salt-tolerant plants for food, fodder and fuel from world flora. In: Lieth H, Al Masoom AA (eds) Towards the rational use of high salinity tolerant plants (Vol 2): agriculture and forestry under marginal soil water conditions, tasks for vegetation science 28. Kluwer Academic, Dordrecht, pp 295–304

Ahmad R, Khan D, Ismail S (1985) Growth of *Azadirachta indica* and *Melia azedarach* in coastal sand using highly saline water for irrigation. Pakistan J Bot 17(2):229–233

Ahmad R, Ismail S, Khan D (1987) Saline agriculture at coastal sandy belt. Final Research Report of Coordinated Research Program on Saline Agriculture. Karachi University/PARC, p 183

Aksoy U, Anac S, Anac D, Can HZ (1997) The effect of ground water salinity on satsuma mandarins: preliminary results. Acta Hortic 449:629–633

Aragues R, Puy J, Royo A, Espada JL (2005) Three yield field response of young olive trees (*Olea europaea* L. cv Arbequina) to soil salinity: trunk growth and leaf ion accumulation. Plant Soil 271:265–273

Armitage FB (1984) Irrigated forestry in arid and semi-arid lands: a synthesis. IDRC, Ottawa, p 160

Aronson JA (1989) HALOPH-a data base of salt tolerant plants of the world. Office of Arid Land Studies, University of Arizona, Tucson, p 77

Aronson JA, Pasternak D, Danon A (1988) Introduction and first evaluation of 120 halophytes under seawater irrigation. In: Whitehead EE, Hutchinson CF, Timmermann BN, Varady RG (eds) Arid lands- today and tomorrow. Westview Press, Colorado, pp 737–746

Aslam Z, Mujtab M, Alshtar J, Waheed R, Malik KA, Naqvi M (1993) Biological methods for economically utilizing salt-affected soils in Pakistan. In: Davidson N, Galloway R (eds) Productive use of saline land. ACIAR proceedings No. 42, Australian Centre for International Agricultural Research, Canberra, Australia, pp 29–31

Bandara G, Morris J, Stackpole D, Collopy J (2002) Soil profile monitoring at irrigated plantation sites in northern Victoria: progress report 2001. CFTT Report No. 2002/025, Forest Science Centre, Melbourne

Banuls J, Primo-Millo E (1995) Effects of salinity on some Citrus scion-rootstock combinations. Ann Bot 76:97–102

Barrett-Lennard EG (2003) Saltland pastures in Australia-a practical guide (Second Edition). State of western Australia and CRC for plant-based management of dryland salinity, South Perth, Western Australia, p 176

Bernstein L (1980) Salt tolerance of fruit crops. USDA Inform Bull 292

Bielorai H, Shalhevet J, Levy Y (1983) The effect of high sodium irrigation water on soil salinity and yield of mature grapefruit orchard. Irrig Sci 4:255–266

Bielorai H, Shalhevet J, Levy Y (1985) Grapefruit response to variable salinity in irrigation water and soil. In: Frenkel H, Meiri A (eds) Soil salinity, two decades of research in irrigated agriculture, vol 1. Van Nostrand Reinhold Co, pp 321–330

Boman B (2000) Effect of saline irrigation water on citrus production. University of Florida, Indian River Research and Education Center, Fort Pierce

Carter CT, Grieve CM (2008) Mineral nutrition, growth and germination of *Antirrhinum majus* L. (Snapdragon) when produced under increasing saline conditions. Hortic Sci 43:710–718

Carter CT, Grieve CM, Poss JA (2005) Salinity effects on emergence, survival and ion accumulation of *Limonium perezii*. J Plant Nutr 28:1243–1257

Carter CT, Grieve CM, Poss JA (2006) Response of Matthiola incana to irrigation with saline waste waters. Hortic Sci 42:119–123

Cassaniti C, Romano D, Hop MECM, Flowers TJ (2013) Growing floricultural crops with brackish water. Env Exp Bot 92:165–175

Chaturvedi AN (1984) Firewood crops in areas of brackish water. Indian Forester 110(4):364–366

Chaturvedi AN (1985) Fuelwood farming in degraded lands in the Gangetic plains. UP Forest Bulletin 50, Government of Uttar Pradesh, Lucknow

Dagar JC (2003) Biodiversity of Indian saline habitats and management & utilization of high Salinity tolerant plants with industrial application for rehabilitation of saline areas. In: Alsharhan AS, Wood WW, Goudie AS, Fowler A, Abdellatif EM (eds) Desertification in the third millennium. Swets & Zeitlinger Publishers, Lisse, pp 151–172

Dagar JC (2014) Greening salty and waterlogged lands through agroforestry systems for livelihood security and better environment. In: Dagar JC, Singh AK, Arunachalam A (eds) Agroforestry systems in India: livelihood security & environmental services-advances in agroforestry, vol 10. Springer, New Delhi/Dordrecht/New York, pp 273–332

Dagar JC, Singh G (2007) Biodiversity of saline and waterlogged environments: documentation, utilization and management. National Biodiversity Authority, Chennai, p 76

Dagar JC, Tomar OS, Kumar Y, Yadav RK (2004) Growing three aromatic grasses in different alkali soils in semi-arid regions of northern India. Land Degrad Dev 15:143–151

Dagar JC, Kumar Y, Tomar OS (2005) Performance of ornamental and medicinal periwinkle under saline environment. Indian J Hortic 62(2):175–180

Dagar JC, Tomar OS, Kumar Y, Bhagwan H, Yadav RK, Tyagi NK (2006) Performance of some under-explored crops under saline irrigation in a semi-arid climate in northwest India. Land Degrad Dev 17:285–299

Dagar JC, Tomar OS, Minhas PS, Singh G, Jeet-Ram (2008) Dryland biosaline agriculture -Hisar experience. Tech Bull 6. CSSRI, Karnal, p 28

Dagar JC, Rao GG, Shukla YK, Sharma HB (2009) Performance of three flower-yielding plants in different sodic soils. Indian J Hortic 66(3):404–409

Dagar JC, Yadav RK, Ahamad S (2012) *Euphorbia antisyphilitica*: a potential petro-crop for degraded calcareous soils and saline water irrigation in dry regions of India. J Soil Salinity Water Qual 4(2):86–91

Dagar JC, Tomar OS, Minhas PS, Mukesh K (2013) Lemongrass (*Cymbopogon flexuosus*) productivity as affected by salinity of irrigation water, planting method and fertilizer doses on degraded calcareous soil in a semi-arid region of northwest India. Indian J Agric Sci 83(7):734–738

Dagar JC, Yadav RK, Minhas PS, Tomar OS, Gajender, Lal K (2015) Fruit-based agroforestry systems for saline water irrigated semi-arid hyperthermic camborthids regions of north-west India. Agrofor Syst DOI 10.1007/s10457-015-9889-4 (in print)

Davidson N, Galloway R (eds) (1993) Productive use of saline land. ACIAR proceedings No. 42. Australian Centre for International Agricultural Research, Canberra, p 124

Gallagher JI (1985) Halophytic crops for cultivation at seawater salinity. Plant Soil 89:323–336

Ghassemi F, Jakeman AJ, Nix HA (1995) Salinisation of land and water resources: human causes, extent, management and case studies. CABI Publishing, Wallingford

Glenn EP, O'Leary JW, Watson MC, Thompsom TL, Kuehl RO (1991) *Salicornia-bigelovii* Trr – an oilseed halophyte for seawater irrigation. Science 251:1065–1067

Grieve CM, Poss JA, Shouse PJ (2008a) Modeling growth of Matthiola incana in response to saline waste waters differing in nitrogen level. Hortic Sci 43(6):1787–1793

Grieve CM, Wu L, Rollins L, Harivandi A (2008b) Tolerance to landscape plants of salinity and of specific ions. In: (Chapter V) A comprehensive literature review on salt management guide for landscape irrigation with recycled water in coastal southern California. http://www.salinitymanagement.org

Gucci R, Tattini M (1997) Salinity tolerance in olive. Hortic Rev 21:177–214

Gupta RK, Singh NT, Sethi M (1994) Groundwater quality for irrigation in India (Map 1: 6 million scale), Bull 19. Central Soil Salinity Research Institute, Karnal

Gupta RK, Tomar PS, Minhas PS (1995) Managing salty soils and waters for afforestation. Tech Bull No. 7/95. CSSRI, Karnal, p 23

Gururaja-Rao G, Nayak AK, Chinchmalatpure AR (2003) *Salvadora persica*: a life support species for salt affected black soils. Tech Bull 1/2003. CSSRI, Karnal, p 44

Gururaja-Rao G, Nayak AK, Chinchmalatpure AR, Nath A, Babu VR (2004) Growth and yield of *Salvadora persica* a facultative halophyte grown on saline black soil (vertic haplustept). Arid Land Res Manage 18:51–61

Hades A, Frankel H (1982) Infiltration as affected by long term use of sodic saline water for irrigation. Soil Sci Soc Am J 46:524–529

Hanjra SH, Rasool S (1993) Potential of *Atriplex* as forage for livestock in Pakistan. In: Davidson N, Galloway R (eds) Productive use of saline land. ACIAR proceedings No. 42, Australian Centre for International Agricultural Research, Canberra, Australia, pp 68–70

Hendricks RC, Bushnell DM (2008) Halophytes energy feed stocks: back to our roots. In: The 12th international symposium on transport phenomena and dynamic of rotation machinery, Honolulu, Hawaii, 17–22 Feb 2008, ISROMAC12-2008-20241, pp 1–11

Hillel D (2000) Salinity management for sustainable agriculture. The World Bank, Washington, DC

Hoffman GJ, Shannon MC, Mass EV, Grass L (1988) Rubber production of salt-stressed guayule at various plant populations. Irrig Sci 9:213–226

Hussain A, Gul P (1993) selection of tree species suitable for saline and waterlogged areas in Pakistan. In: Davidson N, Galloway R (eds) Productive use of saline land. ACIAR proceedings No. 42, Australian Centre for International Agricultural Research, Canberra, Australia, pp 53–55

Jain BL, Goyal RS, Muthana KD (1983) Performance of some tree species in relation to irrigation with saline waters. Ann Arid Zone 22:233–238

Jain BL, Muthana KD, Goyal RS (1985) Performance of tree species in salt-affected soils in arid regions. J Indian Soc Soil Sci 33:221–224

Jaradat AA (2003) Halophytes for sustainable farming systems in the Middle East. In: Alsharhan AS, Wood WW, Goudie AS, Fowler A, Abdellatif EM (eds) Desertification in the third millennium. Swets & Zeitlinger Publishers, Lisse, pp 187–204

Kamiab F, Talaie A, Javanshah A, Khezri M, Khalighi A (2012) Effect of long-term salinity on growth, chemical composition and mineral elements of pistachio (*Pistacia vera* cv Badami-Zarand) rootstock seedlings. Ann Bio Res 3(12):5545–5551

Kefu Z, Zi-Yi C, Shou-Jin F, Giang HX, Hai F, Zeng LF, Harris PJC (1995) Halophytes in China. In: Khan MA, Ungar IA (eds) Biology of salt tolerant plants. Department of Botany, University of Karachi, Karachi, pp 284–293

Klein L, Ben-Tal Y, Lavee S, David I (1994) Saline irrigation of cv Manzanillo and Uovo De piccione trees. Acta Hortic 356:176–180

Le Houerou HN (1994) Forage halophytes and salt-tolerant fodder crops in the Mediterranean basin. In: Squires VR, Ayou AT (eds) Halophytes as a resources for livestock and for rehabilitation of degraded land. Kluwer, Dordrecht, pp 123–137

Leith H, Lohmann M, Guth M, Menzel U (2000) Cash crop halophytes for future halophytes growers. Institute of Environmental System Research, University of Osnabreuck, Osnabreuck, p 32

Levy Y, Syvertsen J (2004) Irrigation water quality and salinity effects in citrus trees. Hortic Rev 30:37–82

Liu XZ, Wang CZ, Su Q, Lib CK (2012) The potential resource of halophytes for developing bio-energy in China coastal zone. Herald J Agric Food Sci Res 1:44–51

Malik KA, Aslam Z, Naqui M (1986) Kallar grass: a plant for saline land. Nuclear Institute for Agriculture and Biology, Faisalabad, p 93

Marcar NE, Khanna PK (1997) Reforestation of salt-affected and acid soils. In: Nambier EKS, Brown AG (eds) Management of soil, nutrients and water in tropical plantation forests. ACIAR monograph 43, Canberra, Australia, pp 481–524

Marcar NE, Morris J (2005) Plantation productivity in saline landscapes. In: Nambiar EKS, Ferguson J (eds) New forests: wood production and environmental services. CSIRO, Collingwood, pp 51–74

Marcar NE, Crewford DF, Leppert PM (1993) The potential of trees for utilization and management of salt-affected land. In: Davidson N, Galloway R (eds) Productive use of saline land. ACIAR proceedings No. 42, Australian Centre for International Agricultural Research, Canberra, Australia, pp 17–22

Marcone MF (2003) *Batis maritime* (Saltwort/Beachwort): a nutritious, halophytic, seed bearings, perennial shrub for cultivation and recovery of otherwise unproductive agriculture land affected by salinity. Food Res Int 36:123–130

Mass EV (1986) Salt tolerance in plants. Appl Agric Res 1:12–26

Meena RL, Ambast SK, Gupta SK, Chinchmalatpure AR, Sharma DK (2014) Performance of Fennel (*Foeniculum vulgare* Mill.) as influenced by saline water irrigation and organic input management in semi-arid conditions. J Soil Salinity Water Qual 6(1):52–58

Minhas PS (1996) Saline water management for irrigation in India. Agric Water Manage 30:1–24

Minhas PS (2012) Sustainable management of brackish-water agriculture. Soil water and agronomic productivity. In: Lal R, Stewart BA (eds) Adv Soil Sci. 19:289–327

Minhas PS, Bajwa MS (2001) Use and management of poor quality waters in rice-wheat production system. J Crop Prod 4:273–306

Minhas PS, Gupta RK (1992) Quality of irrigation water: assessment and management. ICAR, New Delhi, p 123

Minhas PS, Samra JS (2004) Waste water use in Peri-urban agriculture: impacts and opportunities. Tech Bull 1/2004. Central Soil Salinity Research Institute, Karnal, India, p 78

Minhas PS, Singh YP, Tomar OS, Gupta RK, Gupta Raj K (1996) Effect of saline irrigation and its schedules on growth, biomass production and water use by *Acacia* and *Dalbergia* on a highly calcareous soil. J Arid Environ 36:181–192

Minhas PS, Singh YP, Tomar OS, Gupta RK, Gupta Raj K (1997) Saline water irrigation for the establishment of furrow planted trees in northwestern India. Agrofor Syst 35:177–186

Minhas PS, Yadav RK, Lal K, Chaturvedi RK (2015) Effect of long-term irrigation with wastewater on growth, biomass production and water use by *Eucalyptus tereticornis* planted at variable stocking density. Agric Water Manage 152:151–160

Morris J, Bickford R, Collopy J (1994) Tree and shrub performance and soil conditions in a plantation irrigated with saline ground water. Res. Report No. 357. Dept. Conserv and Natural Resource, Victoria, p 37

Munns R (1993) Physiological processes limiting plant growth in saline soil: some dogmas and hypotheses. Plant Cell Environ 16:15–24

NAS (1990) Saline agriculture: salt-tolerant plants for developing countries. National Academy of Science (NAS), Washington, DC, p 143

Niu G, Cabrera RI (2010) Growth and physiological responses to landscape plants to saline water irrigation-a review. Hortic Sci 45(11):1605–1609

O'Connor GA, Elliott HA, Bastian RK (2008) Degraded water reuse: an overview. J Environ Qual 37(5):S157–S168

Panta S, Flowers T, Lane P, Doyle R, Haros G (2014) Halophytes agriculture: success stories. Env Exp Bot 107:71–83

Pearlsteina SL, Felger RS, Glenn EP, Harringtond J, Al-Ghanem KA, Nel-sona SG (2012) NyPa (*Distichlis palmeri*): a potential grain crop for saltwater irrigation. J Arid Environ 82:60–70

Picchioni GA, Miyamoto S (1990) Salt effects on growth and ion uptake of pistachio rootstock seedlings. J Am Soc Hortic Sci 115(4):647–653

Qadir M, Ahmad N, Qureshi RH, Ilyas M (1995) Role of salt tolerant plants to reduce salination and sodification. In: Khan MA, Ungar IA (eds) Biology of salt tolerant plants. Department of Botany, University of Karachi, Karachi, pp 371–379

Qureshi RH, Aslam M, Rafiq M (1993a) Expansion in the use of forage halophytes in Pakistan. Davidson N, Galloway R (eds) Productive use of saline land. ACIAR proceedings No. 42, ACIAR, Canberra, Australia, pp 12–16

Qureshi RH, Nawaz S, Mahmood T (1993b) Performance of selected tree species under saline-sodic field conditions in Pakistan. In: Lieth H, Al Masoom AA (eds) Towards the rational use of high salinity tolerant plants (Vol 2): agriculture and forestry under marginal soil water conditions, tasks for vegetation science 28. Kluwer Academic, Dordrecht, pp 259–269

Rashid A, Khattak JK, Khan MZ, Iqbal MJ, Akbar F, Khan P (1993) Selection of halophytic forage shrubs for the Peshawar Valley, Pakistan. In: Davidson N, Galloway R (eds) Productive use of saline land. ACIAR proceedings No. 42, ACIAR, Canberra, Australia, pp 56–61

Reddy MP, Shah MT, Patolia JS (2008) *Salvadora persica*, a potential species for industrial oil production in semiarid saline and alkali soils. Ind Crop Prod 28:273–278

Rogers AL (1985) Foliar salt in Eucalyptus species. Aust For Res 15:9–16

Schofield NJ (1992) Tree planting for dryland salinity control in Australia. Agrofor Syst 20:1–23

Shillo R, Ding M, Pasternak D, Zaccai M (2002) Cultivation of cut flower and bulb species with saline water. Sci Hortic 92:41–54

Skimina CA (1992) Recycling water, nutrients and waste in the nursery industry. Hortic Sci 27:968–971

Stevens RM, Sweeney SM, Meissner AP, Frahn WA, Davies G (1999) Survival, growth and water use of a range of tree species irrigated with saline drainage water. Aust For 62:97–105

Tanwar BS (2003) Saline water management for irrigation. Work Team Paper on Use of Poor Quality Water for Irrigation (WT-PQW). International Commission on Irrigation and Drainage (ICID), New Delhi, p 123

Tewari JC, Ram M, Roy MM, Dagar JC (2014) Livelihood improvements and climate change adaptations through agroforestry in hot arid environments. In: Dagar JC, Singh AK, Arunachalam A (eds) Agroforestry systems in India: livelihood security & environmental services-advances in agroforestry, vol 10. Springer, pp 155–184

Theiveyanathan S, Benyon RG, Polglase P, Falkiner RA, Myers BJ, Rushdy El-Gendy (2000) Irrigation management of tree plantations in the southern Murray-Darling Basin. In: GJ Connellan GJ (ed) Proceedings of the conference water — essential for life, Irrigation Association of Australia, Melbourne, Australia, 23–25 May 2000, pp 52–61

Thorburn PJ, Walker GR, Jolly ID (1995) Uptake of saline groundwater by plants: an analytical model for semi-arid and arid areas. Plant Soil 175:1–11

Tomar OS, Gupta RK (1985) Performance of some forest tree species in saline soils under shallow and saline water table conditions. Plant Soil 87:329–335

Tomar OS, Minhas PS (1998) Afforestation of salt-affected soils. In: Tyagi NK, Minhas PS (eds) Agricultural salinity management in India. Central Soil Salinity Research Institute, Karnal, pp 453–471

Tomar OS, Minhas PS (2002) Performance of some ornamental winter annual flowering species under saline irrigation. Indian J Hortic 59(2):201–206

Tomar OS, Minhas PS (2004a) Relative performance of aromatic grasses under saline irrigation. Indian J Agron 49(3):207–208

Tomar OS, Minhas PS (2004b) Performance of medicinal plant species under saline irrigation. Indian J Agron 49(3):209–211

Tomar OS, Minhas PS, Sharma VK, Gupta RK (2003a) Response of nine forage grasses to saline irrigation and its schedules in a semi-arid climate of north-west India. J Arid Environ 55:533–544

Tomar OS, Minhas PS, Sharma VK, Singh YP, Gupta RK (2003b) Performance of 31 tree species and soil condition in a plantation established with saline irrigation. For Ecol Manage 177:333–346

Tomar OS, Dagar JC, Minhas PS (2010) Evaluation of sowing methods, irrigation schedules, chemical fertilizer doses and varieties of *Plantago ovata* Forsk to rehabilitate degraded calcareous lands irrigated with saline water in dry regions of north western India. Arid Land Res Manage 24:133–151

Wang L, Zhao ZY, Zhang K, Tian CY (2012) Oil content and fatty acid composition of dimorphic seeds of desert halophytes *Suaeda aralocaspica*. Afr J Agric Res 7(12):1910–1914

Watson MC, O'Leary JW, Glenn EP (1987) Evaluation of *Atriplex lentiformis* (Torr.) S. Wats. and *Atriplex nummularia* Lindl. As irrigated forage crops. J Arid Environ 13:293–303

Weber DJ, Gul B, Khan MA, Williams T, Wayman P, Warner S (2001) Comparision of vegetable oil from seeds of native halophytic shrubs. In: McArthur ED, Fairbanks DJ (eds) Proceedings of the Shrubland ecosystem genetics and biodiversity. RMRS-P-21, USDA Forest Service, Ogden UT, Rocky Mountain Research Station, USA, pp 287–290

Weissbein S, Wiesman Z, Ephrath Y, Silberbush M (2008) Vegetative and reproductive response of olive cultivars to moderate saline water irrigation. Hortic Sci 43(2):320–327

Wiesman Z, Itzhak D, Ben-Dom N (2004) Optimization of saline water level for sustainable Bornea olice and oil production in desert conditions. Sci Hortic 100:257–266

Yadav JSP (1991) Problems and potentials of reforestation of salt affected soils in India. Report of Regional W/E Development Programme in Asia (GCP/RAS/III/NET). Bangkok, FAO Field Document 14, p 56

Yajun B, Xiaojing L, Weiqiang L (2003) Primary analysis of four salt tolerant plants growing in Hai-He plain, China. In: Leith H, Mochtchenko M (eds) Cash crop halophytes: recent studies. Kluwer, London, pp 135–138

Zarrouk M, El Almi H, Ben Youssef N, Sleimi N, Smaoui A, Ben Miled D, Abdelly C (2003) Lipid composition of seed of local halophytes: *Cakile maritime*, *Zygophyllum album* and *Crithmum maritimum*. In: Leith H, Mochtchenko M (eds) Cash crop halophytes: recent studies. Kluwer, London, pp 121–124

Zeinadini A, Etesami H, Nagad-Mashizi SG (2009) Effect of the developing saline water on the quality of irrigation waters and soils physicochemical characteristics of pistachio cultivation of Sirjan area. World Appl Sci J 6(6):748–753

Potential of Wastewater Disposal Through Tree Plantations

R.K. Yadav, P.S. Minhas, Khajanchi-Lal, and J.C. Dagar

Abstract

The adverse effects of irrigation with treated and untreated wastewater include health risks due to pathogens, salts, nutrients, and toxic elements that contaminate the food chain and the environment. However, the areas especially those afflicted by water scarcity can afford for recycling and reuse of wastewater in tree plantations as an effective and sustainable strategy. This is associated with high water, nutrient, and pollutant (metal) assimilation capacity of tree plantations. In woody species, wood, bark, and roots form important sinks for biologically available metals. Since these tissues are slow to enter the decomposition cycle, accumulated metals remain immobilized for considerably longer periods. Urban plantations and green areas along with nonedible crops like aromatic grasses and floriculture crops further offer many economic, social, recreational, and biodiversity conservation benefits. Nevertheless, as a caution, it is stated that the deep-rooted perennials have the ability, but are not guaranteed, to profligate wastewater disposal under all the soil and climatic conditions. Therefore, the regulatory mechanisms must be evoked to control loading rates for safe disposal of wastewater and protection of groundwater from being contaminated.

Introduction

Increasing supplies of water to meet the expanding demand of nonagricultural sectors are resulting in generation of huge volumes of wastewater in commensuration with rapidly increasing population, urbanization, improved living standards, and changing dietary habits. In most of the countries, urban drainage and disposal systems are common and the wastewaters consist of effluents from residential, institutional, commercial and

R.K. Yadav (✉) • J.C. Dagar
Central Soil Salinity Research Institute, Karnal 132 001, Haryana, India
e-mail: rk.yadav@icar.gov.in; dagarjc@gmail.com

P.S. Minhas
National Institute of Abiotic Stress Management, Baramati, Pune 413115, Maharashtra, India
e-mail: minhas_54@yahoo.co.in

Khajanchi-Lal
Water Technology Centre, Indian Agricultural Research Institute, New Delhi 110012, India
e-mail: khajanchi@iari.res.in

J.C. Dagar, P.S. Minhas (eds.), *Agroforestry for the Management of Waterlogged Saline Soils and Poor-quality Waters*, Advances in Agroforestry 13, DOI 10.1007/978-81-322-2659-8_10

industrial establishments and urban runoff water arising from rains or storms. These contain a variety of contaminants that require suitable treatment before use in agriculture. After appropriate treatment, wastewater can be recycled for irrigation of crops and aquaculture, landscape, urban and industrial uses, recreational and environmental uses, and artificial groundwater recharge, but irrigation continues to be the most prevalent practice. About 20 million hectares of agricultural land are irrigated with treated and untreated wastewater throughout the world (Raschid-Sally and Jayakody 2008). Since most of the low- to middle-income developing countries do not have sufficient resources to treat wastewaters, raw or diluted wastewaters are mainly used for irrigation of crops or disposed either in surface water bodies or on land. Despite substantial benefits of wastewater use in agriculture through supplementing nutrients and water resources and supporting the livelihood of millions of smallholder farmers, use of untreated wastewater directly or polluted water from rivers and streams poses potential risks of pathogen and toxic chemical contamination of crops and to the public health and environment in the form of soil and groundwater pollution (Minhas and Samra 2004; Drechsel and Evans 2010; Drechsel et al. 2010).

As an alternative, irrigation of tree species grown for fuel and timber with wastewater is another approach, which can help in overcoming health hazards associated with sewage farming (Braatz and Kandiah 1998; Thawale et al. 2006). Developing and enlivening the green belts around the cities with forest trees under wastewater irrigation can further revive the ecological balance and improve environmental quality by self-treatment of wastewater through land application and forest irrigation. Tree plantations are often expected to use water at higher rates than the shorter vegetation. This is because of greater aerodynamic roughness of tree plantations, clothesline effect in tree rows, and deeper rooting system for accessing water down to several meters of soil. Thus, the systems of agroforestry, which have come to be known as HRTS (high-rate transpiration systems), are the land application systems based upon the transpiration capacity of tree species that promote the treatment of wastewater through the renovating capability of a living soil filter enabling recycling and reuse of wastewater and conservation of nutrient energy into biomass and thereby bringing multiple benefits to society such as fuel wood, environmental sanitation and eco-restoration. However, to ensure optimal plantation growth and environmental protection, loading rates of wastewater and its constituents should match the water and nutrient requirements of proposed plantations. Keeping above in view, some of the interventions for minimizing the adverse effects of wastewater irrigation, especially for the safer disposal of wastewater in urban plantations and landscapes for greening urban areas, have been discussed here.

Management Interventions for Risk Reduction Using Wastewater

The risks of using untreated or partially treated wastewater in agriculture can be reduced through wastewater treatment and nontreatment options or a combination of both (WHO 2006). These include the following:

Water Quality Improvements

The first and foremost step for improvement in wastewater quality is primary treatment. This is simply a sedimentation process in which organic and inorganic solids are allowed to settle and then removed. The process reduces the biological oxygen demand (BOD) by 25–50 %, the total suspended solids by 50–70 %, and the oil and grease contents by 55–65 %. Some organic nitrogen, phosphorus, and heavy metals are also removed. Primary-treated effluents may be of acceptable quality for irrigation (Ayers and Westcot 1985) of trees, orchards, vineyards, fodder crops, and some processed food crops.

Secondary treatment can be implemented using methods such as waste-stabilization ponds, constructed wetlands, infiltration–percolation, and upflow anaerobic sludge blanket reactors (Mara 2003). Storing reclaimed water in reservoirs improves microbiological quality and provides peak-equalization capacity, which increases the reliability of supply and improves the rate of reuse (Qadir et al. 2010). Constructed wetlands also serve as habitat for wildlife and anthropogenic wastewater discharge and treatment and stabilize other related ecological disturbances. Aquatic plants such as *Typha latifolia*, *Phragmites karka*, *Eichhornia crassipes*, *Salvinia molesta*, *Pistia stratiotes*, *Scirpus tabernaemontani*, *Colocasia esculenta*, and *Azolla filiculoides* established in wetlands can also be used for paper pulp. The wetland acts as a biofilter, removing sediments and pollutants such as heavy metals from the water. Groundwater recharge with deep percolation through soil aquifer treatment (SAT) can remove microorganisms, provided that soil properties are appropriate and the process is properly managed (Asano and Cotruvo 2004).

Human Exposure Control

Protective measures such as wearing of gloves, boots, and mask, washing hands properly, and changing irrigation methods can reduce farmers' exposure. The sprinklers should not be used for irrigation. Public awareness campaigns are also important in minimizing the transmittance of diseases through wastewater use.

Phytoremediation

Phytoremediation is an emerging technology using selected plants to clean up the contaminated environment from hazardous contaminants to improve the environment quality. Heavy metals (HMs), the most potential contaminants in wastewater, cannot be degraded biologically but only transformed from one oxidation state to another or form organic complexes. Among the 300 accessions of 30 plant species, *Brassica* (*juncea, napus, rapa*) exhibited moderately enhanced Zn and Cd accumulation (Ebbs et al. 1997). The fern *Pteris vittata* was capable of accumulation of As to the extent of only 0.7 mg g^{-1} dry weight of plant, and aquatic *Azolla caroliniana* and terrestrial *Populus nigra* could accumulate 0.2 mg As g^{-1} dry weight of plant root (Tangahu et al. 2011). Some species have shown hyper-accumulation of different metals such as *Brassica campestris, B. carinata, B. juncea*, and *B. nigra* >100 mg Pb g^{-1} dry weight and *B. napus, B. oleracea*, and *Helianthus annuus* >50 mg Pb g^{-1} dry weight and *B. juncea* >1 mg Hg g^{-1} dry weight but with every chance of food chain contamination. However, if cultivated for fuel wood, trees as component of agroforestry can serve the purpose of phytoremediation. Similarly, Lal et al. (2008) observed that cut flowers such as marigold (*Tagetes erecta*), chrysanthemum (*C. indicum*), and gladiolus (*Gladiolus grandiflorus*) have shown promise in Cd-contaminated environment while *Jasmine sambac, Jasmine grandiflorum*, and *Polianthes tuberosa* are the other ornamental and cut flower species suitable for urban greening and avenue culture with wastewater irrigation (Augustine 2002). Lal et al. (2013) observed that lemon grass (*Cymbopogon flexuosus*) could be successfully grown using primary-treated municipal wastewater for achieving higher productivity without contamination of the end product – the essential oil. Lemon grass could also accumulate heavy metals such as Cd, Cr, Ni, and Pb from wastewater used for irrigation containing these metals acting as phytoremedial agent.

More than 400 plant species have been identified as metal hyper-accumulators (Reeves 2003; Lone et al. 2008; Table 1) which include either high-biomass plants such as willow (*Salix* spp.)

Table 1 Number of plant species that are reported to have hyper-accumulation traits (metal concentration >1000 mg kg^{-1} dry weight)

Metal	Number of species	Metal	Number of species
As	04	Pb	14
Cd	01	Se	20
Co	34	Zn	04
Cu	34	Hg	01
Ni	>320		

Source: Reeves (2003), Lone et al. (2008), and Tangahu et al. (2011)

or those that have low biomass but high hyperaccumulating characteristics such as species of *Thlaspi* and *Arabidopsis*.

Farm-Level Wastewater Management

Improved wastewater irrigation management at farm level includes suitable practices such as crop selection, irrigation management, and other soil-based interventions. Such interventions can reduce potential health and environmental risks to only some extent. However, global surveys indicate that three-fifth of diluted or untreated wastewater is used for raising vegetables and cereals (Raschid-Sally and Jayakody 2007). Thus, there are always chances of food chain contamination as many of the vegetables are consumed raw and also contain metals beyond the permissible levels. A safer alternative could be the production of urban forestry, agroforestry, and avenue and roadside plantations grown for fuel and timber and also the other crops where the economic component is nonedible and thus do not come directly in the food chain.

When choosing irrigation methods, farmers should consider the quality of water supply to manage use-associated potential health and environmental implications due to pathogenic and metal contamination of crops. Furrow irrigation, especially subsurface drippers, provides higher health protection to farmers and consumers as compared to flooding (Minhas and Samra 2004). An additional possibility is the cessation of irrigation, prior to harvest, to allow pathogens' natural die-off.

Soil-based interventions without the production of edible plants are important, particularly in the case when wastewater is contaminated with heavy metals, which usually accumulate in surface soil layers. For moderate levels of metals and metalloids in wastewater, there is no particular management needed if the soils are calcareous; however, there can be a problem in acidic soils, which require lime treatment, and when irrigating with wastewater containing elevated levels of sodium, soil structure deterioration may occur and we require application of a calcium source such as gypsum (Qadir et al. 2010). Care has also to be taken regarding detrimental effects of salts, nitrates, metals, and pathogens reaching groundwater; the shallower the water table, the more the danger.

Harvest and Post-harvest Interventions

These interventions involve the process of harvest, post-harvest cleaning, and handling during transport, marketing, storage, and preparation in kitchens. Minhas et al. (2006) gave details of these processes and also suggested to harvest cereal and fodder crops above a certain height from the ground to minimize pathogens. They also advocated the introduction of low-cost and relatively safer practices such as washing and post-harvest handling methods to reduce the pathogenic load of wastewater-irrigated crops.

Urban Plantations for Wastewater Disposal

Use of wastewater for irrigation is an age-old practice; however, large-scale controlled irrigation in established sewage farms for disposal and prevention of pollution in surface water bodies dates back to only the last century. Although crops were raised on these farms, the crop productivity was a secondary consideration. The El-Gabal El-Asfar sewage farm (Egypt) was established in 1911 covering 200 ha of tree plantations to dispose Cairo City's untreated wastewater and was expanded to 1260 ha in the mid-1980s with conversion of forests to *Citrus* along with the production of cereals and vegetables (Braatz and Kandiah 1998). Pioneering work on the application of treated municipal wastewater on forest lands as a means of purification and groundwater recharge was carried out in Central Pennsylvania, USA, during 1963 to 1977 (FAO 1978). Effluent irrigation on tree plantations in Victoria, Australia, commenced in 1973 (Myers et al. 1996). These provide the benchmark potential productivity of wastewater-irrigated 14-year-old *Eucalyptus grandis* and *E. saligna* in terms of mean annual increment (MAI) in wood volume of 41 and 31 $m^3ha^{-1}year^{-1}$ (Baker 1998; Duncan et al. 1998). Wastewater-irrigated 4-year-old *E. globulus* plantations at 1333 and 2667 stems ha^{-1} stocking density also produced annual volume growth and MAI of 126 and 91 m^3 ha^{-1} and 27 and 36 m^3 ha^{-1}, respectively, while the corresponding MAI values for *E. grandis* were 19 and 26 m^3 ha^{-1}. Hopmans et al. (1990) also observed MAI of 33 $m^3ha^{-1}year^{-1}$ for *E. saligna* at 1500 stems ha^{-1} and 31 m^3ha^{-1} for *E. grandis*. This indicates that short-rotation coppicing of these two species can serve as a suitable option for wastewater disposal (Boardman 1996).

Untreated sewage was used to irrigate *Acacia salicina*, *Eucalyptus camaldulensis*, and *Tamarix aphylla* (Armitage 1985); mixed hardwood stand consisting mainly of oaks (*Quercus* spp.), red pine (*Pinus resinosa*), and white spruce (*Picea glauca*) (Braatz 1996) in Australia; and *neem* (*Azadirachta indica*) and date palm (*Phoenix dactylifera*) in UAE forestry plantations for urban greening. In Murray–Darling Basin (Australia), area under wastewater-irrigated tree plantations increased from 500 to 1500 ha in > 60 variable-size effluent sites (CSIRO 1995). Most of these studies were aimed on handling the disposal of wastewater problem, however, to utilize the nutrient potential of wastewater; economic gains are also considered in recent plantations. For example, in Egypt, Omran et al. (1998) observed better growth of orange trees; increase in water and nutrient availability through effluent application influenced the growth of trees such as *Pinus radiata* (Sheriff et al. 1986), *Eucalyptus grandis* (Stewart et al. 1990), *Populus deltoides*, *E. tereticornis*, *Leucaena leucocephala* (Das and Kaul 1992), *Casuarina glauca*, *E. camaldulensis*, *Tamarix aphylla* (El-Lakany 1995), *Hardwickia binata* (Paliwal et al. 1998), *Acacia nilotica* trees (Singh and Bhati 2004), *Dalbergia sissoo* (Singh and Bhati 2005), olive tree *Olea europaea* (Aghabarati et al. 2008), *Casuarina equisetifolia* (Kumar and Reddy 2010), *Pinus eldarica* (Tabari et al. 2011), and *E. tereticornis* (Minhas et al. 2015). The availability of sufficient quantity of nutrition in wastewater induced luxuriant growth of the most of these plantations. Tabari et al. (2011) observed in afforested *Pinus eldarica* (15-year-old) stands that trees showed better growth in the field irrigated using municipal water than in plots irrigated with well water, as indicated by the increased diameter at breast height (DBH), basal area, and standing volume of trees in wastewater-irrigated fields (Table 2).

Table 2 Growth parameters of 15-year-old *Pinus eldarica* when irrigated with wastewater and well water

Irrigation type	DBH (cm)	Height (m)	Basal area (cm^2)	Standing volume (m^3)
Wastewater	17.95	10.04	264.2	0.139
Well water	13.50	9.02	135.0	0.65

Source: Tabari et al. (2011)
DBH diameter at breast height

Selecting Appropriate Tree Plantations

Selection of plantation species for urban agroforestry or greening with beneficial use of wastewater will depend on the prevailing environmental conditions for which they are planned. However, after due consideration of given local climate and soils and wastewater quality and quantity, the following important traits should be considered in species selection:

- Fast growth, although it is relative to the quality of the wood or other products produced. Generally, fast growth is suitable for pulpwood and materials for panel products but not for sawn wood products because of the often lower density of fast-grown wood. Some species (e.g., *Populus* and *Eucalyptus*) that traditionally have been grown fast in plantations can be managed for saw wood by lengthening rotations, aggressive thinning, and early pruning. However, for wastewater use, urban tree species should have the following characteristics:

- Tolerance to soil conditions, i.e., reaction, salinity, metal load, and excess water.
- Tolerance to climatic conditions like temperatures, insolation, and wind conditions.
- Ease of propagation, including a reliable seed supply if seedlings are used.
- Evergreeness; this allows the plantation to utilize higher quantities of wastewater.

In Egypt, substantial volumes of wastewater generated in cities and villages are used in forest plantations (MSEA 2006). Zalesny et al. (2015) recommended some plantation species, based on their suitability, growth potential, use, and economic value, for wastewater irrigation. These include pine (*Pinus* spp.), eucalyptus (*Eucalyptus* spp.), and poplar (*Populus* spp.) for pulpwood or sawn wood, mahogany (*Khaya ivorensis*) and teak (*Tectona grandis*) for high-value products, and beech wood (*Gmelina arborea*) for only pulpwood. *Salix* has an excellent capacity to take up metals such as cadmium and cesium (Cs-137) from the soil and could be used for environmental protection. Cesium and potassium have been found to compete in the metabolism of the plant, and thus uptake of cesium could be increased by reducing potassium fertilization. Salt tolerance will also be an important criterion for potentially saline effluent disposal on urban sites and environments, while water use is the main consideration in controlling groundwater pollution. Information on salt tolerance of different plantation crops is available in chapters "Global Perspectives on Agroforestry for the Management of Salt-affected Soils and Agroforestry to Rehabilitate the Indian Coastal Saline Areas" of this book. *Eucalyptus* species are generally considered to be effective for wastewater utilization purposes. *Eucalyptus camaldulensis* is a hardy tree that grows under a wide range of climatic conditions and soil types has been found quite successful. Some provenances of the species even tolerate saline water and soil conditions quite well. *Acacia nilotica*, *Dalbergia sissoo*, and *Tecomella undulata*, *Populus*, and *Tamarix* are other species that have performed quite well in plantations under excess soil moisture conditions (Bhutta and Chaudhry 2000). In addition to considerations of quantity and composition of wastewaters, economic benefits accruing from urban plantations is another major factors that decides the choice of plantation species.

There are many species of trees adapted to urban and suburban growing conditions, such as *Leucaena leucocephala*, that provide high-quality fodder for livestock. Similarly, a large percentage of urban dwellers, especially the poor, use firewood as their primary cooking and heating fuel and depend on nearby green areas for their source of wood. Urban greening can provide sustainable fuel wood plantings to meet the needs of these urban residents. Fruits, nuts, and fiber are some of the other forest products that could be harvested from wastewater-irrigated urban and suburban plantations and green areas. Most trees that provide these products are found in private lots and gardens. Generally, the ornamental value is the main consideration of selecting suitable horticultural plantation species for greening public urban areas as these are less subject to damage and theft.

In the peri-urban areas of Hubli in Karnataka state of India, all farmers bordering the wastewater *nalla* (small channel) engage in less water-requiring wastewater-irrigated agroforestry plantations on their private properties which reduces exposure to wastewater. In some areas, the main wastewater-irrigated agroforestry land uses are orchards and agrisilviculture which consists of spatially mixed tree–crop combinations. The two important tree species are sapota (*Achras zapota*) and guava (*Psidium guajava*) and other common species are coconut, mango, areca nut, and teak. Species grown on farm boundaries included neem (*Azadirachta indica*), tamarind (*Tamarindus indica*), *Eucalyptus* spp., poplar (*Populus deltoides*), *Acacia* spp., coconut (*Cocos nucifera*), and teak (*Tectona grandis*). About 20–25 % yield advantage has been observed from wastewater irrigation in comparison to tube well water-irrigated fields (Bradford et al. 2003).

Regulating Wastewater Use Disposal in Urban Plantations

Wastewater-irrigated urban forestry/agroforestry plantations have been recognized as a strategy to use urban wastewater, while also rehabilitating and greening wastelands. The availability of permanent streams of wastewater has enabled urban farmers to diversify their cropping practices. Spatial distribution of plantations, flowers, or agroforestry systems results from a combination of availability and composition of wastewater, labor, soil type, area, and its landscape within urban or peri-urban areas. As such, present scenario of wastewater use in close urban and peri-urban areas of developing countries includes adoption of a year-round intensive vegetable system. Further away from the cities, less intensive farming systems are practiced, without consideration of adverse effects of wastewater irrigation. However, the wastewater still offers advantages in terms of early-season irrigation and increasing growth and production from green plantations, flower crops, and fruit trees in agroforestry systems.

Plants such as *Eucalyptus*, poplar (*Populus* spp.), pine (*Pinus* spp.), bamboo (*Bambusa arundinacea*), acacia (*Acacia mangium*), neem (*Azadirachta indica*), and Indian rose wood (*Dalbergia sissoo*) which have high-rate transpiration system (HRTS) can be effectively utilized as safer alternative for beneficial disposal of wastewater. Such plants can transpire higher quantum of wastewater than the potential evapotranspiration possible from the site soil matrix alone. The higher wastewater use in plantations is due to the combination of deeper rooting, extended growing seasons, and higher inputs of radiant energy because of lower albedos as compared to herbaceous covers or crop lands. Khanzada et al. (1998) monitored the water use of *Acacia nilotica*, *A. ampliceps*, and *Prosopis pallida* on 3–5-year-old plantation sites with contrasting soil and water quality conditions in the Indus Valley in Pakistan. *A. nilotica* used the maximum water, which varied from 1248 to 2225 mm depending upon plantation growth conditions. Under tropical semiarid conditions of Northwest India, Minhas et al. (2015) monitored the transpiration rates of wastewater-irrigated *Eucalyptus* plantations to range between 418–473, 1373–1417, and 1567–1628 mm during 7–10 years of planting under low (163), recommended (517), and high (1963 stems per ha) stocking density (per ha), respectively (Fig. 1). When compared to reference evapotranspiration computed using the Penman method (ETref), the sap flow for recommended density varied between 0.87–1.23 × ETref with an average 1.03 × ETref. Similarly, at Wagga Wagga, Myers et al. (1996) observed that *P. radiata* and *Eucalyptus grandis* attained comparable water use and *Eucalyptus* at closed canopy had the maximum daily water use rate of< 8 mm d^{-1} and varied between 0.84–0.93 × Epan (open pan evaporation). Thus, overall water use by trees varies with specific site conditions defining soil type, evaporative demands, and even the depth to groundwater and its salinity (details in chapter "Use of Tree Plantations in Water Table Drawdown and Combating Soil Salinity"). Under favorable conditions (sandy and deep soils, shallow water table of good quality, cooler climate), trees may draw soil water at about 0.8 × Epan, but these may reduce to about 0.2 × Epan under less optimal conditions (clayey and shallow soils, saline/deeper water table, hot and dry summer, etc.). Nevertheless, their major advantage in wastewater use can be viewed in terms of year-round water with drawls unless being deciduous. Also the exact amount of nutrients taken up by *Eucalyptus* depends upon climate and plantation vigor (Rockwood et al. 1996, 2004). To avoid the groundwater contamination, due to wastewater use in plantations, the wastewater application should be regulated as per the evapotranspiration and nutrient use potential of the site plantations. Nutrients present in the wastewater should be used by the plants and partly retained in the soil matrix without affecting the soil ecosystem.

Heavy Metal Recycling Potential

To tackle the limitations of conventional wastewater treatment systems and avoid food chain

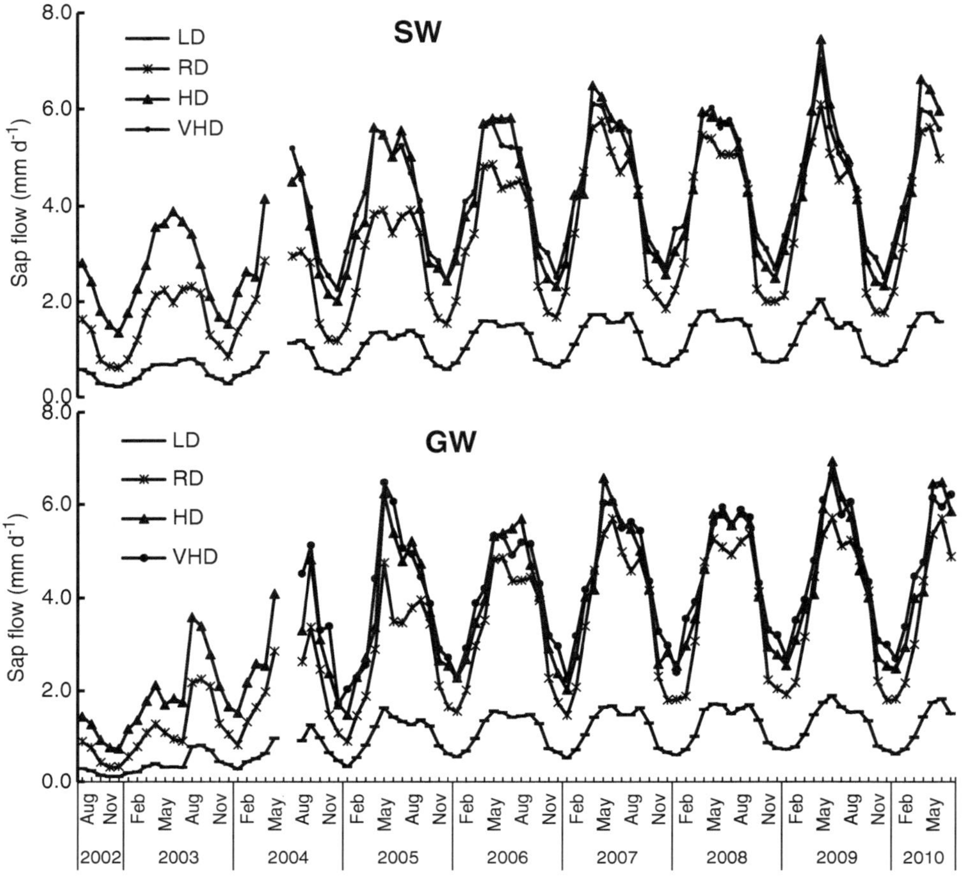

Fig. 1 Monthly average of mean daily sap-flow values per hectare (mm d^{-1}) for groundwater-irrigated (GW) and sewage water-irrigated (SW) *Eucalyptus* plantations (LD, RD, HD, and VHD denote low (163), recommended (517), high (1993), and very high density (3520 stems ha^{-1})) (Source: Minhas et al. 2015)

contamination due to use in agriculture, alternative low-cost, eco-friendly methods need to be evolved for safer disposal and desired level of treatment. Phytoremediation, a cost-effective "green" technology, mainly relies on nutrients, salts, and metal-accumulating plants to remove polluting metals from soil or water (Salt et al. 1998; Reeves 2003; Lone et al. 2008). A list of about 400 terrestrial plants species, having 100–1000 times more accumulation potential for one or more heavy metals (HMs) than those normally accumulated by plants grown under the same conditions, has been prepared by Hooda (2007). In comparison to food arable crops, wastewater irrigation in plantations is a relatively safer, cost-efficient, and environmentally sound way to treat and dispose wastewater (Armitage 1985). On wastewater-irrigated soils, *Acacia nilotica*, *Dalbergia sissoo*, and *Acacia modesta* accumulated relatively higher HMs than several bushes and grass species. HM concentrations in these species varied as per the composition of wastewater in the order of Fe > Zn > Cr > Pb > Ni > Cd > As. All the species exhibited higher HM composition in the root as compared to shoot (Irshad et al. 2015). Though there is lack of reports on symptoms of HM toxicities in tree species, it indicates their tolerance mechanisms to withstand higher HM concentrations than agricultural crops (Riddell-Black 1993, 1994). It has been observed that even those trees which are not selected for metal tolerance generally survive in metal-contaminated soil but with reduced growth rate (Dickinson et al. 1992). Beneficial effects of organic load in wastewater and sludge on tree growth processes have been found to far outweigh any adverse impacts of the added metals. Prolonged sewage irrigation markedly increased the amount of Fe, Zn, Mn, Cu, Pb, and Ni in the leaves and fruits of *Citrus* and olive without adversely affecting their growth and accumulation of metals beyond the safe limits in fruits (Khalil 1990; Maurer et al. 1995; Aghabarati et al. 2008). Therefore, use of low-strength wastewater did not pose any threat to *Citrus* and olive trees and consumers from heavy metal accumulation. Batarseh et al. (2011) found the accumulation being independent of the heavy metal concentration in the wastewater, suggesting a selective uptake of the metals by the olive plants. Also the trend of heavy metal transfer from soil to olive fruits and leaves was almost the same, showing a consistency of transfer. Dinelli and Lombini (1996) observed that metal concentrations were generally higher in the early vegetative growth stage, due to a relatively high nutrient uptake compared to growth rate. This was followed by a period of vigorous growth, which diluted the concentrations until the flowering stage, in which the minimum values for almost all elements were obtained. Several tree species grown on sludge-amended spoil had the highest concentrations of Cd, Cu, Ni, and Zn in root tissues. Wood and bark are important sinks for biologically available metals, with additional sink tissue being formed each growing season. These tissues are slow to enter the decomposition cycle; accumulated metals can, therefore, be immobilized in a metabolically inactive compartment for a considerable period of time (Lepp 1996). Massive root systems of trees upon establishment bind the soil thus promoting soil stabilization. Moreover, addition of litter to the surface quickly leads to an organic cover over the contaminated soil. In addition, transpiration by the trees reduces downward and lateral flow of water in the soil and thus reduces the amounts of heavy metals transferred to groundwater and surface water. Deep-rooting plants could reduce the highly toxic Cr(VI) to Cr(III), which is much less soluble and, therefore, less bioavailable (James 2001). It could be because organic products of root metabolism, or resulting from the accumulation of organic matter, could act as reducing agents (Pulford and Watson 2003). Proper management of municipal effluent irrigation and periodic monitoring of soil and plant quality parameters are required to ensure successful, safe, long-term municipal effluent irrigation.

Wastewater Treatment and Nutrient Removal

A major concern with wastewater irrigation is the fate of excess from plantation sequestration potential of nitrogen (N) and phosphorus (P) in the environment as these cause pollution of surface water and groundwater (Duncan et al. 1998). While P will not usually be a problem in short to medium term, particularly on soils with high P adsorption capacity, there is always potential risk for leaching of N (as nitrate) to groundwater. Fast-growing trees initially accumulate significant amounts of N, but the net N requirement declines after canopy closure stage of plantation when recycling sets in via decomposition of litter and internal translocation. Therefore, removal of N and other nutrients by plantation can be maximized by growing trees in short rotations. However, short rotations (<6 years) may compromise water use and also limit the potential products to biomass for fuel rather than higher-value wood. Across *Eucalyptus* species, aboveground elemental uptake (mainly N, P, K) by coppiced or original trees generally increase in proportion to biomass accumulation with > half of N being in foliage. Duncan et al. (1998) observed that under high-N-strength (15 mg L^{-1}) wastewater irrigation, maximum amount of N was sequestered in the potentially harvestable biomass of 3-year short rotations; while longer rotations (12 years) could achieve a similar balance of inputs through assimilation of N equivalent to supply from only lower-strength (5 mg L^{-1}) effluents. Another serious issue associated with wastewater irrigation is the management of metallic pollution. The tree species for urban greening should be selected on the basis of nature of elements present in wastewater to be used. *Acacia*, *Mimosa*, *Anadenanthera*, and *Salix* are efficient in absorbing Cd; *Eucalyptus* and mangroves are efficient in Pb accumulation; *Genipa americana* is efficient in Cr absorption; and *Salix viminalis* can remove up to 20 % Cd and 5 % Zn.

Sewage treatment plants are established in large cities only, and as such, there are no or a very few wastewater treatment facilities available in small and medium urban settlements of most of third-world countries. Apparently for these areas, decentralized natural treatment systems such as urban plantations with green areas and constructed wetlands could prove to be cost-effective and environment-friendly as an alternative wastewater treatment. Urban plantations in combination with the constructed wetlands (CWL) remove higher nitrogen (66–73 %) and phosphate (23–48 %) compared to un-vegetated wetlands (Juwarkar et al. 1995). Similarly, treatment performance and removal efficiencies of various types of CWL for BOD, TSS, and nutrients (total P and total N and NH_4-N) was compiled by Vymazal (2010). Under sewage irrigation and compost mulch application, *Eucalyptus* biomass yields were found to increase more than twice in comparison to those of *Populus* after 3 years of growth at Orlando (USA), revealing better performance of *Eucalyptus* both in terms of environmental and economic implications. Rockwood et al. (2004) observed that, under sewage irrigation, *Eucalyptus* can reduce N and P leaching by 75 %. Relative concentrations of N, P, and K in *Eucalyptus* plant tissues were reported to be in order of foliage > stem bark > branches > stem wood. HRTS plantations could remove N by 60–76.2 %, whereas the removal of phosphate was comparatively less than nitrogen and it ranged from 17.7 to 70.3 %. N removal efficiency of *Casuarina equisitifolia* was more as compared with *Dendrocalamus strictus*. In addition to N removal, these plantations also reduced wastewater biological oxygen demand (BOD) with removal efficiency ranging from 80.0 to 94.3 % (Thawale et al. 2006). Urban greening in the form of agroforestry system plantations has many advantages, which include sink potential for water and air pollutants, aesthetics, and biomass generation for energy.

Wastewater Treatment with Urban Plantations

The ponds, rivers, and wetlands with plantations as part of natural treatment of wastewater also serve for recreation, wildlife habitat, aesthetics, and educational use. Wetlands, one of the most biologically diverse ecosystems and a resource for tertiary wastewater treatment, increase habitats for flora and fauna in and along the waterways. The biological functions and physical aeration occurring in the wastewater during the passage of wastewater in the waterways remove many of the toxic effluents from the wastewater (ICLEI 1995). These plantations with wetlands in urban park systems are the low-cost wastewater treatment facilities for low-income cities. There are several alternatives to wastewater treatment and disposal that can be incorporated in green areas. As such wastewater can be used to irrigate urban and suburban agriculture and forests, horticultural projects (flowers for export), city landscaping and parks, and tree farms. All of these options provide for a safe and productive means of wastewater disposal (Braatz 1993). This reuse of wastewater not only recharges the aquifer but also reduces the demand on scarce freshwater resources. Controlled recycling wastewater into urban plantations and green park areas or forested, farmed and degraded lands may also be more economical than finding ways to dispose of it somewhere else.

Benefits of Wastewater Irrigation in Urban Plantations

The potential benefits of wastewater irrigation in tree plantations include relatively safer and low-cost treatment and disposal; augmenting nutrient and water supplies; environmental services such as climate improvement, soil enrichment, biodiversity improvement, and carbon sequestration, hence mitigating climate change; and livelihood security through various products such as timber, fuel wood, food, and employment. Some of these benefits are discussed here in brief.

Safe, Low-cost Treatment and Disposal

The cost of conventional methods of treatment of entire wastewater will be very high, prohibitively so for most developing countries. As a result, these countries depend on other forms of relatively cheaper disposal and treatment options. Among these, use in urban tree plantations and greening areas can be one such alternative. However, long-term sustainability of wastewater irrigation in tree plantations also depends on site-specific soil, climate, species, application techniques, and sociopolitical environment. There should be balance between wastewater disposal rates and evapotranspiration and nutrient-/pollutant-carrying capacity of the plantations grown at the site. Controlled application of wastewater in forestry plantations at 2.5 cm per week effectively filtered out excess N, P, and other constituents and made it acceptable for crop production and even drinking. Availability of nutrients from wastewater also improved tree growth by 80 to 186 % (Braatz and Kandiah 1998). In many cases, wastewater when passed through created wetlands was purified enough for irrigation of arable crops and agroforestry. Thus, low-strength municipal wastewater can be recycled through urban forestry plantation ecosystems with the benefits of increasing tree growth, restoring the water quality, and recharging groundwater reserves.

Livelihood Source

Wastewater use in urban forestry plantations supports livelihoods of urban poor in many parts of the world. It is a common reality in urban and peri-urban areas of more than three-fourth of the cities of developing nations in Asia, Africa, and Latin America. The majority of urban poor in these cities have an urgent need for improvement in quality of life with employment opportunities, provision of shelter, potable water, and recreation. In these regions, care should be taken to design urban plantations and green areas to supplement these needs and improved quality of life. An important aspect of urban greening is the jobs for poor, skilled and unskilled, laborers. Urban greening projects are often labor intensive and provide both initial jobs, such as soil preparation, planting, etc., as well as more permanent employment in the form of maintenance and management of plantations and green areas. Project managers of forestry components of an urban greening program in Mexico City have estimated that the program needed 3380, 3700, 800, and 100 workers to produce and transport plants, for working in the plantations, for management, and for protection and surveillance, respectively, in existing green areas (IDB 1992). In addition to basic amenities, urban green space also satisfies diverse basic human needs as food, fuel, and shelter from trees and shrubs, because tree products, if sold, provide direct cash benefits and, if used within the household, they provide indirect cash benefits by freeing cash income for other uses. Trees themselves can improve existing savings/investments, secure tenure, or increase property value. As such, urban greening has many indirect benefits in terms of conservation of land and environment, controlling floods and erosion, saving energy, and providing habitat for wildlife in addition to recreation and health and other material benefits. However, in this chapter, our main focus remains on more safer and beneficial use of wastewater through plantations and urban green areas.

Urban plantations can provide significant material benefits in areas where poles, firewood, and fodder are in high demand. Tree species that produce poles for fence posts are highly valued, especially in arid regions where low-cost fencing materials are scarce. Poles are also used in construction, furniture making, and crafts. There are many species of trees adapted to urban and suburban growing conditions, such as *Leucaena leucocephala*, that provide high-quality fodder for livestock. Similarly, a large percentage of urban dwellers, especially the poor, use firewood as their primary cooking and heating fuel and depend on nearby green areas for their source of wood. Urban green park areas developed using wastewater can provide sustainable fuel wood plantings to meet the needs of these urban residents. Most trees that provide fruits, nuts, and fiber can be grown in private gardens.

Environmental Benefits

Urban plantations and green areas provide some direct and other indirect benefits related to improvement in quality of life. In addition to direct benefits such as fuel wood, food, fodder, and poles, these improve air, water, and land resources and provide also a safer outlet for disposal of urban wastes which help in the improvement of health, recreation, environmental education, aesthetics, and enhancement of landscape, especially for the urban poor. Plantations also help in controlling erosion, urban water supplies, and habitats for wildlife. Depending on management objectives of urban plantations, the focus is quite different in developed cities and relatively poorer urban dwellings; however, multipurpose urban plantations are beneficial in all conditions. Urban plantations and green areas should be designed on the basis of needs and desires of local populations so that these can serve maximum possible benefits. Overall, the term urban agroforestry including urban greening using plantations involves the management of urban and peri-urban plantation in a planned, integrated, and systematic manner to achieve the maximum environmental, social, and economic well-being of the urban society.

Carbon Sequestration

Tree plantations offer additional advantage of mitigating a predicted increase in atmospheric carbon concentration through their potential to absorb more carbon efficiently. *Eucalyptus* plantation can play an important role as carbon sinks and contribute significantly to the removal of CO_2 from the atmosphere (Prasad et al. 2012). During the process of photosynthesis, the atmospheric CO_2 is utilized by the leaves to produce photosynthetic pathway compounds, which get stored either in the roots or bole. The carbon absorption by tree plantations in a given area varies with plantation age corresponding to variations in growth as well as plantation density. Carbon absorption is also expected to increase with better tree growth caused by essential plant nutrients supplied through sewage irrigation. Minhas et al. (2015) recorded that rate of increments in stock volumes of wastewater-irrigated *Eucalyptus tereticornis* plantations increased with plantation density and age, and for densities <2000 stems ha^{-1}, it peaked during the 6th year of growth as compared to earlier in higher densities. The overall carbon temporal sequestration potential of different densities of wastewater and tube well water-irrigated *Eucalyptus* plantations varied from 19.9 Mg ha^{-1} for wastewater irrigation in 3rd year of growth to 351 Mg ha^{-1} (Table 3). Thus the wastewater irrigation in plantations can help in increasing the carbon sequestration potential of urban plantation.

Improvement in Climate and Energy Savings

While air pollution indices in many cities in more developed countries have dropped over the years, air pollution levels have been also rising in cities throughout much of Latin America and the Caribbean. Carter (1993) reported that average level of particulate suspension in the atmosphere of Mexico City increased from 615 % between 1974 and 1990. Those most affected by such detrimental air contaminants are children, the elderly, and poor people with respiratory prob-

Table 3 Temporal changes in carbon sequestration potential of sewage-irrigated (SW) and groundwater-irrigated (GW) *Eucalyptus* (Mg ha^{-1})

Plantation age (years)	Stocking density (stems ha^{-1})							
	165		520		1990		6530	
	SW	GW	SW	GW	SW	GW	SW	GW
3	19.9	21.0	117.0	113.4	83.2	82.3	150.4	144.2
7	41.7	39.6	264.7	253.8	156.2	151.5	193.4	181.1
10	52.0	46.6	351.0	328.6	237.6	229.1	214.8	196.4

Source: Minhas et al. (2015)

lems. Therefore, in these cities, an aggressive and multifaceted approach to combating pollution is all the more urgent. Growing plantations and developing green areas reduce air pollution and also improve city beautification. Air pollution is directly reduced when dust and smoke particles are trapped by the plantations. In addition, plants absorb toxic gases, especially vehicle exhausts, which are a major component of urban smog (Nowak et al. 1996). The temperature moderating effect of urban plantations can reduce temperature extremes and thus reduce the smog formation arising with rise in temperature (Kuchelmeister 1991). Carbon dioxide, a major component of air pollution and greenhouse effect, can also be reduced through photosynthesis and reducing heat island effect with urban greening and plantations.

Urban plantations influence climate in two distinct manners, depending on the size, spacing, and design of plantations and green areas, first directly through effect on human comfort and second indirect effect on the energy budget of buildings in cities where air-conditioning is used. Plantations increase human comfort by influencing the degree of solar radiation, air movement, humidity, and air temperature and providing protection from heavy rains. Plantations and other vegetated areas also have an important impact on the energy budgets of buildings and, in turn, of entire cities. Plantations have been found to reduce the average air temperature in buildings by as much as 5 °C (Akbari et al. 1992). Studies in Chicago suggest that an increase of 10 % plantation cover can reduce the total energy requirements of the city by an equal extent (McPherson et al. 1994). Urban plantations also supply renewable energy in the form of fuel wood and other substitutes of fossil fuels. Treating wastewater in plantations eliminates the need for major sewage treatment plants that need fuel for their operation. Similarly, organic municipal solid waste serves as composted fertilizer, mulch for green areas, and animal feed, thereby reducing the energy and transport.

Social Benefits

Although difficult to quantify, the benefits of urban plantations and greening to human health are considerable. Urban plantations and green parks improve air quality, contribute toward aesthetically pleasing and relaxing environment, and thus have positive impacts on health in terms of decrease in respiratory illnesses and reduction in stress. Urban forests provide a connection between people and their natural environment that would otherwise be missing in a city. This connection is important for everyday enjoyment, productivity, and general mental health of workers (Nowak et al. 1996). Plantations also reduce ultraviolet light exposure thereby lowering the risks of harmful health effects such as skin cancer and cataracts (Heisler et al. 1995).

Green areas provide recreational sites, especially for lower-income residents who tend to frequent city parks more than wealthier citizens because of financial constraints and restrictions on leisure time. The urban poor generally have few affordable options for recreation and thus place a high value on green areas. Parks and other green areas also provide educational opportunities to learn about the environment and natural processes.

The aesthetic value of plantations and green areas, though not considered as important as food and shelter, is also very meaningful to urban residents. Vegetation reduces sun glare and reflection, complements architectural features, and tones down the harshness of large expanses of concrete. Rehabilitating lands with vegetation are often more attractive and cost effective than constructing buildings. Aesthetically pleasing green areas help in enhancing the properties values. For example, the vegetated beautification of Singapore and Kuala Lumpur has been adjudged as a major factor in attracting huge foreign investment and their rapid economic growth (Braatz 1993). Similarly, focused urban greening along roadways and railway lines in the Black Country district of England, a region of polluted lands, helped to attract huge investments (Jones 1995). The range of benefits that urban greening provides is both practical and comprehensive and addresses many of the social, environmental, and economic problems most cities face. Though urban plantations and green park areas are not the panacea for every urban problem, nonetheless these can significantly improve many of them and create a much more desirable environment to live.

Conclusions

Wastewater has enormous irrigation, nutrient, and labor employment potential, which is likely to increase with urbanization. Therefore, the wastewater needs to be considered as a resource rather than a menace by the urban planners and policy makers, especially in freshwater-scarce urban situations. However, wastewater also contains salts, pathogens, heavy metals, and other pollutants. Therefore, the benefits of wastewater use can be offset by the associated adverse health and environmental impacts in the long run, especially in developing countries where large volumes of raw or diluted wastewater are used in high-value vegetables, food grains, and fodder crops in peri-urban agriculture.

To overcome the hazards associated with wastewater use in agriculture, the chapter emphasizes on low-cost appropriate alternative measures like urban plantations and green park areas and some guidelines for selection of suitable species. Urban plantations and green areas should be designed on the basis of needs and desires of local populations so that these can serve maximum possible benefits to all residents. Plantations have the potential to improve air, water, and land resource quality, moderate the extreme high and low temperatures, and control floods and erosion with additional advantages such as creating habitats for wildlife, recreational activities, soothing environment, and above all aesthetic value of the cities.

To make the urban plantations and green park areas a safer and viable alternative for wastewater use, the plantation species should include fast-growing, high-rate transpiration multipurpose trees tolerant to salts and waterlogging and generate regular income. Tree plantations such as *Eucalyptus*, poplar (*Populus* spp.), pine (*Pinus* spp.), *Melaleuca* spp., bamboo (*Bambusa arundinacea*), acacia (*Acacia mangium*), neem (*Azadirachta indica*), and Indian rose wood (*Dalbergia sissoo*) which have high-rate transpiration system can be effectively utilized as safer alternative for beneficial disposal of wastewater. But to avoid contamination of natural resources, wastewater disposal rate needs to be regulated depending upon the plantation transpiration rate, tolerance to salts, and uptake of toxic substances.

Since the plantation transpiration capacity and their water requirement also decrease due to low evaporative demand during winter and rainy seasons, therefore, the provision of alternative storage and soil aquifer treatment needs to be developed in integration with constructed wetland with urban plantations and green park areas. But as a caution, regulatory mechanisms are required to be evoked to control loading rates since these are not as profligate consumers of water.

References

Aghabarati A, Hosseini SM, Maralian H (2008) Heavy metal contamination of soil and olive trees (*Olea europaea* L.) in suburban areas of Tehran, Iran. Res J Environ Sci 2:323–329

Akbari H, Davis S, Dorsano S, Huang J, Winnett S (1992) Cooling our communities: a guidebook on tree planting and light-colored surfacing. United States Environmental Protection Agency, Washington, DC

Armitage FB (1985) Irrigated forestry in arid and semi-arid lands: a synthesis. IDRC, Ontario

Asano T, Cotruvo JA (2004) Groundwater recharge with reclaimed municipal wastewater: health and regulatory considerations. Water Res 38:1941–1951

Augustine DS (2002) Eco-friendly management of solid and liquid wastes of farm, urban and industries. Lecture notes of training at TNAU, Coimbatore, p 213

Ayers RS, Westcot DW (1985) Water quality for agriculture. Irrigation and Drainage Paper 29, Review, FAO, Rome

Baker TG (1998) Tree growth in irrigated plantations at Wodonga and Kyabram-unpublished paper. Centre for Forest Tree Technology, University of Melbourne, Melbourne

Batarseh MI, Rawajfeh A, Ioannis KK, Prodromos KH (2011) Treated municipal wastewater irrigation impact on olive (*Olea europaea* L.) trees at Al-Tafilah, Jordan. Water Air Soil Pollut 217:185–196

Bhutta MN, Chaudhry MR (2000) Biological control of waterlogging. In: Proceedings of the eighth ICID international drainage workshop, vol IV(V), New Delhi, India, pp 33–45

Boardman R (1996) Opportunities and constraints for plantations irrigated with effluent from Adelaide. In: Proceedings of the Australian Water and Wastewater Association Conference, Sydney, May 1996, pp 287–293

Braatz S (1993) Urban forestry in developing countries: status and issues. In: Kollin C (ed) Proceedings of the sixth national urban forest conference. American Forests, Minnesota

Braatz K (1996) The use of municipal wastewater for forest and tree irrigation-Unasylva. 185:9

Braatz S, Kandiah A (1998) The use of municipal wastewater for forest and tree irrigation. FAO Corporate Document Repository -Unasylva No. 185-Forest Influences. http://www.fao.org/docrep/w0312E/w0312e09.htm

Bradford A, Brook R, Hunshal CS (2003) Wastewater irrigation in Hubli–Dharwad, India: implications for health and livelihoods. Environ Urban 15:157–170

Carter J (1993) The Potential of Urban Forestry in Developing Countries: a Concept Paper. Forestry Department of the Food and Agriculture Organization of the United Nations, Rome

CSIRO (1995) Effluent irrigated plantations: design and management. CSIRO Technical Paper No. 2, Canberra

Das DC, Kaul RN (1992) Greening wastelands through wastewater. National Wasteland Development Board, New Delhi

Dickinson NM, Turner AP, Watmough SA, Lepp NW (1992) Acclimation of trees to pollution stress: cellular metal tolerance traits. Ann Bot 70:569–572

Dinelli E, Lombini A (1996) Metal distributions in plants growing on copper mine spoils in Northern Apennines, Italy: the evaluation of seasonal variations. Appl Geochem 11:375–385

Drechsel P, Evans AEV (2010) Wastewater use in irrigated agriculture. Irrig Drain Syst 24:1–3

Drechsel P, Scott CA, Raschid-Sally L, Redwood M, Bahri A (2010) Wastewater irrigation and health: assessing and mitigating risk in low-income countries. Earthscan-International Development Research Centre (IDRC)-International Water Management Institute (IWMI)

Duncan M, Baker T, Wall G (1998) Wastewater irrigated tree plantations: productivity and sustainability. Paper presented in 61st annual water industry engineers and operators' conference civic centre-Shepparton, 2–3 Sept 1998, pp 18–26

Ebbs SD, Lasat MM, Brady DJ, Cornish J, Gordon R, Kochian IV (1997) Phytoextraction of cadmium and zinc from a contaminated soil. J Environ Qual 26(5):1424–1430

El-Lakany MH (1995) Urban and peri-urban forestry in the Near East region: a case study of Cairo. FAO Forestry Department Paper

FAO (1978) Municipal recycling in forest ecosystems. In: Proceedings of the 8th world forest congress, Jakarta, Rome, Food and Agriculture Organization (FAO), Rome

Heisler GM, Grant RH, Grimmond S, South C (1995) Urban forests -cooling our communities? In: Kollin C, Barratt M (eds) Proceedings of the seventh national urban forest conference. American Forests, Washington, DC, pp 31–34

Hooda V (2007) Phytoremediation of toxic metals from soil and waste water. J Environ Biol 28:367–376

Hopmans P, Stewart HTL, Flinn DW, Hillman TJ (1990) Growth, biomass production and nutrient accumulation by seven tree species irrigated with municipal effluent at Wodonga, Australia. For Ecol Manage 30:203–211

ICLEI (1995) Multi-functional park design and management. Case Study 27. International Council for Local Environmental Initiatives (ICLEI), Durban, South Africa. Toronto

IDB (1992) Ecological conservation project for the Mexico City Metropolitan area. Internal Bank document. Inter-American Development Bank (IDB), USA

Irshad M, Ahmad S, Pervez A, Inoue M (2015) Phytoaccumulation of heavy metals in natural plants thriving on wastewater effluent at Hattar industrial estate, Pakistan. Int J Phytoremediat 17:154–158

James BR (2001) Remediation-by-reduction strategies for chromate-contaminated soils. Environ Geochem Health 23(175–17):9

Jones N (1995) The black country urban forestry initiative: a United Kingdom case study. In: Proceedings of the seventh national urban forest conference. American Forests, Washington, DC

Juwarkar AS, Thawale PR, Juwarkar AA, Khanna P (1995) A case study of environmental problems in the use of wastewater for crop irrigation. In: National workshop on health, agriculture and environmental aspects of wastewater use. UNEP-WHO-NEERI

Khalil ME (1990) Accumulation of some nutrients and heavy metals in Abu Rawash area, Giza Governorate. M.Sc. thesis, Faculty of Agriculture, Moshtohor, Zagazig University, Egypt

Khanzada AN, Morris JD, Ansari R, Slavich PG, Collopy JJ (1998) Groundwater uptake and sustainability of *Acacia* and *Prosopis* plantations in southern Pakistan. Agric Water Manage 36:121–139

Kuchelmeister G (1991) Peri-urban multipurpose forestry in development cooperation, experience, deficits, and recommendations. Commission of European Communities Contract Article B946/90-19

Kumar YA, Reddy VM (2010) Effects of municipal sewage on the growth performance of *Casuarina equisetifolia* (Forst. & Forst.) on sandy soil of east coast at Kalapakkam (Tamil Nadu, India). Appl Ecol Environ Res 8(1):77–85

Lal K, Minhas PS, Shipra, Yadav RK (2008) Extraction of cadmium and tolerance of three annual cut flowers on Cd- contaminated soils. Bioresour Technol 99:1006–1011

Lal K, Yadav RK, Kaur R, Bundela DS, Khan MI, Chaudhary M, Meena RL, Dar SR, Singh G (2013) Productivity, essential oil yield, and heavy metal accumulation in lemon grass (*Cymbopogon flexuosus*) under varied wastewater-groundwater irrigation. Ind Crops Prod 45:270–278

Lepp NW (1996) Uptake, mobility and loci of concentrations of heavy metals in trees. In: Glimmerveen I (eds) heavy metals and trees. Proceedings of a discussion meeting, Glasgow. Institute of Chartered Foresters, Edinburgh, pp 68–82

Lone MI, He Z-l, Stoffella PJ, Yang X (2008) Phytoremediation of heavy metal polluted soils and water: progresses and perspectives. Revi J Zhejiang Univ Sci B 9(3):210–220

Mara D (2003) Domestic wastewater treatment in developing countries. Earthscan, London

Maurer MA, Davies FS, Graetz DA (1995) Reclaimed wastewater irrigation and fertilization of mature 'Redblush' grapefruit trees on spodosols in Florida. J Am Soc Hortic Sci 20:394–402

McPherson EG, David JN, Rowan AR (eds) (1994) Chicago's urban forest ecosystem: results of the Chicago urban forest climate project. Northeast Forest Experiment Station, Radnor

Minhas PS, Samra JS (2004) Wastewater use in Peri-urban agriculture: impacts and opportunities. Bulletin No. 2/2004. Central Soil Salinity Research Institute, Karnal, p 78

Minhas PS, Yadav RK, Sharma N, Joshi PK (2006) Prevalence and control of pathogenic contamination in some sewage irrigated vegetable, forage and grain crops. Bioresour Technol 97:1174–1179

Minhas PS, Yadav RK, Lal K, Chaturvedi RK (2015) Effect of long-term irrigation with wastewater on growth, biomass production and water use by Eucalyptus (*Eucalyptus tereticornis* Sm.) planted at variable stocking density. Agric Water Manage 152:151–160

MSEA (2006) Ministry of State for Environmental Affairs (MSEA) Annual Report (Egypt)

Myers BJ, Theiveyanathan S, O'Brien ND, Bond WJ (1996) Growth and water use of effluent-irrigated *Eucalyptus grandis* and *Pinus radiata* plantations. Tree Physiol 16:211–219

Nowak DJ, Dwyer JF, Childs G (1996) The benefits and costs of urban greening. Draft document to be published in Proceedings of urban greening seminar held at Mexico City December 2–4, edited by Krishnamurthy, L. and J. RenteNascimento. Universidad Autonoma de Chapingo, Mexico

Omran M, Waly TM, Abd-Elmaim EM, El-Nashir BMB (1998) Effect of sewage irrigation on yield, tree components and heavy metals accumulation in Naval Orange trees. Biol Wastes 23:7–24

Paliwal K, Kurunlchamy KSTK, Ananthavalli M (1998) Effect of sewage water irrigation on growth parameter, biomass and nutrient accumulation in *Hardwickia binata* under nursery condition. Bioresour Technol 68:105–110

Prasad JVNS, Srinivas K, Srinivasa Rao C, Ramesh C, Venkatravamma K, Venkateswarlu B (2012) Biomass productivity and carbon stocks of farm forestry and agroforestry systems of leucaena and eucalyptus in Andhra Pradesh, India. Curr Sci 103(5):536–540

Pulford ID, Watson C (2003) Phytoremediation of heavy metal-contaminated land by trees: a review. Environ Int 29:529–540

Qadir M, Wichelns D, Sally LR, McCornick PG, Drechsel P, Bahri A, Minhas PS (2010) The challenges of wastewater irrigation in developing countries. Agric Water Manage 97:561–568

Raschid-Sally L, Jayakody P (2007) Understanding the drivers of wastewater agriculture in developing countries-results from global assessment. In: Comprehensive assessment research report series. IWMI, Colombo

Raschid-Sally L, Jayakody P (2008) Drivers and characteristics of wastewater agriculture in developing countries: results from a global assessment. Research Report 127. International Water Management Institute, Colombo

Reeves RD (2003) Tropical hyperaccumulators of metals and their potential for phytoextraction. Plant Soil 249(1):57–65

Riddell-Black D (1993) A review of the potential for the use of trees in the rehabilitation of contaminated land. WRC report CO 3467. Water Research Centre, Medmenham

Riddell-Black D (1994) Heavy metal uptake by fast growing willow species. In: Aronsson P, Perttu K (eds) Willow vegetation filters for municipal wastewaters and sludges: a biological purification system. Swedish University of Agricultural Sciences, Uppsala, pp 145–151

Rockwood DL, Pisano SM, McConnell WV (1996) Superior cottonwood and Eucalyptus clones for biomass production in waste bioremediation system. In: proceedings of 7th national bioenergy conference, 15–20 Sept 1996, Nashville, TN, pp 254–261

Rockwood DL, Naidu CV, Carter DR, Rahman M, Spriggs TA, Lin C, Alker GR, Isebrands JG, Segrest SA (2004) Short rotation woody crops and phytoremediation: opportunities for agroforestry. Agrofor Syst 61:51–63

Salt DE, Smith RD, Raskin I (1998) Phytoremediation. Annu Rev Plant Physiol Plant Mol Biol 49:643–668

Sheriff DV, Nambiar EKS, Fife DN (1986) Retranslocation between nutrient status, carbon assimilation and water use efficiency in Pinus radiate D. Don needles. Tree Physiol 2:73–88

Singh G, Bhati M (2004) Soil and plant mineral composition and productivity of *Acacia nilotica* (L.) under irrigation with municipal effluent in an arid environment. Environ Conserv 31(4):331–338

Singh G, Bhati M (2005) Growth of *Dalberia sissoo* in desert regions of western India using municipal effluent and plant chemistry. Bioresour Technol 69:1019–1028

Stewart HTL, Hopmans P, Flinn DW, Hillman TJ (1990) Nutrient accumulation in trees and soil following irrigation with municipal effluent in Australia. Environ Pollut 63:155–177

Tabari M, Salehi A, Mohammadi J (2011) Impact of municipal wastewater on growth and nutrition of afforested *Pinus eldarica* stands. In: Sebastin F, Einschlag G (eds) Waste water- evaluation and management, pp 303–312. www.intechopen.com/books

Tangahu BV, Abdullah SRS, Basri H, Idris M, Anuar N, Mukhlisin M (2011) A review on heavy metals (As, Pb, and Hg) uptake by plants through phytoremediation. International. J Chem Eng vol 2011, Article ID 939161, Hindawi Publishing Corporation, p 31. http://www.hindawi.com

Thawale PR, Juwarkar AA, Singh SK (2006) Resource conservation through land treatment of municipal wastewater. Curr Sci 90:704–711

Vymazal J (2010) Constructed wetlands for wastewater treatment. Water 2:530–549

WHO (2006) Guidelines for the safe use of wastewater-excreta and grey water. Volume 2. Wastewater use in agriculture, Geneva

Zalesny RS Jr, Stanturf JA, Evett SR, Kandil NF, Soriano C (2015) Opportunities for woody crop production using wastewater in Egypt. I. Afforestation strategies. Int J Phytorem 13:102–121

Tree Plantations in Saline Environments: Ecosystem Services, Carbon Sequestration and Climate Change Mitigation

S.R. Gupta, J.C. Dagar, and Mukesh-Kumar

Abstract

Nearly one billion hectares of arid and semiarid areas of the world are salt-affected and remain barren due to salinity or water scarcity. These lands can be utilized by adopting appropriate planting techniques and integrating trees with tolerant crops, forage grasses, oil-yielding crops, and aromatic and medicinal plants. Biosaline agroforestry provides various ecosystem services such as the improved soil fertility, carbon sequestration, and biomass production. Provisioning services relating to biomass production have been fairly well studied in different biosaline agroforestry. Tree plantations and agroforestry enrich the soil in organic matter and exert a considerable ameliorative effect on soil properties. The soil microbial biomass serves as a useful indicator of soil improvement under salt stress. Arbuscular mycorrhizal fungi colonized the roots of grasses in silvopastoral systems on saline and sodic soils, the dominant AM fungal species being *Glomus* and *Acaulospora*. By integrating trees with the naturally occurring grassland systems on highly sodic soils, the soil organic carbon content increased from 5.3 Mg ha^{-1} (in sole grass) to 13.6, 10.9, and 14.2 Mg ha^{-1}, when *Dalbergia sissoo, Acacia nilotica*, and *Prosopis juliflora* trees were introduced with grass. The soils of biosaline agroforestry could store 25.9–99.3 Mg CO_2 ha^{-1} in surface 0.3 m soil. Maintaining the stores and sink of carbon in agroforestry could play a key role in climate change mitigation as well as help in adapting to changing environmental conditions.

S.R. Gupta (✉)
Department of Botany, Kurukshetra University,
Kurukshetra 136119, Haryana, India
e-mail: sgupta2002158@gmail.com

J.C. Dagar
Central Soil Salinity Research Institute,
Karnal 132001, Haryana, India
e-mail: dagarjc@gmail.com

Mukesh-Kumar
Water Technological Center, Indian Agricultural
Research Institute, Pusa, New Delhi 110012, India
e-mail: mukesheco2007@rediffmail.com

J.C. Dagar, P.S. Minhas (eds.), *Agroforestry for the Management of Waterlogged Saline Soils and Poor-quality Waters*, Advances in Agroforestry 13, DOI 10.1007/978-81-322-2659-8_11

Introduction

The ecosystem services are the benefits that the natural environment provides to humanity. The Millennium Ecosystem Assessment defined four categories of ecosystem services, i.e., provisioning services, regulating services, cultural services, and supporting services that contribute to human well-being (MEA 2005). According to the World Economic Forum, 60 % of the earth's ecosystem services have been degraded in the past 60 years. The Convention on Biological Diversity, the Millennium Development Goals, and other international agreements have clearly indicated relationship between biodiversity conservation, poverty alleviation, and human well-being (MEA 2005; TEEB 2010; Sachs et al. 2009; Turner et al. 2012). The main services from tree plantations and agroforestry systems include habitat provision; pollination and seed dispersal; clean water; flood mitigation; biodiversity; oxygen production; nutrient cycling and soil fertility improvement; genetic resources for crops; spiritual, cultural, recreational, and tourism values (Jose 2009) besides carbon sequestration (Montagnini and Nair 2004; Nair et al. 2009); and climate change mitigation and adaptation (Nair 2012). It has been well recognized that services and benefits provided by agroforestry practices occur over a range of spatial and temporal scales (Izac 2003; Jose 2009).

Current CO_2 levels (396 ppm) are higher than at any time, at least the last 800,000 years (170–300 ppm), indicating a pronounced human impact (IPCC 2013). Averaged over all land and ocean surfaces, temperatures have warmed 0.85 °C (0.65–1.06 °C) over the period 1880–2012 (IPCC 2013). The ultimate goal of climate change mitigation is to reduce emissions of greenhouse gases, whereas adaptation aims to develop strategies to reduce the negative impacts (Nair 2012). The linkages between mitigation and adaptation in forestry sector have been discussed by several workers (Klein et al. 2007; Ravindranath 2007; Duguma et al. 2014). The soil carbon pool, which consists of both organic and inorganic carbon, also plays an important role in the global carbon cycle by sequestering carbon. Soil carbon sequestration is an important factor in the greenhouse gas emissions balance and is strongly related to site conditions (e.g., soil structure, initial soil carbon content, climate), structure of agroforestry, and soil management (Montagnini and Nair 2004; Nair et al. 2009). The improvement of soil carbon in tree plantations on salt-affected soils offers substantial global greenhouse gas mitigation potential. Management practices that favor carbon sequestration in agroforestry tends to enhance resilience in the face of climate variability and enhancing longer-term adaptation to changing climates (FAO 2010). This chapter gives an overview of tree plantations and agroforestry systems of salt-affected and waterlogged areas, ecosystem services of agroforestry, carbon sequestration in biosaline agroforestry, and the role of tree plantations in climate change mitigation.

Agroforestry Systems for Salt-affected Soils

Several approaches including chemical amendments, tillage operations, crop-assisted interventions, tree plantations, and phytoremediation have been used to reclaim sodic and saline-sodic soils (Gupta and Abrol 1990; Minhas and Sharma 2003; Qadir et al. 2007a). The plant-assisted approach for bioamelioration of saline and sodic soils has been used by several workers (Kumar and Abrol 1984; Mishra et al. 2002). Phytoremediation helps in improving plant nutrient availability in the soil and increases soil aggregate stability and porosity as a result of root activity in sodic soils (Qadir et al. 2007a).

The sodic soils usually have a narrow spectrum of flora and are characterized by low-species diversity and single-species dominance (Sinha et al. 1988). *Desmostachya bipinnata,*

Table 1 Some examples of agroforestry systems on salt-affected soils in India

Saline environments	Occurrence	Agroforestry system, role of trees	Study areas, reference
High soil sodicity with calcareous hard pans underlain with good quality fresh groundwater	Haryana, UP, Bihar, Punjab	Temporary agroforestry system, from silvi-agro to agro-halophytic trees to remediate soil + conventional agroecosystem, energy plantation	Lucknow, BIOSAFOR (2011), (Singh et al. 2013)
Waterlogged saline soils with water table <2 m	Haryana, Rajasthan, Punjab	Permanent agroforestry system: trees for bio-drainage (prevention); agro and pasture with salt-tolerant species	Sampla, Haryana, BIOSAFOR (2011)
Saline-sodic topsoil, sodic subsoils, waterlogged, slight	Haryana	Existing agroforestry system with alley cropping	Puthi, Haryana, Kumar (2012)
Sodic soil, calcite layer; low soil permeability and impeded drainage	Haryana	Tree plantations and silvopastoral agroforestry systems	Bichhian, Haryana, Kaur et al. (2002a)
Calcareous soils irrigated with saline water	Haryana	*Acacia nilotica/Salvadora persica* silvopastoral system	Hisar, Haryana, Kumari (2008)

Sporobolus marginatus, and *Diplachne (Leptochloa) fusca* were prevalent in sodic soils of the Indo-Gangetic plains (Rana and Parkash 1987; Sinha et al. 1988). For the different sodic grassland communities (pH 8.9–10.10), species richness, species diversity, and species evenness decrease with sodicity (Sinha et al. 1988; Neeraj et al. 2004). Grasses such as *Cynodon dactylon, Sporobolus arabicus, Imperata cylindrica*, and *Aeluropus lagopoides* dominate the saline or saline arid habitats of the Salt Range in Pakistan (Chaudhary et al. 2001). Some examples of bio-saline agroforestry systems for salt-affected areas in India are summarized in Table 1. The different kinds of agroforestry systems have been developed for high soil sodicity with calcareous hard pans + fresh groundwater (GW); high soil sodicity + sodic GW; permanent waterlogged saline soils; temporary waterlogged saline soils; saline or neutral soil; and saline groundwater or aquifer. Agroforestry systems range from silvo-agro to agro-halophytic trees to remediate soil, agro-silvo-aqua-pasture, trees for bio-drainage, permanent agroforestry system, agro-silvo-pasture for bio-drainage, permanent agro-silvopastoral, and silvopastoral systems.

Agroforestry systems on salt-affected soils or biosaline agroforestry systems may be an alternative land use option as some tree species are less susceptible to soil salinity/sodicity, and their cultivation can help regenerate these soils (Bell 1999; Lambert and Turner 2000; Singh 1995; Wicke et al. 2013). Examples of species tolerant to soil salinity, soil sodicity, or both are *A. nilotica, E. camaldulensis, E. tereticornis*, and *P. juliflora* (Marcar and Crawford 2004). *P. juliflora* can grow on a variety of soils including saline flats and on shifting sand dunes and coastal sand and can grow in waterlogged conditions and is tolerant to high salinity. Several studies focused on sodic soils found agroforestry systems to be an economically viable land use option (Bell 1999; Kaur et al. 2002a; Marcar and Crawford 2004; Masters et al. 2007; Qadir and Oster 2002; Zhang et al. 2004).

The integration of salt-tolerant trees with naturally growing grasses has been reported to be a viable land use option for improving the biological productivity and fertility of highly sodic soils (Kaur et al. 2002a, b). Monoculture plantations on degraded lands could improve the soil condition and enrich the species diversity of herbaceous plants (Singh et al. 2002). Vegetation of salt waste lands has been found to ameliorate soil conditions and improve soil biological activity (Tripathi and Singh 2005). Creation of new forests on barren land has contributed significantly to soil amelioration in the degraded sodic soil of the Indo-Gangetic plains.

Ecosystem Services of Biosaline Agroforestry

The ecosystem services are the benefits that the natural environment provides to humanity. The Millennium Ecosystem Assessment (MEA 2005) highlighted the condition of ecosystem and ecosystem services and distinguished four broad categories of ecosystem services, i.e., provisioning, regulating, cultural, and supporting services (Fig. 1). The provisioning services describe the processes that yield foods, fibers, fuels, water, biochemical, medicinal plants, pharmaceuticals, and genetic resources. The cultural services comprise a set of largely non-material benefits of the environment including recreation and tourism and the spiritual, religious, aesthetic, and inspirational well-being. The regulating services are the benefits obtained from the regulation of ecosystem processes and include erosion control or soil stabilization, water purification, waste treatment, air quality maintenance, climate regulation, hydrological flows, and natural hazard protection. The supporting services are those that are necessary for the production of all other ecosystem services, and these differ from provisioning, regulating, and cultural services for their indirect impacts.

Ecosystem services depend on ecosystem structure and processes and provide a link between ecology and economics. Structural components in ecosystems include trees, crops, soil, topography, and animals. Ecosystem processes include the flow of water, animal life cycles, photosynthesis, nutrient cycles, and others. These structural components and ecosystem processes support ecosystem functions such as soil accumulation, habitat creation, and buffers to flooding. Ecosystem functions generate benefits to people called ecosystem goods and services (De Groot et al. 2002). For conservation of biodiversity, it is important to show the relationship between biodiversity and ecosystem services and the importance and value of ecosystem services provided by sites important for biodiversity. Ecosystem services are generated as a result of interaction and exchange between biotic and abiotic components of ecosystems. Since most of ecosystem services are not part of commercial market, they are often given little weight in policy decisions. The assessment of ecosystem services requires evaluation of changes in ecosystem processes and structures. Trees in agroforestry can be used for timber, pulp, firewood, fodder, honey and other products (e.g., leaf oils and tannins), shelter and shade, wind, water and water table erosion control, wildlife corridors, and aesthetics. Physical, chemical, and biological soil properties are easily measurable parameters that can serve as good indicators of ecosystem services.

PROVISIONING SERVICES	REGULATING SERVICES	CULTURAL SERVICES
Products obtained from the agroforestry systems • Food • Timber • Fuel wood • Fiber • Medicinal plants • Genetic resources	Benefits obtained from regulation of ecosystem processes • Climate regulation • Water quality • Carbon sequestration • Bioremediation • Biological control • Pollination • Soil Quality • Nutrient mineralization	Non material benefits that contribute to wider needs and desires of society • Recreation and tourism • Existence value • Aesthetic values • Educational values
SUPPORTING SERVICES Necessary for the production of all other ecosystem services ♦ Nutrient cycling ♦ Primary production ♦ Soil fertility ♦ Ecosystem resilience		

Fig. 1 Major ecosystem services provided by biosaline agroforestry (Based on MEA 2005)

Provisioning Services

Agroforestry practices can provide significant amounts of timber and fuelwood on marginal salt lands. Some provisioning services are more readily appropriated by human society, which gives economic and social benefits. If the trees are used as fodder or as shade trees, they can be planted scattered over the fields that are used as pastures but can also be planted in blocks or as timber belts. The trees can be directly browsed by cattle or the leaves and pods can be harvested to feed cattle.

Prosopis juliflora is a multipurpose tree used as timber, fuelwood, charcoal, animal feed, and also for reclamation of waste lands, sand dunes, and salt-affected soils. It is a hardy and resilient tree, adaptable to all frost-free semiarid climatic regions (Hocking 1993). The generally crooked stems and branches of *P. juliflora* make good fuelwood with a calorific value of 18.9–19.9 MJ kg^{-1} and can be used for making good charcoal (Qureshi and Barrett-Lennard 1998). Charcoal from *P. juliflora* wood is used extensively. In India, a planting of 2500 plants ha^{-1} gave 13 Mg ha^{-1} from cut side branches after 40 months of growth. Singh and Singh (1992) reported that the biomass production in 10 years was 260 Mg ha^{-1} in a high density plantation (10,000 trees ha^{-1}) of *P. juliflora* per hectare on a high alkaline soil. The annual increment in a plantation with 5000 plants ha^{-1} under the same conditions was 28 Mg ha^{-1} (Singh and Singh 1992). Pod yields of *P. juliflora* ranged from 2 to 8 Mg ha^{-1} under optimum conditions (Kuiper and Probos 2005).

Kaur et al. (2002a) estimated bole and branch wood production in different silvopastoral systems of *P. juliflora*, *D. sissoo*, and *A. nilotica* that were used for the rehabilitation of sodic soils. In the silvopastoral agroforestry systems of *A. nilotica* with *D. bipinnata*, *D. sissoo* with *D. bipinnata*, and *P. juliflora* with *D. bipinnata*, the bole wood that can be used as timber was 4.62–9.78 Mg ha^{-1}, and branch biomass production varied between 4.16 and 20.82 Mg $ha^{-1}year^{-1}$ (Kaur et al. 2002a). Timber and fuelwood biomass in clonal *E. tereticornis* plantation in different spacing in shallow water table areas after 4 years showed timber production of 11.17–28.65 Mg ha^{-1} (Kumar 2012). More than a billion people in developing countries rely on wood for cooking and heating. Fuelwood can be obtained from salt-tolerant trees including *Prosopis*, *Tamarix*, *Acacia*, *Casuarina*, *Pithecellobium*, *Parkinsonia*, and *Salvadora* (Ladeiro 2012). The salt-adapted tree species like *Prosopis juliflora*, *Pongamia pinnata*, *Populus euphratica*, and *Tamarix* spp. could provide good quality wood. In coastal areas, the mangroves species of *Rhizophora*, *Ceriops*, *Avicennia*, and *Aegiceras* are good fuelwoods and also contribute to charcoal production.

Soil Enrichment and Bio-amelioration

Incorporation of organic matter into soils through root growth and litterfall and their decomposition leads to reduction in sodicity and thus higher infiltration and leaching of reaction products. Most of salt-affected soils, being low in carbon, can provide for environmental benefits not only in terms of carbon accumulations and ameliorations of soils but also their direct benefits through their effects on the greenhouse gas carbon dioxide.

Amelioration of Soil Properties

A. nilotica and *E. tereticornis* plantations were noted to have a considerable ameliorative effect on soil properties when planted on alkali soil; the soil organic carbon content increased to about double the initial value for *Eucalyptus* but tripled under *Acacia* plantation (Gill and Abrol 1993). Twenty-year-old plantations of *P. juliflora*, *A. nilotica*, *E. tereticornis*, *A. lebbeck*, and *Terminalia arjuna* could ameliorate the alkali soil by adding organic matter through litter to the extent that arable crops could be grown successfully on the soil (Singh and Gill 1992). On the alkali soil, *Tamarix articulata* ameliorated the soil by inducing the maximum reduction of sodicity and pH values followed by *P. juliflora* and *A. nilotica* (Dagar et al. 2001).

Prosopis cineraria grows successfully in highly saline (ECe > 15 dS m^{-1}) and alkaline soils (pH up to 9.8). It is highly drought tolerant; its taproot can reach groundwater at 20-m depth. It is an excellent fuel (calorific value of sapwood: 20.9 MJ kg^{-1}) and also gives high-quality charcoal (5,000 kcal kg^{-1}) (Qureshi and Barrett-Lennard 1998). Due to its deep root system, mono-layered canopy, and ability to fix atmospheric nitrogen, *P. cineraria* is extensively used as an agroforestry tree throughout arid and semi-arid India. The tree has a boosting effect on the yield of crops growing in its vicinity. The crops draw their moisture and nutrients from the top 0.5–0.6 m of soil, while the tree gets its nutrients from a deeper horizon. In addition, the tree provides shade to crops during summer (Dutton et al. 1992). The tree is favored for agroforestry as it fixes large amounts of nitrogen and does not affect growth of plants under its canopy. Aggarwal et al. (1993) found that soil nitrogen, phosphorus, and potassium were higher under *P. cineraria* canopies than in open fields. The biomass of *Pennisetum typhoides* was three times higher when grown in *P. cineraria* soil than in open soil. *P. typhoides* grown in *P. cineraria* soils was always at least two times larger than when grown in soil from outside of *P. cineraria* canopies regardless of nutrient additions; the effects being neutral to positive effects on the species beneath them (Aggarwal et al. 1993). The improvement in soil properties and biological activity increased with age of *P. juliflora* and *D. sissoo* (Mishra and Sharma 2003; Tripathi and Singh 2005). Trees such as *T. articulata*, *A. nilotica*, *P. juliflora*, and *E. tereticornis* and several other tree species enhanced the organic carbon contents of soils by more than 3.5 g kg^{-1} in about 10 years of their growth (Tomar et al. 2003).

The energy plantation of *Jatropha curcas* improved soil fertility and decrease soil sodicity after 6 years of its growth at Banthra Research Station, Lucknow, northern India (Singh et al. 2013). Soil amelioration potential of *Jatropha curcas* on soil properties was significant when compared to initial soil properties at 0–15-cm soil depth. Soil bulk density, pH, electrical conductivity (EC), and exchangeable sodium percentage (ESP) decreased. There was a significant increase in soil organic carbon (SOC), nitrogen (N), phosphorus (P), and enzyme activities (dehydrogenase, glucosidase, and protease) beneath the canopy of *Jatropha curcas* than outside canopy (Singh et al. 2013).

Soil Microbial Biomass

Salinity under arid conditions has harmful influence on the size and activity of soil microbial biomass and on biochemical processes essential for maintenance of soil quality (Yuan et al. 2006). The tree plantations and silvopastoral agroforestry systems raised on sodic soils improved soil carbon and microbial activity through input of organic matter from aboveground and belowground parts of the plants (Kaur et al. 2002a, b). Plant cover through its effects on quantity and quality of organic matter influence the levels of soil microbial biomass. In salt-affected soils, the size and dynamics of soil labile carbon pool has been shown to vary with land use type and tree species (Kaur et al. 2000). Microbial biomass carbon was related to plant biomass carbon as well as the flux of carbon in net primary productivity. Nitrogen mineralization rates were found greater in silvopastoral systems compared to sole grass system. Soil organic matter was linearly related to microbial biomass carbon, soil N, and nitrogen mineralization rates.

Wong et al. (2008) observed an increase in soil microbial biomass with the increase in salinity because of the availability of more soluble carbon in the soil. In sodic soils, soil organic carbon can be rapidly lost due to disintegration of soil aggregates and desorption of carbon from clays (Wong et al. 2008). Soil organic carbon and biological properties were improved after the establishment of a tree plantation of *Casuarina equisetifolia* and *Pithecellobium dulce* with undergrowth of native grasses and forbs in salt-affected soils in Mahasarakam Province, Thailand (Naingoo et al. 2012). Total soil organic carbon was significantly higher in the tree plantation zones than in the fallow soil. Soil microbial biomass was higher under tree plantations (164.4–228.5 mg C kg^{-1} soil) as compared to fallow soil (99.4 mg C kg^{-1} soil).

Nitrogen Cycling

N fraction in vegetation on sodic soils was 0.32 and 0.29 of the soil in *P. juliflora* + *D. bipinnata* and *P. juliflora* + *S. marginatus* silvopastoral system, respectively. N returned in litterfall varied from 0.0 75 to 0.14 Mg $ha^{-1}year^{-1}$, while the turnover of fine root biomass returned 0.019 to 0.037 Mg N $ha^{-1}year^{-1}$. Thus, total N returned to soil was 0.11–0.177 Mg N $ha^{-1}year^{-1}$. N sequestration in the system was 0.046–0.10 Mg N $ha^{-1}year^{-1}$, which was 28–36 % of total uptake of the nitrogen in the agroforestry system (Kaur et al. 2002a). *P. juliflora* can also fix atmospheric nitrogen through symbiosis with cowpea-type *Rhizobium*. The roots also form mycorrhizal associations with *Glomus* fungi. Plants with both *Rhizobium* and mycorrhizal associations show significantly higher nitrogen fixation rates than those lacking the mycorrhiza (www.worldagroforestry.org).

Arbuscular Mycorrhizal Fungal Diversity

At Bichhian in northeastern India, *Desmostachya bipinnata* and *Sporobolus marginatus* have been integrated with trees in the silvopastoral systems on sodic soils. These salt-adapted grasses are characterized by moderately high diversity of arbuscular mycorrhizal (AM) fungi in their rhizosphere (Jangra et al. 2011). A total of 27 species of AM fungi belonging to six genera, i.e., *Acaulospora*, *Entrophospora*, *Gigaspora*, *Glomus*, *Sclerocystis*, and *Scutellospora* have been identified. In *Acacia nilotica* and *Salvadora persica* silvopastoral system on saline-sodic soils at Hisar, northwestern India, the AM root colonization in various grass species varied from 47.8 to 71.2 % (Kumari 2008). In the agrohorticultural system of *Carissa carandas* along with barley (*Hordeum vulgare*), some 23 species of mycorrhizal fungi belonging to *Glomus*, *Acaulospora*, and *Gigaspora* were identified. The species of *Glomus* and *Acaulospora* dominated the AM fungi in both agrohorticultural and silvopastoral systems possibly having a significant role in salt adaptation (Jangra et al. 2011; Kumari 2008).

AM fungi could survive and colonize roots of salt-adapted plants under extreme saline-sodic soil conditions (Garcia and Mendoza 2007). Besides, a large amount of carbon is found in tissues of mycorrhizal fungi, which could be long lived in the soil (Treseder and Allen 2000). For example, chitin, which is not readily decomposed (Gooday 1994), can constitute up to 60 % of fungal cell walls (Muzzarelli 1977). Arbuscular mycorrhizal fungi are also the sole producers of glomalin, a potentially recalcitrant glycoprotein (Wright et al. 1996; Wright and Upadhyaya 1999). Thus, mycorrhizal fungi could sequester increased amounts of carbon in living, dead, and residual hyphal biomass in the soil (see Treseder and Allen 2000) and may play key role in soil carbon sequestration.

Carbon Sequestration as an Indicator of Ecosystem Services

The agroforestry systems are important for carbon storage in the soil-plant system, which is an indicator of regulatory ecosystem services. The various pools of carbon in the agroforestry systems include aboveground woody biomass, belowground biomass, litterfall, ground floor litter, and soil carbon. The extent of carbon sequestration in agroforestry systems depends on the amount and quality biomass input from trees and non-trees components of the system as well as on soil properties such as soil structure and their aggregates (Nair 2012).

Carbon Sequestration in Plant Biomass

Agroforestry has potential to store carbon in plant biomass and wood products, besides being a self-reliant and risk proof system required for sustainable land management, microclimate moderation, and rehabilitation of degraded land and carbon sequestration (Nair 2007; Alam et al. 2010; Palsaniya et al. 2010). Deep-rooted trees are important for incorporating the SOC pool in the subsoil and enhancing soil structure. The trees on salt-affected soils have the potential for carbon sequestration by increasing soil carbon and plant biomass production (Bhojvaid and Timmer 1998; Garg 1998; Kaur et al. 2002a, b). Bhojvaid and Timmer (1998) reported the annual rate carbon sequestration to be 1.4 Mg C ha^{-1} $year^{-1}$ over a 30-year period of *Prosopis* plantation. Glenn et al. (1992) estimated that 0.6–1.2 giga-tones of C per year could be assimilated annually by halophytes on saline soils; evidence from decomposition experiments suggested that 30–50 % of this carbon might enter long-term storage in soil. Thus, halophytes adapted to saline soils could play an important role in soil carbon sequestration (Glenn et al. 1993).

In southwestern Australia, the rates of C sequestration in biomass of *E. globules* over a 10-year period ranged from 3.3 to 11.5 Mg C ha^{-1} $year^{-1}$on a large-scale watershed, the rates of C sequestration being high (Harper et al. 2005, 2007). In silvopastoral agroforestry systems on sodic soils at Bichhian, northeastern India, the total carbon storage was 1.18–18.55 Mg C ha^{-1}, and carbon input in net primary production varied between 0.98 and 6.50 Mg C $ha^{-1}year^{-1}$ (Kaur et al. 2002a) under different grass-tree combinations. Biomass and carbon sequestered by 5-year-old and 4-month-old clonal *E. tereticornis* on waterlogged soils at Puthi, Hisar, northwestern India, was 15.5 Mg ha^{-1} (Jeet-Ram et al. 2011). The *Eucalyptus*-based agroforestry on waterlogged soils showed soil carbon storage of 15.8 Mg C ha^{-1} compared to baseline of the cropland; the net carbon sequestration amounted to 4.4 Mg C ha^{-1} over a period of 4 years (Table 2).

Biomass production from biosaline agroforestry in waterlogged, saline-sodic soils in India was found to sequester 6 Mg CO_2 eq ha^{-1} over the 15-year lifetime of the plantation (Wicke et al. 2013). The emissions from agrochemical and fossil fuel use could be compensated by carbon sequestration in belowground biomass and soil. The economic value of the carbon sequestration is small and ranges between 0.003 and 0.046 kV ha^{-1} depending on the carbon credit price assumed.

Table 2 Biomass and carbon sequestered by 5-year-old and 4-month-old clonal *E. tereticornis* on waterlogged soils at Puthi, Hisar, northwestern India

Tree component	Dry biomass (Mg ha^{-1})	C (%)	C-sequestered (Mg ha^{-1})
Timber	22.1	47.0	10.4
Fuelwood	0.8	43.5	0.3
Twigs and leaves	1.1	43.9	0.5
Roots	8.9	48.0	4.3
Total	32.9	–	15.5

Source: Jeet-Ram et al. (2011)

Soil Carbon Sequestration in Tree-based Systems

The soil organic matter is an important indicator of soil quality and the ecosystem services. Improving SOC Pool is an important strategy of reclaiming salt-affected soil. The goal is to create a positive ecosystem C budget. Soil C sequestration in agroforestry systems on salt-affected soils has been studied by several workers (Table 3). Soil carbon sequestration potential in 0.3-m soil layer ranged from 6.84 to 27.09 in some agroforestry systems and grassland systems of salt-affected soils. In *P. juliflora* + *D. bipinnata* and *P. juliflora* + *S. marginatus* silvopastoral systems on sodic soils, the soil carbon pool was 13.4 Mg C ha^{-1}, *P. juliflora* + *D. bipinnata*, and 9.6 Mg C ha^{-1}, *P. juliflora* + *S. marginatus* (Table 3). The total organic carbon in surface 0.3 m soil was 6.8, 21.1 and 20.1 Mg C ha^{-1} for the native grassland, the *A. nilotica* + *C. ciliaris* silvopastoral systems, and *S. persica* system, respectively. The integration of trees with forage grasses improved SOC (Kumari 2008).

Table 3 Carbon sequestration (The values within brackets are in eqCO_2 in surface 0.3 m soil) under selected agroforestry systems on salt-affected soils in India

Agroforestry system	Nature of soil/location	C-sequestered (Mg ha^{-1})	Reference
P. juliflora + *D. bipinnata* silvopastoral	Sodic soils, Bichian, northwestern India	13.4 (49.2)	Kaur et al. (2002b)
P. juliflora + *S. marginatus* silvopastoral	-do-	9.6 (35.2)	Kaur et al. (2002b)
Eucalyptus plantation*	Sodic soil of UP India	24.5–27.1 (90.0–99.3)	Mishra et al. (2003), Lal (2009)
Baseline*	Sodic soil of UP India	20.7 (75.9)	Mishra et al. (2003), Lal (2009)
Acacia nilotica silvopastoral	Saline-sodic calcareous soils, Hisar, India	21.1 (77.6)	Kumari (2008)
S. persica silvopastoral	-do-	20.1 (73.9)	Kumari (2008)
Grassland	-do-	6.8 (25.9)	Kumari (2008)
Eucalyptus clonal-based agroforestry	Waterlogged soil Puthi, Haryana, India	15.8 (58.0)	Kumar (2012)
Cropland	-do-	11.3 (41.6)	Kumar (2012)
J. curcas, energy plantation**	Sodic soil, Banthra Lucknow, India	10.4–7.6 (38.2–28.0)	Singh et al. (2013)

*Soil depth 1.5 m; **Soil depth 0.15 m

Garg (1998) showed increase in SOC pool from about 10 Mg ha^{-1} to>45 Mg ha^{-1} after a 5-year period under *Acacia nilotica* and about 40 Mg ha^{-1} under *Dalbergia sissoo*. The establishment of mesquite in a sodic field increased the soil organic carbon from 11.8 to 13.3 Mg C ha^{-1} in the first 5 years, increasing to 34.2 Mg C ha^{-1} by year 7 (Bhojvaid and Timmer 1998). It reached 54.3 Mg C ha^{-1} in a 30-year period, indicating the average annual rate of increase in soil organic carbon to be about 1.4 Mg ha^{-1} (Bhojvaid and Timmer 1998). Mishra and Sharma (2003) reported that with the increase in age of *P. juliflora* and *D. sissoo* plantations from 3 years to 9 years on sodic soils, there was an increase in the levels of soil organic matter. There was a measurable increase in SOC pool to 150-cm depth after planting *Eucalyptus* on sodic soils (Mishra et al. 2003). The rate of soil carbon sequestration was 1.1 to 1.5 Mg C ha^{-1} year^{-1}.

Trees have also been grown in association with salt-tolerant grasses through agroforestry systems. Experiments conducted in the Indo-Gangetic plains showed that growing mesquite (*P. juliflora*) and other perennials are an effective strategy for increasing the SOC pool in salt-affected soils. Establishing mesquite on an alkaline soil in northwestern India increased its SOC concentration over a 74-month period from 0.18 to 0.43 % in surface 0.15 m soil and from 0.13 to 0.29 % in 0.15–0.30-m soil (Singh and Dagar 2005). Establishing mesquite in association with Kallar grass (*L. fusca*) increased SOC concentration over a 74-month period from 0.19 to 0.58 % in surface 0–0.15 m soil compared with 0.12 to 0.36 % in 0.15–0.30-m soil depth. Change in bulk density and organic carbon concentration and organic carbon pool of soil under less than 4-year *Eucalyptus* plantations at Puthi research site of waterlogged areas was 32.63 to 29.02 Mg C ha^{-1} (Kumar 2012).

In the Western Australian wheat belt, the restoration of native eucalypt forests for managing degraded agricultural landscapes is a critical part of managing dryland salinity and rebuilding biodiversity. Harper et al. (2012) conducted two 26-year-old reforestation experiments with four *Eucalyptus* species (*E. cladocalyx*, *E. occidentalis*, *E. sargentii*, and *E. wandoo*) compared with agricultural field. SOC storage (to 0.3-m depth) ranged between 33 and 55 Mg ha^{-1}, with no statistically significant differences between tree species and adjacent farmland (Harper et al. 2012). In contrast, the reforested plots contained additional carbon in the tree biomass (23–60 Mg ha^{-1}) and litter (19–34 Mg ha^{-1}), with the greatest litter accumulation associated with *E. sargentii*. Litter represented between 29 and 56 % of the biomass carbon and the protection or utilization of this litter in fire-prone, semiarid farmland will be an important component of carbon management (Harper et al. 2012). These workers emphasized the importance of considering litter in reforestation carbon accounts.

Soil carbon sequestration is a process in which CO_2 is removed from the atmosphere and stored in the soil carbon pool. Soil carbon sequestration due to the conversion of CO_2 from air found in soil into inorganic forms such as secondary carbonates in arid and semiarid regions is also important (Lal and Kimble 2000, Lal 2008). The soluble salts that occur in soils consist mostly of various proportions of anions, such as sulfate, chloride, and biocarbonate, and the cations such as calcium, sodium, and magnesium. Schlesinger (1985) calculated that the input rate of SIC in Aridisols was 0.24 g C m^{-2} $year^{-1}$ in the Mojave Desert, and the accumulation rate of secondary carbonates ranged between 0.12 and 0.42 g C m^{-2} $year^{-1}$. The soil inorganic carbon represents a large proportion of total soil carbon in Indian soils with long turnover time (Bhattacharyya et al. 2008). The soil inorganic carbon represents a large proportion (67.8 %) of total soil carbon in sodic soils, mainly in lower depths (Gupta et al. 2015). In *Prosopis juliflora* and *Eucalyptus tereticornis* plantations on reclaimed sodic soils at Saraswati Reserved Forest, northwest India, the soil inorganic carbon (Mg IC ha^{-1}) stocks in the tree plantations were 1.561–2.458 (0.30–0.45-m soil), 4.242–5.252 (0.45–0.60-m soil), and 18.596–16.901 (0.60–1.00-m soil) (Jangra 2009). This study indicated that soil inorganic carbon (SIC) at increasing soil depth provided greater potential for carbon sequestration. The soil inorganic carbon pool has been considered to improve soil physical properties, as well as improve total carbon sequestration in the soils (Pal et al. 2000; Bhattacharyya et al. 2004). Recently, soil-dissolved inorganic carbon (SDIC) and soil inorganic carbon (SIC) in saline and alkaline soil profiles up to 9-m soil depth from six profiles in the southern Gurbantongute Desert, China, have been analyzed by Wang et al. (2013). This study showed that deep-layer soil contained considerable inorganic carbon, with more than 80 % of the soil carbon stored below 1 m, in the form of soil-dissolved inorganic carbon or soil inorganic carbon. Thus, it is important to understand the role of inorganic carbon in soil carbon sequestration so as to optimize management strategies for carbon sequestration.

Climate Change Mitigation and Adaptation

Rising temperatures due to climate warming may lead to higher evapotranspiration and aridity, resulting in higher concentration of salts in soil. The three principal mechanisms of salinization, i.e., salt accumulation, seepage, and wind deposition are likely to be affected by climate change. It is known that the accumulation of salts in soils often has negatively affects soil properties and processes, including nutrient-holding capacity, nutrient dynamics, bulk density, soil structure, and porosity. It is pertinent to point out that reliance on using saline waters seems inevitable for irrigation because of increasing pressure on freshwater resources (Bouwer 2002; Qadir et al. 2007a, b). The environmental changes in the Aral Sea Basin in Central Asia basin have occurred on large scale due to human activity in recent times; other examples of such degradation exist elsewhere in other parts of the world (see Qadir et al. 2014).

Climate change mitigation refers to activities that reduce GHGs in the atmosphere or enhance the storage of GHGs in ecosystems (IPCC 2007). Adaptation to climate change is focused on actions that reduce or eliminate the negative effects of climate change or take advantage of the positive effects (IPCC 2007, 2014). Addressing soil and water salinity involves either reducing salinity (mitigation) or adapting to salinity (adaptation) by using innovative soil and water management technologies and practices for salt-affected soils and saline water. The various cost-efficient mitigation measures in biosaline agroforestry include enhancing resource use efficiency, increasing plant productivity, carbon sequestration in plant biomass and soil, and creating positive carbon budget (Wicke et al. 2013). Impacts of agroforestry systems on nitrous oxide (N_2O) and methane (CH_4) emissions need to be considered with particular reference to salt-affected soils. Studies on these GHGs in agroforestry systems are limited (Schoeneberger et al. 2012). Integrating agroforestry into agricultural operations reduces N_2O emissions by eliminating nitrogen (N) application on the part occupied by trees. Additionally, emissions may be further reduced through tree uptake of excess nitrogen (Bergeron et al. 2011; Schoeneberger et al. 2012). The integration of tree components in silvopastoral system could result in diminished N_2O emissions due to increased nutrient use efficiency.

Klein et al. (2007) have argued that adaptation and mitigation measures must be addressed simultaneously if the risks associated with climate change are to be reduced. Agroforestry practices illustrate the linkages between mitigation and adaptation as these systems can promote soil carbon sequestration while improving ecosystem function and resilience to climate extremes by enriching soil fertility and soil water retention (Smith et al. 2007). In agroforestry systems, due emphasis is placed on systems integrity and functioning, involvement of multi-stake holders, and using a multifunctional approach. The various enabling conditions for synergy between mitigation and adaptation measures have been discussed with special reference to developing countries by Duguma et al. (2014), which seem to be pertinent in the context of tree plantations on salt-affected soils.

The woody biomass component (bole and branches) represents the major portion of easily observed and measured new carbon in the agroforestry systems. For example, in *Prosopis juliflora* with *Desmostachya bipinnata* silvopastoral systems on highly sodic soils, aboveground woody biomass carbon in bole and branches comprised 82 % of the total biomass carbon in 6-year-old systems (Kaur et al. 2002a). While delivering other production and natural resource services, biosaline agroforestry has the potential to address climate change mitigation and adaptation needs on salt-affected and waterlogged soils (Wicke et al. 2013). Studies have shown that including trees in the agricultural production system can help remove excess water, thereby reducing waterlogging and lowering the groundwater tables (see Wicke et al. 2013). To develop adaptation strategies, it is necessary to understand the main factors that arise from the changing climatic conditions and their effects on the agroforestry systems of salt-affected soils.

The Intergovernmental Panel on Climate Change (IPCC 2014) has emphasized adopting a climate-resilient pathway for development by combining flexibility, innovativeness, and effectiveness in mitigating and adapting to climate change. The salt-affected soils represent the stressed environment; the climate change effects will be relatively severe, which requires considerations for comprehensive approach to soil, water, and agroforestry management with a focus on the amelioration and rehabilitation of salt-affected lands, improved productivity per unit of water used, and environmental protection so as to promote adaptation on a long-term basis. As climate change is a growing threat to development in saline environments, the agroforestry systems must enhance resilience to effects of climate change by reducing impact of extreme weather events, providing habitat diversity, greater structural and functional diversity, and diversified production.

Conclusions

Soil carbon sequestration in salt-affected soils is critical to formulate management strategies for their ecological restoration and for improving soil productivity so as to meet increasing demands of human society for food, fodder, biomass energy, and industrial products. Implementing appropriate management practices to build up soil carbon stocks in grasslands could lead to considerable mitigation, adaptation, and development benefits. Carbon sequestration also provide associated ecosystem co-benefits such as increased soil water holding capacity, better soil structure, improved soil quality and nutrient cycling, and reduced soil erosion. Under waterlogged conditions in saline environments, tree-based systems can reduce water table by pumping out excess water more rapidly than only cropland systems. Agroforestry systems can offer greater economic stability and reduce risk under climate change by creating more diversified systems with greater income distribution over time.

The sparse research conducted to date for climate change mitigation and adaptation through agroforestry has mainly focused on carbon sequestration in woody biomass. However, the complexity in agroforestry that provides its potential for meeting GHG objectives and providing the resiliency for attaining the production and the other ecosystem services, we need to work out the goals to be achieved under different climatic and soil conditions. Agroforestry has the potential to affect numerous production and ecosystem services such as aesthetics, recreation, microclimate, carbon sequestration, natural pest control, pollination, water quality, soil erosion protection, and biodiversity enhancement that will be impacted by climate change. Thus, climate change-integrated tools along with ecosystem functioning and services will be needed to ensure sustainable agroforestry in saline environments.

References

Aggarwal RK, Kumar P, Raina P (1993) Nutrient availability from sandy soils underneath *Prosopis cineraria* (Linn.) Macbride compared to adjacent open site in an arid environment. Indian For 199:321–325

Alam A, Kilpelannen A, Kellomaki S (2010) Potential timber and energy wood production and carbon stock in fuelwood under varying thinning regimes and climatic scenarios. Bioenerg Res 3:362–372

Bhattacharyya T, Pal DK, Chandran P, Mandal C, Ray SK, Gupta RK, Gajbhiye KS (2004) Managing soil carbon stocks in the Indo-Gangetic Plains, India. Rice–Wheat Consortium for the Indo-Gangetic plains, New Delhi, p 44

Bhattacharyya T, Pal DK, Chandran P, Ray SK, Mandal C, Telpande B (2008) Soil carbon storage capacity as a tool to prioritize areas for carbon sequestration. Curr Sci 95:482–493

Bell DT (1999) Australian tree for rehabilitation of waterlogged and salinity damaged landscape. Aust J Bot 47:697–716

Bergeron M, Lacombe S, Bradley RL, Whalen J, Cogliastro A, Jutras MF, Arp P (2011) Reduced soil nutrient leaching following the establishment of tree-based intercropping systems in eastern Canada. Agrofor Syst 83:321–330

Bhojvaid PP, Timmer V (1998) Soil dynamics in an age sequence of *Prosopis juliflora* planted for sodic soil restoration in India. For Ecol Manag 106:181–193

BIOSAFOR Biosaline (Agro) Forestry (2011) Remediation of saline wastelands through production of renewable energy, biomaterials and fodder. Deliverable D24: final activity report: evaluation of results. European Commission FP6 Project STREP Contract no 032502

Bouwer H (2002) Integrated water management for the 21st century: problems and solutions. J Irrig Drain Eng 28:193–202

Chaudhary AA, Hameed M, Ahmed R, Hussain A (2001) Phytosociological studies in ChhumbiSurela Wild Life Sanctuary, chakwal, Pakistan. Species diversity. Int J Agric Biol 3:369–374

Dagar JC, Singh G, Singh NT (2001) Evaluation of forest and fruit trees used for rehabilitation of semiarid alkali/sodic soils in India. Arid Soil Res Rehabil 15:115–133

De Groot RS, Wilson M, Boumans R (2002) A typology for the description, classification and valuation of ecosystem functions, goods and services. Ecol Econ 41:393–408

Duguma LA, Peter AM, van Noordwijk M (2014) Climate change mitigation and adaptation in the land use sector: from complementarity to synergy. Environ Manag 54:420–432

Dutton RW, Powell M, Ridley RJ (eds) (1992) Prosopis species: aspects of their value, research and development. In: Proceedings of the prosopis symposium, held by CORD, University of Durham, UK

FAO (2010) Global forest resources assessment 2010. Main report. FAO Forestry Paper 163, Rome

Garcia IV, Mendoza RE (2007) Arbuscular mycorrhizal fungi and plant symbiosis in a saline-sodic soil. Mycorrhiza 17:167–174

Garg VK (1998) Interaction of tree crops with a sodic soil environment: potential for rehabilitation of degraded environments. Land Degrad Dev 9:81–93

Gill HS, Abrol IP (1993) Afforestation and amelioration of salt-affected soils in India. In: Davidson N, Galloway R (eds) The productive use of saline land. Proceedings of a workshop held in Perth, Western Australia. ACIAR Proceedings No. 42, pp 23–27

Glenn EP, Pitelka LF, Olsen MW (1992) The use of halophytes to sequester carbon. Water Air Soil Poll 64:251–263

Glenn EP, Squires V, Olsen MW, Frye R (1993) Potential for carbon sequestration in drylands. Water Air Soil Poll 70:341–355

Gooday GW (1994) Physiology of microbial degradation of chitin and chitosan. In: Ratledge C (ed) Biochemistry of microbial degradation. Kluwer, Dordrecht, pp 279–312

Gupta RK, Abrol IP (1990) Salt-affected soils: their reclamation and management for crop production. Adv Soil Sci 11:223–288

Gupta SR, Jangra R, Dagar JC (2015) Carbon pools and fluxes in grassland systems on sodic soils of northern India. In: Singh AK, Dagar JC, Arunachalam A, Gopichandran R, Shelat KN (eds) Climate change modelling, planning and policy for agriculture. Springer, New Delhi, pp 119–140

Harper RJ, Beck AC, Ritson P, Hill MJ, Mitchell CD, Barett DJ, Smetter KRJ, Mann SS (2007) The potential of greenhouse sink to underwrite improved land management. Ecol Eng 29:329–341

Harper RJ, Smettem KRJ, Tomlinson RJ (2005) Using soil and climatic data to estimate the performance of trees, carbon sequestration and recharge potential at the catchment scale. Aust J Exp Agric 45:1389–1401

Harper RJ, Okom AEA, Stilwell AT, Tibbett M, Dean C, George SJ, Sochacki SJ, Mitchell CD, Mann SS, Dods K (2012) Reforesting degraded agricultural landscapes with Eucalypts: effects on carbon storage and soil fertility after 26 years. Agric Ecosyst Environ 163:3–13

Hocking D (ed) (1993) Trees for drylands. International Science Publisher, New York, p 370

IPCC (2007) In: Parry M, Canziani O, Palutikof J, van der Linden P (eds) Climate change 2007: impacts, adaptation and vulnerability. Contribution of working group II to the fourth assessment report of the intergovernmental panel on climate change. Cambridge University Press, Cambridge, UK/New York, 976 pp

IPCC (2013) In: Stocker TF, Qin D, Plattner GK, Tignor M, Allen SK, Boschung J, Nauels A, Xia Y, Bex V, Midgley PM (eds) Climate change 2013: the physical science basis. Contribution of working group I to the fifth assessment report of the intergovernmental panel on climate change. Cambridge University Press, Cambridge, UK/New York, 1535 pp

IPCC (2014) Summary for policymakers. In: Field CB (ed) Climate change 2014: impacts, adaptation, and vulnerability. Part A: global and sectoral aspects. contribution of working group II to the fifth assessment report of the intergovernmental panel on climate change. Cambridge University Press, Cambridge, UK/New York, pp 1–32

Izac AMN (2003) Economic aspect of soil fertility management and agroforestry practices. In: Scroth G, Sinclair F (eds) Tree crop and soil fertility: concept and research methods. CABI, Wallingford, UK, p 464

Jangra R (2009) Carbon dynamics in land use systems on a reclaimed sodic soil. Ph.D. thesis, Botany Department, Kurukshetra, University Kurukshetra

Jangra R, Gupta SR, Kumar R, Bhalla E (2011) Soil respiration, microbial biomass, and mycorrhizal diversity in sodic grassland ecosystems in northwestern India. American-Eurasian J Agric & Environ Sci 10:863–875

Jeet-Ram, Dagar JC, Lal K, Singh G, Toky OP, Tanwar VS, Dar SR, Chauhan MK (2011) Biodrainage to combat waterlogging, increase farm productivity and sequester carbon in central command areas of northwest India. Curr Sci 100:1673–1680

Jose S (2009) Agroforestry for ecosystem services and environmental benefits. Agroforest Syst 76:1–10

Kaur B, Gupta SR, Singh G (2000) Soil carbon, microbial activity and nitrogen availability in agroforestry systems on moderately alkaline soils in northern India. Appl Soil Ecol 15:283–294

Kaur B, Gupta SR, Singh G (2002a) Carbon storage and nitrogen cycling in silvi-pastoral systems on a sodic soil in northwestern India. Agrofor Syst 54:21–29

Kaur B, Gupta SR, Singh G (2002b) Bioamelioration of a sodic soil by silvopastoral system in northwestern India. Agrofor Syst 54:13–20

Klein RJT, Huq S, Denton F, Downing TE, Richels RG, Robinson JB, Toth FL (2007) In: Parry ML, Canziani OF, Palutikof JP, van der Linden PJ, Hanson CE (eds) Inter-relationships between mitigation and adaptation. Climate change 2007: impacts, adaptation and vulnerability. Contribution of working group II to the fourth assessment report of the intergovernmental panel on climate change. Cambridge University Press, Cambridge, UK, pp 745–777

Kuiper L, Probos O (2005) Biosaline (agro)forestry: a literature review. Biomass upstream stuurgroep, p 17

Kumar A, Abrol IP (1984) Studies on the reclaiming effect of Karnal-grass and para-grass grown in a highly sodic soil. Indian J Agri Sci 54:189–193

Kumar M (2012) Waterlogging control and carbon sequestration through biodrainage. PhD thesis, Kurukshetra, University Kurukshetra, p 253

Kumari C (2008) Soil carbon and plant diversity in agroforestry systems of calcareous soils under saline water irrigation. M.Phil dissertation, Botany Department, Kurukshetra University, Kurukshetra

Ladeiro B (2012)saline agriculture in the 21st century: using salt contaminated resources to cope food requirements. J Bot. Volume 2012, Article ID 310705, 7 pages. doi:10.1155/2012/310705

Lal R (2008) Carbon sequestration. Phil Trans R Soc Lond B 63:815–830

Lal R (2009) Carbon sequestration in saline soils. J Soil Saline Water Qual 1:30–49

Lal R, Kimble JM (2000) Pedogenic carbonates and the global carbon cycle. In: Lal R, Kimble JM, Eswaran H, Stewart BA (eds) Global climate change and pedogenic carbonates. CRC Press, Boca Raton, pp 1–14

Lambert MJ, Turner J (2000) Change in organic carbon in forest plantation soils in eastern Australia. For Ecol Manage 133:231–247

Marcar N, Crawford D (2004) Trees for saline landscapes. Rural Industries Research and Development Corporation (RIRDC), Canberra

Masters DG, Benes SE, Norman HC (2007) Biosaline agriculture for forage and livestock production. Agr Ecosyst Environ 119:234–248

MEA (Millennium Ecosystem Assessment) (2005) Ecosystems and human well-being: biodiversity synthesis. World Resources Institute, Washington, DC

Minhas PS, Sharma OP (2003) Management of soil salinity and alkalinity problems in India. Journal of Crop Production 7:181–230

Mishra A, Sharma SD, Khan GH (2002) Rehabilitation of degraded sodic lands during a decade of *Dalbergia sissoo* plantation in Sultanpur district of Uttar Pradesh, India. Land Degrad Dev 13:375–386

Mishra A, Sharma SD, Khan GH (2003) Improvement in physical and chemical properties of sodic soil by 3, 6, and 9 years old plantation of *Eucalyptus tereticornis* - Biorejuvenation of sodic soil. For Ecol Manage 184:115–124

Mishra A, Sharma SD (2003) Leguminous trees for the restoration of degraded sodic wasteland in eastern Uttar Pradesh, India. Land Degrad Dev 14:245–261

Montagnini F, Nair PKR (2004) Carbon sequestration: an underexploited environmental benefit of agroforestry systems. Agrofor Syst 61:281–295

Muzzarelli RAA (1977) Chitin. Pergamon, New York

Nair PKR (2007) The coming of age of agroforestry. J Sci Food Agric 87:1613–1619

Nair PKR (2012) Climate change mitigation: a low-hanging fruit of agroforestry. In: Nair PKR, Garitty D (eds) Agroforestry: the future of global land-use. Springer, Dordrecht, pp 31–68

Nair PKR, Kumar BM, Nair VD (2009) Agroforestry as a strategy for carbon sequestration. J Plant Nutr Soil Sci 172:10–23

Neeraj, Gupta SR, Malik V, Kaur B, Neelam (2004) Plant diversity, carbon dynamics and soil biological activity in tropical successional grassland systems at Kurukshetra. Intern J Ecol Environ Sci 30:285–298

Naingoo A, Iwai CB, Saenjan P, Topark-Ngarm B (2012) Soil carbon dynamics during the amelioration of salt-affected areas in northeast Thailand. Intern J Environ Rural Develop 3–1:120–125

Pal DK, Dasong GS, Vadivelu S, Ahuja RL, Bhattacharyya T (2000) Secondary calcium carbonate in soils of arid and semiarid regions of India. In: Lal R, Kimble JM, Eswaran H, Stewart BA (eds) Global change and pedogenic carbonate. CRC Press, Boca Raton, pp 149–185

Palsaniya DR, Tewari RK, Singh R, Yadav RS, Dhyani SK (2010) Farmer -agroforestry land use adoption interface in degraded agroecosystem of Bundelkhand region, India. Range Manag Agrofor 31:11–19

Qadir M, Oster JDC (2002) Vegetative bioremediation of calcareous sodic soils: history, mechanisms, and evaluation. Irrig Sci 21:91–101

Qadir M, Oster JD, Schubert S, Noble AD, Sahrawat KL (2007a) Phytoremediation of sodic and saline- sodic soils. Adv Agron 96:197–247

Qadir M, Quillérou E, Nangia V, Murtaza G, Singh M, Thomas RJ, Drechsel P, Noble AD (2014) Economics of salt-induced land degradation and restoration. Nat Resour Forum 38:282–295

Qadir M, Sharma BR, Bruggeman A, Choukr-Allah A, Karajeh F (2007b) Non-conventional water resources and opportunities for water augmentation to achieve food security in water scarce countries. Agric Water Manage 87:2–22

Qureshi RH, Barrett-Lennard EG (1998) Saline agriculture for irrigated land in Pakistan: a handbook. Monograph No. 50. Australian Centre for International Agricultural Research, Canberra, p 142

Rana RS, Parkash V (1987) Floristic characterization of alkali soils in north-western India. Plant Soil 99:447–451

Ravindranath NH (2007) Mitigation and adaptation synergy in forest sector. Mitigat Adapt Strat Global Change 12:843–853

Sachs JD, Baillie JEM, Sutherland WJ (2009) Biodiversity conservation and the millennium development goals. Science 325:1502–1503

Schlesinger WH (1985) The formation of caliche in soils of the Mojave Desert, California. Geochimica et Cosmochimica Acta 49:57–66

Schoeneberger MM, Bentrup G, de Gooijer H, Soolanayakamahally R, Sauer T, Brandle J, Zhou X, Current D (2012) Branching out: agroforestry as a climate change mitigation and adaptation for agriculture. J Soil Water Cons 67:128–136

Singh AN, Raghubanshi AS, Singh JS (2002) Plantation as a tool for mine spoil restoration. Curr Sci 82:1436–1441

Singh G (1995) An agroforestry practice for the development of salt lands using *Prosopis juliflora* and *Leptochloa fusca*. Agrofor Syst 29:61–75

Singh G, Dagar JC (2005) Greening sodic soils: Bichhian model. Technical Bulletin 2/2005. Central Soil Salinity Research Institute, Karnal, p 51

Singh G, Gill HS (1992) Ameliorative effect of tree species on characteristics of sodic soils at Karnal. Indian J Agric Sci 62:142–146

Singh G, Singh NT (1992) Mesquite for the revegetation of salt lands. Central Soil Salinity Research Institute, Karnal, p 24

Singh K, Singh B, Tuli R (2013) Sodic soil reclamation potential of *Jatropha curcas*: a long-term study. Ecol Eng 58:434–440

Sinha A, Rana RS, Gupta SR (1988) Phytosociological analysis of some natural grassland communities of sodic soils. Trop Ecol 29:136–145

Smith P, Martino D, Cai Z, Gwary D, Janzen H, Kumar P, McCarl B, Ogle S, O'Mara F, Rice C, Scholes B, Sirotenko O (2007) Agriculture. In: Metz A (ed) Climate change 2007: mitigation. contribution of working group III to the fourth assessment report of the intergovernmental panel on climate change. Cambridge University Press, Cambridge, UK/New York

TEEB (2010) In: Pushpam-Kumar (ed) The economics of ecosystems and biodiversity: ecological and economic foundations. Earthscan, London/Washington

Tomar OS, Minhas PS, Sharma VK, Singh YP, Gupta RK (2003) Performance of 31 tree species and soil conditions in a plantation established with saline irrigation. For Ecol Manag 177:333–346

Treseder KK, Allen MF (2000) Mycorrhizal fungi have a potential role in soil carbon storage under elevated CO_2 and nitrogen deposition. New Phytol 147:189–200

Tripathi KP, Singh B (2005) The role of revegetation for rehabilitation of sodic soils in semi-arid sub-tropical forest, India. Rest Ecol 13:29–38

Turner WR, Brandon K, Brooks TM, Gascon C, Gibbs HK, Lawrence KS, Mittermeir RA, Selig ER (2012) Global biodiversity conservation and the alleviation of poverty. Bioscience 62:85–92

Wang Y, Wang Z, Li Y (2013) Storage/turnover rate of inorganic carbon and its dissolvable part in the profile of saline/alkaline soils. PLoS ONE 8(11), e82029. doi:10.1371/journal.pone.0082029

Wicke B, Edward SWM, Akanda R, Stille L, Singh RK, Awan AR, Mahmood K, Faaij APC (2013) Biomass production in agroforestry and forestry systems on salt affected soils in South Asia: exploration of the GHG balance and economic performance on three case studies. J Environ Manage 127:324–334

Wong VNL, Dalal RC, Greene RSB (2008) Salinity and sodicity effects on respiration and microbial biomass of soil. Biol Fert Soils 44:943–953

Wright SF, Franke-Snyder M, Morton JB, Upadhyaya A (1996) Time-course study and partial characterization of a protein on hyphae of arbuscular mycorrhizal fungi during active colonization of roots. Plant Soil 181:193–203

Wright SF, Upadhyaya A (1999) Quantification of arbuscular mycorrhizal fungi activity by the glomalin concentration on hyphal traps. Mycorrhiza 8:283–285

Yuan BC, Li ZZ, Liu H, Gao M, Zhang YY (2006) Microbial biomass and activity in salt affected soils under arid conditions. Appl Soil Ecol 35:319–328

Zhang JF, Xing SJ, Li JY, Makeschin F, Song YM (2004) Agroforestry and its application in amelioration of saline soils in eastern China coastal region. For Stud China 6:27–33

Synthesis and Way Forward: Agroforestry for Waterlogged Saline Soils and Poor-quality Waters

P.S. Minhas and J.C. Dagar

Abstract

The development of salt-affected soils has been considered to be an adjunct of irrigated agriculture since time immemorial. The social, economic, and environmental costs being high for the on- and off-farm reclamation techniques, agroforestry is now emerging as a potential tool not only for arresting salinity but also for other environmental services like adaptation to climate change, sequestration of carbon, and restoration of biodiversity. Recent research and developmental efforts, though experimentally in small plots or under microsite conditions in catchments, have demonstrated that trees can be successfully established through appropriate site preparation, careful species selection, post-planting care especially during the earlier and other critical stages of development. The multienterprise models involving the integration of agroforestry with other enterprises like fisheries/dairying should improve productivity vis-à-vis income generation. Special commitment would be required by governments on the insurance, legal, and institutional arrangements and also the collaboration between ecologists and land managers for the large-scale promotion of agroforestry on salt-degraded lands.

The multitude benefits of forestry for fulfilling the economic, environmental, and social expectations of any country are well established. However, a notable development is the importance given to agroforestry since the 1990s by governments, landholders, and other stakeholders, especially in the developing countries. Though it is still considered to be in the developmental stage, as a land management system, the combination of trees and shrubs/agricultural crops can help in achieving the targets of sustainable agriculture, diversifying farm income, reducing trade deficits, developing timber and other related industries, and generating employment opportunities.

P.S. Minhas (✉)
National Institute of Abiotic Stress Management,
Baramati, Pune 413115, Maharashtra, India
e-mail: minhas_54@yahoo.co.in

J.C. Dagar
Central Soil Salinity Research Institute,
Karnal 132 001, Haryana, India
e-mail: dagarjc@gmail.com

J.C. Dagar, P.S. Minhas (eds.), *Agroforestry for the Management of Waterlogged Saline Soils and Poor-quality Waters*, Advances in Agroforestry 13, DOI 10.1007/978-81-322-2659-8_12

Moreover, the major role of agroforestry is now emerging in the domain of environmental services for its potential to reduce land degradation, improve biodiversity, mitigate the climate change, and sequester carbon.

Whereas land degradation is concerned, virtually all global land resources are afflicted by degradation processes to some degree, and as a consequence of a steadily increasing population and thus the demand for food as well as the impact of current episode of climate change, the situation is becoming more acute. The major land degradation processes fall into two categories: intrinsic and extrinsic. Among the intrinsic processes, the development of salt-affected soils has been considered to be an adjunct of irrigated agriculture since time immemorial for the reason that their maxima persisted in irrigated areas. For example, the earliest knowledge of connection between irrigation and soil salinity relates to the Euphrates and Tigris valleys where irrigation-induced soil salinity brought about the decline of a highly progressive Mesopotamian civilization. The same was later realized about the irrigated agriculture for the Indus civilization. Of course, the latest one is the Aral Sea in Kazakhstan and Uzbekistan where bureaucratic incompetence and ecosystem mismanagement while diverting the Amu Darya and Syr Darya rivers in the 1930s for cotton irrigation resulted in widespread salinization. The other one is the development of dryland salinity especially in Australia where the large-sale clearing of forests has afflicted over 2.5 Mha of agricultural landscape and the prognosis is worsening for salinity and waterlogging over next 30–50 years. Additionally, coastal salinity is the consequence of periodical inundation with tidal water and is due to the high water table and its high salt concentrations in lowlands in close proximity to the sea. Moreover, the saline aquifers exit in most of the low-rainfall and water-scarce areas. The ever-increasing exploitation of these aquifers for irrigation along with the wastewaters from urban agglomerations is further leading to salinity and environmental problems. Several on- and off-farm techniques, as well as the use of saline waters for crop production, have been employed to rehabilitate the saline waterlogged soils. However, the social, economic, and environmental costs continue to be the major constraints in their effective utilization. Therefore, as an alternative, the role that agroforestry plays in the viable utilization and maximization of benefits from these degraded land and water resources is the theme of various chapters included in this book.

Recent research and developmental efforts have greatly improved the expectations that agroforestry can deliver in terms of greening the saline waterlogged soils, resource conservation, and also the economic and environmental benefits to meet the challenge of livelihood security of diverse rural populations. The agroforestry on such farms or community lands can produce a broad range of products depending upon the species planted, climatic conditions, site specific characteristics, and the management practices followed. These include timber, biomass for fuel/bioenergy, fodder, aromatic and pharmaceutical compounds, and carbon storage in both above- and belowground parts. Of course, salinity limits the choice of tree species because of their limited tolerance to salinity. Since aridity and salinity go together, the growth rates get further limited under low- and medium-rainfall situations. However, the firewood continues to be the most sought-after product from agroforestry under such harsh environments.

The most research involving tree plantations, specifically low- to medium-rainfall zones (<600 mm), has been conducted in Australia and the Indian subcontinent. The tree species showing promise in Australia include Sargent's mallet (*Eucalyptus sargentii*), *E. occidentalis* (swamp yate) and *E. kondininensis* (stocking gum), and *Casuarina glauca.* Saltbushes (species of *Atriplex, Melaleuca*), blue bush (*Marcaena brevifolia*), and samphire (*Halosarcia* spp.) also performed well. Mesquite (*Prosopis juliflora*), *Tamarix articulata*, *Acacia* (*Acacia nilotica*, *A. tortilis*), *Eucalyptus* (*E. camaldulensis*, *E. tereticornis*), *Casuarina* (*C. equisetifolia*, *C. glauca*), *Parkinsonia aculeata*, *Prosopis cineraria*, and arjun (*Terminalia arjuna*) have been identified in South Asia. In general, the annual wood production of these species ranged between 4 and

15 m^3 ha^{-1}, depending upon the harshness of the salinity situation compared to>30 m^3 ha^{-1} on normal soils. Mesquite (*Prosopis*) species are considered as the tree plantations having the most potential for saline environments in the Indian subcontinent and Arabian countries, while *E. occidentalis*, *E. grandis*, and *Melaleuca halmaturorum* are extremely tolerant to saline swamps in Australia along with the perennial grasses like *Thinopyrum elongatum* (Tall wheat grass), Rhodes grass (*Chloris gayana*), and mixed stands of *Phalaris aquatica*. Since the priority has now shifted to "biofuels," the major focus is on *Jatropha curcas*, *Pongamia pinnata*, and *Ricinus communis* as the "future fuels," while *Azadirachta indica*, *Simarouba glauca*, and *Madhuca indica* also hold promise. Similarly, fruits like date palm (*Phoenix dactylifera*), pistachio (*Pistachio vera*), *Carissa carandas*, *Punica granatum*, *Emblica officinalis*, *Psidium guajava*, *Syzygium cuminii*, and *Ziziphus mauritiana* have been successfully raised on saline soils on a short-term basis. In coastal areas, mangroves need special attention for the protection of coastal lines and conservation of the entire aquatic life along the coasts. They are not only life support systems for coastal wildlife but also act as major sinks for carbon. Some of the associated trees such as *Pandanus* spp., coastal almond (*Terminalia catappa*), cashew nut (*Anacardium occidentale*), *Calophyllum inophyllum*, *Morinda citrifolia*, *Salvadora persica*, and nipa palm (*Nypa fruticans*) are known for their industrial application. On the whole, the agroforestry on saline soils continues to be of the subsistence type for greening with some ecological benefits. It is since the productivity of even the above species is not as satisfactory based on the economic considerations. It is the time to learn from agriculture growth that could gain major advantages from selection, breeding/hybridization, tissue culture, etc. of annual crops. A similar scope exists with tree breeding efforts in terms of the selection of better-performing genotypes/provenances, clonal development, and even hybridization for better and salt-tolerant planting materials.

In addition to salinity/sodicity, several other soil and site factors control the growth of tree plantations in saline soils. These include the lack of accessibility to soil volume (e.g., due to calcite layer below sodic soils) and water (e.g., in regions with summer droughts), nutrient limitations, episodic events (e.g., flooding/waterlogging, droughts), wind throws, etc. Trees may also become susceptible to salinity buildup through highly saline and shallow water tables during their growth. To overcome these constraints, techniques like auger-bore hole to pierce through hardpans of calcite, subsurface planting and irrigation through channels to create niches of better salt and water regime, mounding to alleviate seasonal waterlogging, and scalping to remove surface salts have been developed. Though these techniques have been quite successful in establishing the tree plantations, the growth and health of these plantations is usually governed by salt and water dynamics and agroclimatic conditions during their later stages, e.g., these may become prone to salinity and thus suffer reduced growth rates and possible death after the cessation of the supplemental irrigation. Therefore, more efforts are required to evolve newer planting techniques and the post-planting management for the long-term sustainable growth of tree plantations on salt-affected lands.

A major part of the book is devoted to the role of trees in modifying salt and water dynamics at the field and catchment scale and in aiding in the control of salinity. The main force behind the latter is the water-profligate nature of trees and their deep root systems. Even some trees are phreatophytes and their root systems intercept the saturated zone or capillary fringe above water table. Thereby, a misnomer was created in the 1990s that high-rate transpiration systems (HRTS) with plantations like *Eucalyptus* can act as "biopumps" without any long-term verification. However, the literature world-over now elaborates that the tree plantations are not as profligate water users and rather their transpiration rates are controlled by specific site conditions defining soil type, evaporative demands, and even the depth to groundwater and its salinity. Under favorable conditions (sandy and deep soils, shallow water table of good quality, cooler climate), trees may draw soil water at about 0.8 × Epan, but this may reduce to about 0.2 × Epan under less optimal conditions (clayey and shallow soils, saline/deeper water table, hot and dry summer, etc.). The major advantage of trees in waterlogging control can be viewed in terms of their year-round water withdrawals and that too from rain-recharged soil profiles/the shallow water tables. Plantations, especially of *Eucalyptus* species, have been shown to draw down water table, of course their spatial extent being governed by tree transpiration rates and hydraulic characters of soils. But for trees to be effective, the experiences in Australia and India show that the land requirements for reforestation would be high (10–50 % and even as high as 80 % of total land), the other issue being salt accumulation, once the deeper-rooted trees skim out the water.

When implied for bringing sustainable catchment-scale changes in soil salinity, the impact of broad-scale reforestation on its water balance needs critical evaluations. The results from several field-scale experiments and modeling efforts do show that interactions between the hydrology of an area and the planted trees can help in choosing some of the innovative planting designs, e.g., farm boundary plantations in flatter areas and block plantations along the canals in northwestern India, strategically placed shelterbelts/"walking plantations" in recharge/discharge sites to control dryland salinity in Australia, and even integrating subsurface drainage and tree plantations can be quite successful. In zones of medium to low rainfall on permeable soils in Australia, multiple narrow-belt plantations to capture surface and subsurface runoff from interbelt areas have been utilized to maximize wood production. The experiences gained can be gainfully employed to mitigate and even reverse waterlogging and secondary salinization problems. The tools like GIS (geographical information system) and RS (remote sensing) should help in the prognosis of hot spot areas to be put under plantations and also to prioritize action plans for developing an integrated command framework to control waterlogging and salinity. Similarly, the process-based models (like 3-G) should further help in the prediction of salinity within the basin under the present and afforested conditions and thus the afforestation designs and other management options and priorities. The knowledge of the equilibrium root zone salinity that is likely to develop and the plant types that can grow at the acceptable rates under the prevailing conditions should better match between the specificity of the site, tree species, and other silviculture practices for better control over salinity.

With competitions from other sectors of the economy, freshwaters are becoming scarce for irrigation and allied agricultural activities, and these will have to rely more on marginal-quality water resources like highly brackish groundwater/drainage waters and wastewaters from domestic, commercial, and industrial sectors. This warrants for modifications in existing crop, irrigation, and soil management practices for protecting public health and the environment. Earlier chapters indicate that research efforts have led to the identification of a number of forage grasses and shrubs, aromatic and medicinal species, biofuel crops, and fruit trees suited to be grown in agroforestry systems, which are profitable and suit a variety of agroecosystems. A shift to these is likely to be the key to future agricultural and economic growth in the afflicted areas, but such systems will need to consider three issues: improving the productivity per unit of the marginal water resources, protecting the environment, and involving farmers in the most suitable and sustainable crop-diversifying systems to mitigate any perceived risks. This is specifically the case with wastewaters where the irrigation of high-transpiring forest species has been put forward for their recycling and the reuse and conservation of nutrient energy into biomass, thereby bringing multiple benefits such as fuelwood production, environmental sanitation, and ecorestoration. However, the loading rates the tree plantations can carry continue to be contradictory. The evidences are that these plantations though can act as sites for a year round recycling of sewage but their potential is just about 1.5-fold the agricultural crops when considered on annual basis. This indicates that cautions, rather regulatory mechanism need to be devised to control loading rates for minimizing the risks of groundwater contaminations.

Several of the efforts for the reestablishment of tree plantations, though experimentally in small plots or under microsite conditions in catchments, have demonstrated their role as effective tools in the rehabilitation of salt-degraded lands. The emerging agroforestry options under these environments should not only provide economic or on-farm benefits to farming communities but also help in environment conservation through improvements in salinity/sodicity, the availability status of nutrients, and carbon storage in the post-plantation soil. These have further potential to affect numerous production and ecosystem services such as microclimate modifications, carbon mitigation both via carbon sequestration and biofuels, soil erosion protection, biodiversity restoration, and adaptations to climate change, but among these, the conservation of carbon stocks with agroforestry on salt-affected patches should become a foremost priority. The governmental policies to support such programs as a part of carbon reforestation are essential. Greening of salt-affected lands can also be supported by schemes similar to "REDD" (Reducing Emissions from Degradation and Deforestation) of tropical rainforests in developing countries. The economic analysis of most of the reforestation programs on salt-affected soils in Australia and India shows that these either lead to only small financial gains or are cost neutral. Of course, these estimates do not consider the carbon offsetting and other ecosystem benefits. Thus, a realistic economic assessment of these benefits is fundamental to integrate and justify suitable investment and supportive policies for their reforestation. Moreover, in developing countries, integrated farming systems may provide hope for food security to the smallholders living in these high-risk and harsh conditions of both aridity and salinity. Wherever possible, the multienterprise models involving agroforestry models should be integrated with other enterprises like fisheries/dairying on a pilot scale for higher productivity vis-à-vis income generation and their adoption by the farmers while minimizing the risks. Additionally if large-scale agroforestry is to be promoted on salt-affected lands, special commitment would be required from governments on the insurance, legal, institutional, and even community arrangements. At the same time, the new paradigms of agroforestry in salt-affected lands demand management prescriptions based upon the long-term research collaborations between land managers and ecologists that will lead to more objective analysis and thus understanding of their ecological functioning.

Index

A
Abelmoschus esculentus, 130
Acacia ampliceps, 11, 15, 16, 36, 104, 150, 155, 169
Acacia catechu, 15
Acacia dealbata, 37
Acacia eburnea, 15
Acacia farnesiana, 15, 135, 151, 155
Acacia ferruginea, 135
Acacia leucophloea, 134
Acacia linarioides, 18
Acacia mangium, 136, 169, 176
Acacia mearnsii, 60
Acacia melanoxylon, 13, 60
Acacia modesta, 16, 150, 155, 170
Acacia nilotica, 14–18, 20–26, 36–38, 40, 103, 104, 115–117, 134, 135, 148–151, 153, 155, 167–170, 183, 185–189, 198
Acacia pennatula, 15
Acacia planiformis, 123
Acacia salicina, 60, 167
Acacia saligna, 11, 13, 15, 60
Acacia stenophylla, 16, 60, 61
Acacia tortilis, 15, 16, 18, 150, 151, 153, 155, 198
Acanthus ebracteatus, 124
Acanthus ilicifolius, 124
Acanthus volubilis, 124
Achras zapota, 21, 26, 123, 134, 135, 152, 155, 168
Acrostichum aureum, 124
Adhatoda vasica, 150
Aegle marmelos, 20, 21, 26, 152, 153
Agri-horticultural system, 153
Agro-ecological regions/zones, 122
Agroforestry, 2–4, 6–27, 35, 36, 45, 46, 55, 64, 65, 98, 102–104, 115, 122, 124–126, 129–132, 134, 136–137, 139, 140, 146, 152–153, 155, 157, 164–169, 172–174, 182–192, 197–201
Agroforestry approach, 2, 9, 22, 122
Agroforestry and ecosystem services, 181–192
Agroforestry for adapting climate change, 190–191, 201
Agroforestry for carbon sequestration, 4, 13, 25, 182, 187, 188, 191, 192
Agroforestry for employment/income generation, 201
Agroforestry for energy plantations, 21, 157
Agroforestry for livelihood security, 3, 4, 122
Agroforestry policy, 4
Agroforestry potential, 150, 152–154, 156, 157, 188, 191, 192
Agroforestry practice, 2, 17, 46, 182, 185, 191
Agroforestry research, 58, 62, 68, 75, 102, 106, 116, 118, 129, 139, 186, 189, 192, 198, 200, 201
Agroforestry systems, 2–4, 17, 24, 35, 36, 45, 98, 124, 126, 152, 155, 169, 172, 182, 183, 185–189, 191, 192, 200
Agro-silvo-pastoral systems, 183
Ailanthus excelsa, 15, 23
Albizia amara, 123
Albizia lebbe(c)k, 15, 20–22, 135, 149, 185
Albizia procera, 15, 22
Alkali soils, 7, 16, 20–25, 117, 185
Alley cropping, 183
Allium cepa, 26
Allium sativum, 26
Aloe barbadensis syn. Aloe vera, 156
Anacardium occidentale (Cashew nut), 123, 124, 127, 134, 199
Ananas comosus (pine apple), 150
Andaman and Nicobar Islands, 133
Anethum graveolens, 156
Annona glabra, 133
Annona ramosa, 123
Annona squamosa, 123
Apple, 133, 150
Apricot, 150
Ardisia andamanica, 133
Ardisia solanacea, 133
Areca catechu (Areca nut), 15, 124, 130
Areca triandra, 124, 130
Arthrocnemum indicum, 14, 15, 134, 135, 154
Artocarpus chaplasha, 123
Arundo donax, 117
Asplenium nidus, 124
Atriplex (amnicola, atacamensis, barclayana, canescens, cinerea, glauca, halomus, isatidea, lentiformis, nummularia, paludosa, polycarpa, semibaccata, undulata, etc.), 7–9, 11, 12, 14–17, 19, 67, 103, 104, 117, 118, 134, 135, 139, 153–155, 198
Auger hole technique, 25
Avena sativa (oats), 23

J.C. Dagar, P.S. Minhas (eds.), *Agroforestry for the Management of Waterlogged Saline Soils and Poor-quality Waters*, Advances in Agroforestry 13, DOI 10.1007/978-81-322-2659-8

Avicennia marina, 124, 127, 153
Avicennia officinalis, 124, 127
Azadirachta indica, 15, 16, 21, 22, 25, 26, 37, 40, 137, 149–151, 155, 167–169, 176, 199
Azolla filiculoides, 165

B

Bael tree (*Aegle marmelos*), 20, 21, 26, 152, 153
Bambusa arundinacea, 169, 176
Bambusa bambos, 137
Banana (*Musa* x *paradisiaca*), 124, 130, 133–137
Barley (*Hordeum vulgaris*), 9, 11, 12, 54, 74, 102, 153, 187
Barn yard millet (*Echinochloa-crus-galli*), 14, 16, 135
Barringtonia asiatica, 124
Barringtonia racemosa, 124
Batis maritima, 139, 152, 155
Bauhinia racemosa, 124
Bay Islands, 124, 129, 132
Benefit /cost ratio (B: C ratio), 22
Berseem (*Trifoliun alexandrinum*), 24, 26, 137
Beta vulgaris (Beet root), 139, 155
Betula nigra, 165
Bienertia cycloptera, 14, 135, 154
Bio-amelioration, 146, 185–187
Biodisposal, 115
Biodiversity, 2, 4, 17, 51, 52, 55, 105, 123, 139, 173, 182, 184, 189, 192, 198, 201
Biodiversity conservation, 182
Biodrainage, 3, 34, 35, 110, 114–119, 136
Bioenergy, 109, 198
Biofuels, 139, 155, 199–201
Bio-management of salt-affected soils, 3, 97–106
Biomass production, 4, 15, 19, 21, 25, 59, 105, 117, 154, 185, 188
Bioremediation species, 3, 98, 101–105, 184
Black pepper (*Piper nigrum*), 130
Block plantations, 116, 200
Blue panic (*Panicum antidotale*), 24, 154
Bombax ceiba syn. B. malabaricum, 123
Borassus flabellifer, 15, 123, 137, 155
Bottle gourd (*Lagenaria siceraria*), 137
Boundary plantation, 2, 46, 200
Brachiaria mutica, 16, 24, 25, 135, 139, 154, 155
Brachiaria ramosa, 16
Brachiaria ruziziensis, 134
Brackish water aquaculture, 128, 133
Brassica juncea, 153, 165
Brassica napus, 165
Brassica nigra, 165
Brassica oleracea, 165
Brassica rapa, 165
Brinjal (*Solanum melongena*), 130
Bruguiera cylindrica, 124
Bruguiera gymnorrhiza, 124, 127
Bruguiera parviflora, 124, 127
Butea monosperma, 15, 21

C

Cacao tree (*Theobroma cacao*)
Caesalpinia bonduc, 124
Caesalpinia crista, 60, 124
Cajanus cajan, 18
Callistemon lanceolata, 15
Canal command area, 17, 99, 135
Canavalia maritima, 124
Capparis decidua, 17, 21
Carbon credit, 55, 188
Carbon sequestration, 4, 13, 25, 55, 104, 105, 173–175, 182, 187–192, 201
Carbon stock, 2, 25, 191, 201
Carica papaya, 150
CARI *Dhan*-4, 133
CARI *Dhan*-5, 133
Carissa carandas, 17, 21, 26, 152, 153, 187, 199
Cashew nut tree (*Anacardium occidentale*), 123, 127, 199
Cassava (*Manihot esculenta*), 132
Cassia fistula, 151
Cassia javanica, 151
Cassia senna, 156
Cassia siamea, 15, 21, 22, 149, 151, 155
Castor, 156
Castor (*Ricinus communis*), 27, 124, 156
Casuarina equisetifolia, 15, 16, 20–24, 40, 124, 127, 134–138, 150, 167, 186, 198
Casuarina glauca, 10, 11, 15, 17, 36, 60, 64, 65, 116, 127, 134, 136, 137, 153, 167, 198
Casuarina obesa, 11, 13, 15, 16, 60, 61, 150
Catchment area, 3, 6, 10, 11, 36–38, 40, 41, 52, 53, 55, 56, 62, 67, 103–104, 125, 199–201
Catharanthus roseus, 23, 150, 155
Cenchrus ciliaris, 135, 154, 155, 188
Cenchrus setigerus, 154
Ceriops tagal, 124, 127
Chenopodium album, 138
Chenopodium glaucum, 16, 139, 152
Chloris barbata, 16, 134
Chloris gayana, 7, 12, 13, 24, 25, 103, 106, 134, 139, 155, 199
Chloris virgata, 16
Chrysanthemum indicum, 23, 156, 165
Cinnamomum zeylanicum, 130
Cirrhinus mrigala (carp fish)
Citrus aurantifolia, 150
Citrus limon, 150
Citrus maxima, 150
Citrus sinensis, 150
Citrus spp., 151, 166, 171
Clarias batrachus (Indian malgur fish), 136
Claviceps purpurea (Ergot), 23
Clerodendr(um)on inerme, 15, 123, 127
Climate change, 2, 23, 27, 121, 122, 139, 140, 146, 157, 173, 182, 190–192, 198, 201
Climate change adaptations, 190–191, 201
Climate change mitigation, 182, 190–192

Clitoria ternatea, 124
Clonal *Eucalyptus* (*E. grandis* x *urophylla*), 38
Clonal plantations, 38, 185
Clove *(Syzygium aromatica)*, 130, 133
Cluster bean, 114, 153
Cluster planting, 126
Coast line, 122, 123
Coastal almond/Badam (*Terminalia catappa*), 127, 153, 199
Coastal and island areas, 131
Coastal ecosystems, 129
Coconut *(Cocos nucifera)*, 123, 124, 127, 129, 130, 132, 133, 135, 153, 168
Coix lacryma-jobi, 135
Colocasia esculenta, 132, 165
Command area, 25, 110, 112, 113, 135
Commercial plantation, 137
Condofan pea (*Clitoria ternatea*), 124
Congo signal grass (*Brachiaria ruziziensis*), 134
Cordia rothii, 21, 150, 155
Cordia subcordata, 124
Coriander *(Coriandrum sativum)*, 134
Cotton *(Gossypium arboretum/ hirsutum)*, 19, 38, 102, 114, 198
Cowpea (*Vigna unguiculata*), 23, 137, 187
Crescentia alata, 15
Cressa cretica, 17
CSR-36, 133
Cucurbita maxima, 150
Cumin *(Cuminum cyminum)*, 134
Curcuma domestica syn *C. longa*, 137
Custard apple (*Annona squamosa*), 123
Cut-and-carry system, 25
Cyamopsis tetragonoloba (Cluster bean), 153
Cycus rumphii, 124
Cymbopogon flexuosus, 23, 165
Cymbopogon khasans (Jamrosa), 23
Cymbopogon martini, 23, 156
Cynodon dactylon, 16, 17, 19, 154, 183
Cynometra ramiflora, 124

D

Dalbergia latifolia, 123
Dalbergia monosperma, 124
Dalbergia sissoo, 15, 20–22, 24, 26, 40, 117, 148, 167–170, 176, 185, 186, 189
Date palm, 150, 152, 167, 199
Deep furrow and high ridge technique, 131
Deforestation, 201
Degraded and wastelands, 183
Dendrocalamus strictus, 15, 22, 172
Derris scandens, 124
Desmostachya bipinnata, 26, 139, 155, 182, 185, 187–189, 191
Diagnosis and design, 34, 46, 56, 75, 77, 78, 85, 91, 92, 116, 125, 136, 146, 173–176, 200
Dichanthium annulatum, 16, 17, 25, 134, 155
Dichanthium caricosum, 16
Dichrostachys cinerea, 15
Digitaria ciliaris, 16
Dill (*Anethum graveolens, A. sowa*), 134, 156
Dinanath grass (*Dichanthium annulatum*), 16, 17, 25, 134, 155
Dioscorea alata, 132
Diptotaxis tenuifolia, 138
Dischidia nummularia, 124
Distichlis palmeri, 138, 152
Distichlis spicata, 7, 18, 19
Dolichandrone spathacea, 124
Doruvu technology, 135
Drainage effluents, 2, 6, 13, 102, 103, 106, 114, 146
Dryland agriculture, 189
Dryland agroforestry model, 7

E

Echinochloa colonum, 14, 16, 134
Echinochloa crus-galli, 14, 16, 135
Economics of agroforestry, 6, 12, 14, 18, 19, 21, 23, 27, 98, 115, 183, 192
Ecosystem service, 4, 182, 184, 187, 188, 192, 201
Egyptian clover (*Trifoliun alexandrinum*), 20, 22–24, 26
Eichhornia crassipes, 165
Elaeis guineensis (Oil palm), 123, 135
Eleocharis dulcis, 155
Eleusine coracana (Finger millet/Ragi), 154
Elytrigia elongata, 103, 106
Emblica officinalis, 21, 26, 123, 152, 153, 199
Employment generation, 17, 122, 130, 133
Endocomia macrocoma, 133
Engineering strategies, 101–102
Environmental benefits, 4, 16, 67, 185, 198
Erianthus munja syn. *Saccharum munja*, 117
Eruca sativa, 156
Erythrina indica, 123, 124
Erythrina variegata, 124
Eucalyptus camaldulensis, 10, 11, 13, 15, 16, 19, 36, 37, 41, 45, 59–65, 67, 104, 115–117, 136, 150, 167, 168, 183, 198
Eucalyptus globulus, 10, 11, 13, 36, 59–62, 65, 167
Eucalyptus grandis, 10, 36, 38, 42–44, 59–62, 64–66, 75, 77, 83, 118, 167, 169, 199
Eucalyptus hybrid *(E. botryoides* x *E. tereticornis, E. robusta* x *E. tereticornis)*, 104, 150
Eucalyptus loxophleba, 11, 45, 60
Eucalyptus marginata, 36
Eucalyptus microtheca, 10, 16, 36
Eucalyptus occidentalis, 10, 11, 13, 44, 45, 59, 60, 65, 189, 198
Eucalyptus platypus, 11, 60
Eucalyptus robusta, 13, 60
Eucalyptus saligna, 60, 167
Eucalyptus sargentii, 11, 13, 45, 59, 60, 65, 189, 198
Eucalyptus tereticornis, 10, 15, 16, 20–23, 36, 37, 39–41, 44, 60, 104, 116, 123, 136, 150, 151, 167, 174, 183, 185, 186, 188, 190, 198
Euchlaena mexicana, 134

Euphorbia antisyphilitica, 27, 157
Exchangeable sodium percentage) (ESP), 19–23, 25, 26, 54, 186
Excoecaria agallocha, 124, 127
Exotic species, 16, 150

F
FAO soil classification, 6
Farm forestry, 56
Farming in forests, 3, 122
Farming systems research, 12, 136, 137
Farm-level wastewater management, 166
Fennel *(Foeniculum vulgare)*, 134, 156
Fenugreek (*Trigonella foenum-graecum*), 134
Feronia limonia, 15, 21, 135, 150, 152, 153, 155
Finlaysonia maritima, 124
Floriculture, 156–157
Fodder banks, 135
Fodder trees, 135
Fodder values, 135, 155
Foeniculum vulgare, 156
Forage grasses, 2, 17, 19, 103, 106, 133, 134, 146, 154, 155, 157, 188, 200
Freshwater skimming system, 135
Fruit trees-based agroforestry systems, 22, 23, 25–26, 169
Furrow irrigation, 147, 166
Furrow planting, 14, 156

G
Garlic, 26
German chamomile (*Matricaria chamomilla*), 23
Gliricidia sepium syn. G. maculata, 123, 133, 135, 138
Glycophytes (transitional halophytes), 152
Gmelina arborea, 168
Goose berry, 26, 153
Gram, Chickpea (*Cicer arietinum*), 114
Grapes (*Vitis vinifera*), 150
Grazing lands, 24, 134, 135
Green manure, 133
Grevillea robusta, 21
Guar (*Cyamopsis tetragonoloba*), 153
Guava (*Psidium guajava*), 135, 168
Guazuma ulmifolia, 15, 150, 151
Guettarda speciosa, 124
Guinea grass (*Panicum maximum*), 23

H
Halophytes, 7, 14–19, 122, 134, 135, 138–139, 153, 155, 157, 188
Halopyrum mucronatum, 139, 155
Halopyrum mucronatum, 139, 155
Haloxylon aphyllum, 18
Haloxylon salicornicum, 116
Hardwickia binata, 21, 22, 40, 167
Haritiera kanikensis, 18
Heavy metals recycling potential, 169–171
Helianthus annuus, 165
Heritiera fomes, 124
Heritiera littoralis, 124, 138
Hernandia peltata, 123, 124
Hibiscus tiliaceus, 123, 124, 138
Homegardens, 122, 133
Homesteads, 124, 132–133, 137
Hordium vulgare, 153, 187
Horsfieldia glabra, 133
Horsfieldia irya, 133
Horsfieldia macrocarpa, 133
Household economy, 137, 173
Hoya parasitica, 124
Hybrid napier (*Pennesetum purpureum*), 134
Hydnophyttum formicarum, 124
Hypopthalmichthys molitrix (silver carp), 127–128

I
ICAR, 121, 137
Impact of waterlogging, 113–114
Imperata cylindrica, 183
Improved agroforestry systems, 2–4, 16, 17, 24, 98, 124–126, 145–157, 169, 182–188, 191, 192, 200
Improved Doruvu technique, 135
Indo-Gangetic plains, 183, 189
Infiltration rate, 25, 66, 101, 112, 117, 149
Inga dulce syn. *Pithecellobium dulce*, 21, 151, 186
Integrated farming system, 3, 133, 136, 137, 139, 140, 201
Intercropping, 22, 24, 136, 153
Intsia bijuga, 124
Ipomoea batatas, 132
Ipomoea pes-caprae, 124
Isabgol (*Plantago ovata*), 23, 153, 156
Island ecosystem, 3, 122

J
Jaiphal (*Myristica fragrans*), 133
Jamun (*Syzygium cuminii*), 26, 135
Jasminum grandiflorum, 165
Jatropha curcas, 27, 123, 157, 186, 189, 199
Job's tears (*Coix lacryma-jobi*), 135
Joint Forest Management (agroforestry approach), 2, 9, 22, 122
Jujube-based pastoral system, 150

K
Kallar grass (*Leptochloa fusca*), 20, 24–26, 103, 139, 154, 155, 189
Kandelia candel, 124
Kazungula grass (*Seteria anceps*), 24
Khejri (*Prosopis cineraria*), 14, 15, 21, 116, 135, 150, 153, 155, 186, 198
Kigelia pinnata, 21
Knema andamanicum, 133

Knema conferta, 133
Knema glaucescens, 133
Kochia indica, 17, 134
Kosteletzkya virginica, 139, 152

L
Lagenaria siceraria syn. *L. vulgaris*, 137
Land shaping techniques, 130–133
Lannea coromandelica, 137
Leaf area index, 35, 37, 38, 63, 78, 81
Lemon (*Citrus limon*), 23, 150, 156, 165
Lemon grass (*Cymbopogon flexuosus*), 156
Lepidium sativum, 156
Leptochloa fusca, 14, 16, 17, 20, 24–26, 103, 134, 135, 139, 153–155, 189
Leucaena leucocephala, 14–16, 21, 22, 123, 134, 135, 137, 150, 153, 154, 167, 168, 174
Licuala spinosa, 124
Line planting, 126
Litter fall, 2
Live fence, 173–174
Livelihood security, 3, 4, 18, 122, 130, 145, 173, 198
Liza tade (mullet fish), 129
Lolium multiflorum, 16
Lolium rigidum, 9
Lotus glabra, 19
Lotus ulginosus, 19
Lumnitzera littorea, 124
Lumnitzera racemosa, 124

M
Mace tree (*Myristica fragrans*), 133
Madhuca indica, 137, 199
Maireana brevifolia, 11, 12, 19
Maize (*Zea mays*), 38, 74
Mangifera andamanica, 133
Mangifera camptosperma, 133
Mangifera indica, 21, 123, 133, 137, 150, 199
Mangifera sylvatica, 133
Mango (*Mangifera indica*), 133, 137, 150, 168
Mangroves, 3, 15, 17, 18, 122–129, 138, 139, 153, 172, 185, 199
Manihot esculenta, 132
Manilkara littoralis, 15, 124, 128, 138
Maranta arundinacea, 132
Marginal waters, 200
Matricaria chamomilla, 23, 156
Maytenus emarginata, 15
Medicago sativa, 9, 11, 16, 19, 24, 55
Melia azedarach, 21, 149, 151
Melia dubia, 137
Melilotus denticulata, 24
Mesquite (*Prosopis juliflora*), 24, 103, 189, 198
Microbial carbon, 25
Microclimate, 188, 192, 201
Mimusops elengi, 124
Monoculture, 183
Morinda citrifolia, 123, 124, 130, 133, 155, 199
Moringa (*Moringa oleifera*), 21, 123, 130
Morus alba, 21
MPTs, 124, 134, 138
Mulch, 10, 172, 175
Multi-enterprise farming systems, 137, 201
Multipurpose trees (MPTs), 124, 130, 176, 185
Multistorey cropping, 139
Multistorey tree garden, 122
Musa acuminata, 60, 133
Musa x paradisiaca, 123, 130, 134
Mustard (*Brassica juncea, napus*), 22, 23, 114, 153
Mycorrhiza, 187
Myristica fragrans, 133
Myristica glauca, 133
Myristica prainii, 133
Mysure gum (*Eucalyptus tereticornis*), 104

N
Natural resource, 6, 176, 191
Net primary productivity, 186
Nitrogen fixation, 24, 187
Nitrogen fixing trees, 134
North-eastern region, 187, 188
North-western region, 26, 59, 110, 112, 187–189, 200
Nutmeg (*Myristica fragrans*), 133
Nutrient cycling, 25, 182, 191
Nutritional security, 1

O
Obligate (true) halophytes, 126
Ochrosia oppositifolia, 124
Ocimum sanctum, 156
Oil palm (*Elaeis guineensis*), 123, 135
Okra (*Abelmoschus esculentum*), 130
On-farm research (OFR), 131
Oreochromis urolepis (Tilapia fish), 129
Organic farming, 35
Organic matter, 26, 35, 50, 103, 105, 106, 135, 150, 171, 185, 186

P
Pandanus leram, 123
Panicum antidotale, 24, 154
Panicum coloratum, 154
Panicum laevifolium, 24, 154
Panicum maximum, 16, 24, 134, 154
Panicum turgidum, 139, 153, 155
Panicum virgatum, 155
Papaya (*Carica papaya*), 137, 150
Para grass (*Brachiaria mutica*), 24
Parkinsonia aculeata, 15, 18, 21, 198
Parthenium argentatum, 155
Paspalum conjugatum, 16, 139
Paspalum plicatulum, 16
Paspalum vaginatum, 7, 16, 19

Passiflora edulis (Passion fruit), 150–151
Peach (*Prunus persica*), 150
Pearl millet (*Pennisetum typhoides*), 138, 153, 186
Pennisetum clandestinum, 7, 19
Pennisetum purpureum, 24, 134
Pennisetum typhoides, 138, 153, 186
Phoenix dactylifera, 21, 150, 152, 167, 199
Phoenix paludosa, 124, 133
Phragmites australis, 139, 155
Phragmites karka, 139, 155, 165
Phreatophytes, 115, 126, 199
Pigeon pea (*Cajanus cajan*), 22, 23
Pine apple (*Ananas comosus*), 150
Pinus radiata, 36, 55, 56, 60, 66, 75, 167, 169
Piper nigrum, 130
Pistacia vera (Pistachio nut), 152
Pithecellobium dulce, 21, 151, 186
Plantago ovata, 23, 153, 156
Plantation based cropping systems, 3, 139
Plantation crops, 3, 92, 122, 139, 168
Planting methods with saline irrigation, 14
Polianthes tuberosa, 165
Pomegranate (*Punica granatum*), 20, 26, 134, 135, 150
Pongamia pinnata, 15, 17, 20, 21, 24, 27, 104, 123, 124, 127, 134, 137, 138, 149, 153, 157, 185, 199
Poor quality waters, 2, 4, 98, 99, 146, 157, 197–201
Poplar (*Populus* spp.), 168, 169, 176
Poplar based system, 22
Populus alba, 15, 16, 21, 150
Populus deltoides, 21–23, 103, 167, 168
Populus euphratica, 185
Populus nigra, 21, 165
Populus spp., 168, 169, 176
Portuleca oleracea, 157
Post-harvest interventions, 166
Potato (*Solanum tuberosum*), 23, 76
Premna corymbosa, 124
Prosopis alba, 15, 16, 21, 150
Prosopis articulata, 21
Prosopis chilensis, 16, 150
Prosopis cineraria, 14, 15, 21, 116, 135, 150, 153, 155, 186, 198
Prosopis juliflora, 15–18, 20–22, 24–26, 36, 103, 116, 117, 134, 135, 137, 149–151, 153, 155, 157, 183, 185–191
Prosopis levigata, 21
Prosopis nigra, 21
Prosopis pallida, 21, 104, 169
Prunus amygdalus, 152
Prunus armeniaca, 150
Prunus avium, 150
Prunus domestica, 150
Prunus persica, 21, 150
Psidium guajava, 21, 26, 123–124, 168, 199
Psyllium (*Plantago ovata*), 23, 153, 156
Pterocarpus dalbergioides, 128
Puccinellia ciliata, 7, 11–13, 106
Punica granatum, 20, 21, 26, 124, 134, 135, 150, 199
Pyrus communis, 21, 151

R

Raised and sunken bed technique, 20
Rangeland, 7, 74
Red oil palm (*Elaeis guineensis*), 135
Reforestations, 35, 39, 55, 189, 200, 201
Rhizobium, 24, 187
Rhizophoraceae, 126
Rhizophora apiculata, 124, 127
Rhizophora lamarckii, 124
Rhizophora mucronata, 124, 127
Rhizophora stylosa, 124, 127
Rhizophora x *annamalayana*, 18
Rhizosphere, 187
Rice (*Oryza sativa*), 17, 20, 22, 23, 26, 38, 40, 102, 124–126, 129–133, 136, 137, 188
Ricinus communis, 27, 124, 156, 199
Ridge trenching method of planting, 14
Root biomass, 187
Row planting, 126
Rubber tree (*Hevea brasiliensis*), 155

S

Salicornia bigelovii, 17–19, 138, 152
Saline irrigation, 67, 104, 105, 147–150, 152–157
Saline sodic soils, 15, 154
Saline soils, 3, 7, 13–18, 53, 54, 57, 64, 67, 99, 117, 134, 152, 183, 188, 199
Saline waterlogged soils, 3, 198
Saline waters, 2–4, 9, 17, 18, 34, 42, 44, 59, 61, 63–67, 101, 103, 104, 134, 135, 146, 147, 150, 152, 154, 156, 168, 183, 190, 198
Salix alba, 13
Salix babylonica, 15
Salsola drummondii, 14, 135, 154
Salt-affected soils, 2, 7, 8, 19, 24, 25, 98, 99, 101, 110, 113, 182, 183, 185, 186, 188–191, 198, 201
Salt-excluding halophytes, 57
Salvadora oleoides, 14, 15, 21, 116, 135, 150, 153
Salvadora persica, 14–18, 21, 22, 124, 127, 134, 135, 138, 150, 152, 153, 155, 183, 187–189, 199
Salvinia molesta, 165
Samanea saman, 15, 124
Sand dunes stabilization, 126
Sapindus laurifolia, 21
Sapindus trifoliatus, 134
Sapota (*Achrus zapota*), 26, 134, 135, 151, 168
Scirpus littoralis, 124
Scirpus tabernaemontani, 165
Scyphiphora hydrophyllacea, 124
Sea water intrusion, 3, 98, 122, 123, 136, 146
Seaweeds, 140
Secale cereal, 13, 23
Seepage, 3, 7, 11, 13, 17, 34, 37, 40, 43, 45, 46, 54, 99, 102–104, 112–114, 116, 118, 135, 190
Semiarid tropics, 169
Sesbania cannabina syn. *S. aculeata*, 16, 154
Sesbania grandiflora, 40, 124
Sesbania sesban, 21, 135, 154

Sesuvium portulacastrum, 155
Setaria anceps, 24
Shade trees, 134, 185
Shallow furrow and medium ridge technique, 131
Shelterbelt, 2, 3, 57, 126, 138, 200
Shifting cultivation, 133
Silvopastoral systems, 4, 9, 12, 16–18, 25, 134, 139, 152–154, 157, 183, 185–188, 191
Simarouba glauca, 199
Simmondsia chinensis, 155
Sloping land, 63
Smallholders, 164, 201
Soap nut (*Sapindus trifoliatus*), 134
Social benefit, 185
Social forestry, 197
Sodic soil (land), 24
Sodic water, 101, 150
Sodium absorption ration (SAR), 19, 129
Soil conservation, 27, 164, 198
Soil organic matter, 52, 186, 188, 189
Solanum melongena, 130
Sonneratia alba, 124, 127
Sonneratia apetala, 124
Sonneratia caseolaris, 124, 127
Sonneratia ovata, 124
Sorghum *bicolor* (Sorghum), 22, 23
Species for saline soils, 15, 16, 18
Species for sandy plain, 112
Species for shallow soil areas, 39, 169, 200
Species for sodic soils, 2, 4, 6, 15, 19–26, 52, 102, 105, 106, 154, 182, 183, 185–191, 199
Spices, 3, 122, 130, 133, 134, 139, 156
Sporobolus airoides, 19
Sporobolus marginatus, 16, 26, 183, 187–189
Sporobolus tremulus, 16
Standing biomass, 19, 129
Strip plantation, 42, 45, 116
Submergence area, 113
Subsistence farming, 4
Sub-surface drainage, 6, 34
Sub-surface planting, 13
Sub-surface planting & furrow irrigation system (SPFIS), 147, 148
Sugarcane (*Saccharum officinarum*), 127, 129, 132, 137
Sun flower (*Helianthus annuus*), 165
Sunderbans, 127, 129, 132
Sustainability, 4, 8, 12, 23, 35, 42–45, 64, 74, 99, 103, 104, 106, 137, 173
Syzygium aromaticum, 130
Syzygium cuminii, 20–22, 26, 149, 199
Syzygium samarangense, 124

T

Tabernaemontana crispa, 124
Taca leontopetaloides, 132
Tagetes erecta, 165
Tall wheat grass (*Thinopyrum ponticum* syn. *Elytrigia elongata*), 199
Tamarind (*Tamarindus indica*), 15, 21, 23, 124, 134, 135, 150, 153, 168
Tamarix articulata syn. *T. aphylla*, 167
Tamarix dioica, 117
Tamarix ericoides, 15
Tamarix indica, 0, 135, 150, 153
Tamarix stricta, 135, 150, 153
Tamarix troupii, 15, 17, 134
Tapioca/Cassava (*Manihot esculenta*), 0, 132
Tara-mira (*Eruca sativa*), 156
Taro (*Colocasia esculenta*), 132, 165
Tea (*Camellia sinensis*), 150
Teak (*Tectona grandis*), 137, 168
Tecomella undulata, 117, 168
Tectona grandis, 21, 23, 137, 168
Terminalia arjuna, 15, 20–22, 24, 149, 185, 198
Terminalia catappa, 15, 17, 124, 127, 134, 138, 139, 152, 153, 199
Thespesia populnea, 15, 124, 127, 137, 138
Thinopyrum ponticum, 7, 13, 153
Tournefortia ovata, 124, 138
Traditional agroforestry systems, 135
Trifolium alexandrinum, 24, 26
Trifolium repens, 13
Trifolium resupinatum, 13, 24
Triumfetta repens, 124
Tropics and subtropics, 17
Tuber crops, 132
Turmeric (*Curcuma domestica*), 22, 23, 133, 137
Types of sand dunes, 110
Typha domingensis, 139, 155
Typha latifolia, 165

U

Untreated sewage, 167
Urban plantations, 164, 168, 172–176

V

Vertisol (soil), 24, 25, 155
Vetiveria zizanioides (Vetiver), 16, 23, 25, 26, 156
Vigna marina, 124, 139
Vigna unguiculata, 137
Vitex negundo, 124
Vitex trifoliata, 124
Vitis vinifera, 21, 26, 150
Viviparous seedlings, 126

W

Wastelands, 110, 169
Wastewater disposal, 67, 167, 172, 173, 176
Wastewater irrigation, 164–166, 168–171, 173, 174
Wastewater treatment, 164, 169, 172
Water harvesting, 124, 132

Waterlogged areas, 13, 16, 103, 109, 113–117, 123, 129, 136, 137, 139, 182, 189
Water melon, 136
Water-use efficiency, 38, 91, 93
Wedelia biflora, 124
Wetland, 7, 51, 123, 135, 137, 164, 165, 172, 173, 176
Wetland agroforestry, 7
Wheat (*Triticum aestivum, T. durum, T. dicoccum, T. monococcum*), 11, 19, 20, 22, 23, 26, 38, 40, 52, 76, 114, 117, 137, 189
Windbreaks, 2, 105
Wood apple, 135
Woodlot, 11, 45, 56, 137
Woody perennials, 134

X
Xylocarpus gangeticus, 124
Xylocarpus granatum, 124, 127
Xylocarpus moluccensis, 124

Y
Yams (*Dioscorea* spp.), 132

Z
Zea mays, 155
Ziziphus mauritiana, 15, 21, 26, 117, 134, 135, 150–152, 199
Zygophyllum album, 139, 152
Zygophyllum simplex, 14, 135